# Protective Systems for high temperature applications: from theory to industrial implementation

European Federation of Corrosion Publications
NUMBER 57

# Protective Systems for high temperature applications: from theory to industrial implementation

Edited by
M. Schütze

EUROPEAN FEDERATION OF CORROSION
FÉDÉRATION EUROPÉENNE DE LA CORROSION
EUROPÄISCHE FÖDERATION KORROSION

**Published for the European Federation of Corrosion
by CRC Press
on behalf of
The Institute of Materials, Minerals & Mining**

**CRC Press**
Taylor & Francis Group
Boca Raton  London  New York

CRC Press is an imprint of the
Taylor & Francis Group, an **informa** business

IOm³
The Institute of Materials,
Minerals and Mining

First published in 2011 by Maney Publishing

Published by CRC Press on behalf of the European Federation of Corrosion and The Institute of Materials, Minerals & Mining

2 Park Square, Milton Park, Abingdon, Oxon OX14 4RN
711 Third Avenue, New York, NY 10017, USA

First issued in paperback 2017

*CRC Press is an imprint of the Taylor & Francis Group, an informa business*

ISBN 13: 978-1-138-11636-8 (pbk)
ISBN 13: 978-1-906540-35-7 (hbk)

ISSN 1354-5116

# Contents

v

# European Federation of Corrosion (EFC) publications: Series introduction

The European Federation of Corrosion (EFC), incorporated in Belgium, was founded in 1955 with the purpose of promoting European cooperation in the fields of research into corrosion and corrosion prevention.

Membership of the EFC is based upon participation by corrosion societies and committees in technical Working Parties. Member societies appoint delegates to Working Parties, whose membership is expanded by personal corresponding membership.

The activities of the Working Parties cover corrosion topics associated with inhibition, cathodic protection, education, reinforcement in concrete, microbial effects, hot gases and combustion products, environment-sensitive fracture, marine environments, refi neries, surface science, physico-chemical methods of measurement, the nuclear industry, the automotive industry, the water industry, coatings, polymer materials, tribo-corrosion, archaeological objects and the oil and gas industry. Working Parties and Task Forces on other topics are established as required.

The Working Parties function in various ways, e.g. by preparing reports, organising symposia, conducting intensive courses and producing instructional material, including films. The activities of Working Parties are coordinated, through a Science and Technology Advisory Committee, by the Scientific Secretary. The administration of the EFC is handled by three Secretariats: DECHEMA e.V. in Germany, the Fédération Française pour les sciences de la Chimie (formely Société de Chimie Industrielle) in France, and The Institute of Materials, Minerals and Mining in the UK. These three Secretariats meet at the Board of Administrators of the EFC. There is an annual General Assembly at which delegates from all member societies meet to determine and approve EFC policy. News of EFC activities, forthcoming conferences, courses, etc., is published in a range of accredited corrosion and certain other journals throughout Europe. More detailed descriptions of activities are given in a Newsletter prepared by the Scientific Secretary.

The output of the EFC takes various forms. Papers on particular topics, e.g. reviews or results of experimental work, may be published in scientific and technical journals in one or more countries in Europe. Conference proceedings are often published by the organisation responsible for the conference.

In 1987 the, then, Institute of Metals was appointed as the official EFC publisher. Although the arrangement is non-exclusive and other routes for publication are still available, it is expected that the Working Parties of the EFC will use The Institute of Materials, Minerals and Mining for publication of reports, proceedings, etc., wherever possible.

The name of The Institute of Metals was changed to The Institute of Materials (IoM) on 1 January 1992 and to The Institute of Materials, Minerals and Mining with effect from 26 June 2002. The series is now published by CRC Press on behalf of The Institute of Materials, Minerals and Mining.

**P. McIntyre**
EFC Series Editor
The Institute of Materials, Minerals and Mining, London, UK

EFC Secretariats are located at:

Dr B. A. Rickinson
European Federation of Corrosion, The Institute of Materials, Minerals and Mining,
1 Carlton House Terrace, London SW1Y 5AF, UK

Mr M. Roche
Fédération Européenne de la Corrosion, Fédération Française pour les sciences de la
Chimie, 28 rue Saint-Dominique, F-75007 Paris, France

Dr W. Meier
Europäische Föderation Korrosion, DECHEMA e.V., Theodor-Heuss-Allee 25,
D-60486 Frankfurt-am-Main, Germany

# Volumes in the EFC series

* indicates volume out of print

# Influence of novel cycle concepts on the high-temperature corrosion of power plants*

Bettina Bordenet

*ALSTOM (Switzerland) Ltd., Brown Boveri Strasse 7,
CH-5401 Baden, Switzerland*
*bettina.bordenet@power.alstom.com*

## 1.1    Introduction

The Kyoto Protocol, which came into force in 2005, aims to reduce worldwide emissions of greenhouse gases (GHG). Fossil fuel fired power plants are among the largest and most concentrated producers of $CO_2$ emissions. Additionally, a rapidly growing demand for electricity is projected over the next 30 years [1], especially in highly populated and fast-growing countries such as China and India.

The main strategies to reduce $CO_2$ emissions in the power sector are increased efficiency, and carbon capture and sequestration (CCS). Increased efficiency corresponds to producing more energy for the same amount of fuel, hence effectively reducing $CO_2$ emissions. This can be achieved by worldwide implementation of state-of-the-art power plant technology for new installations and the replacement of old inefficient power plants. The efficiency of conventional power plants can still be increased by further development of the technology, e.g. 700 °C steam power plant.

The long-term strategy for the reduction of $CO_2$ emissions is the recovery and sequestration of $CO_2$ produced during power production. To achieve this goal, several new power plant concepts have been developed in recent years, which are currently under investigation from both engineering and economic points of view. An overview of different public-funded projects in Europe and in the USA on this topic can be found in Ref. 2.

## 1.2    Options for power plants with low $CO_2$ emissions

Three $CO_2$ separation concepts are proposed for fossil fuel fired power plants (Fig. 1.1) [3]:

- *Post-combustion capture*: Carbon dioxide is separated from the exhaust gas coming from a conventional power plant. This technique can be applied to existing power plants by adding the $CO_2$ capture device. The most suitable technology seems to be chemical absorption, but it is very energy demanding and results in a reduction of efficiency.

---

* Reprinted from B. Bordenet: Influence of novel cycle concepts on the high-temperature corrosion of power plants. *Materials and Corrosion*. 2008. Volume 59. pp. 361–66. Copyright Wiley-VCH Verlag GmbH & Co. KGaA.

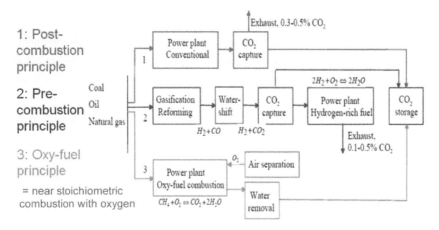

*1.1* Overview of three main options of $CO_2$ separation in fossil fuel fired power plants [3]

For post-combustion capture concepts, the main focus is on the $CO_2$ separation process and on its efficiency increase. The $CO_2$ content in natural gas fired combined-cycle (CC) power plants is rather low at only 4 vol.%, compared to a boiler power plant with ~15 vol.%. Hence, a large amount of exhaust gas must be treated, which leads to the large footprint and high investment cost of the $CO_2$ separation device. Concepts have been developed for CC power plants, which include exhaust gas recirculation to increase the $CO_2$ content in the exhaust [4,5].

- *Pre-combustion fuel decarbonisation*: In this concept, the fuel is converted first to $H_2$ and CO by gasification or reforming and then to $H_2$ and $CO_2$ after the water shift reaction. Afterwards, the $CO_2$ is removed and the $H_2$ is fed as fuel into a power plant with nearly no $CO_2$ in the exhaust.

  For this approach, concepts such as the Integrated Gasification Combined Cycle (IGCC) are investigated [6–9]. One of the main challenges seems to be the combustion of pure $H_2$ in the gas turbine.

- *Oxy-fuel combustion*: Here, the combustion takes place in a nitrogen-free atmosphere. The oxygen is separated from the air and burned in near-stoichiometric conditions with the fuel. Consequently, the exhaust gas is composed mainly of $CO_2$ and $H_2O$. After the condensation of the water, the $CO_2$ can be used directly for example in the enhanced recovery of oil. The oxygen is usually provided by an air separation unit, which is an energy consuming set-up.

  Oxy-fuel firing is under investigation for combined cycles [3,10,11] and boiler applications [10–14]. In most cases, the oxygen is produced by an air separation unit. Some concepts are based on the delivery of $O_2$ through membranes, such as the advanced zero emission power plant (AZEP). For all these cases, the composition of the flue gas is significantly changed compared to conventional power plants.

  Another type of oxy-fuel firing is chemical looping combustion (CLC), which is being evaluated for combined cycle [10] and boiler applications [14–16]. The combustion is split into separate oxidation and reduction reactions by introducing a suitable metal oxide as an oxygen carrier to circulate between the two reactors.

Each of the concepts presented above requires, in general, higher investment and maintenance costs, and exhibits a lower efficiency than a comparable power plant without $CO_2$ capture.

## 1.3    Corrosion risk evaluation for new power plant concepts

A detailed corrosion risk evaluation needs to be performed for each new power plant concept. The gas compositions are different for some components and it must be determined where the materials which have been used so far can withstand the modified environments. In the present study, the focus is put on a relative evaluation in comparison to the standard power plant. The components of a new power plant type that encounter a different environment are identified. Then the materials are evaluated with respect to their corrosion risk and possible changes in corrosion mechanisms are discussed. This approach has been used as a case study for two of these new cycles:

- Combined cycle power plant with exhaust gas recirculation as an example of post-combustion capture cycles
- Pulverised fuel boiler with oxy-fuel firing, which belongs to the group of cycles with oxy-fuel combustion

### 1.3.1    Combined cycle power plant with exhaust gas recirculation (EGR)

Description of a combined cycle power plant with EGR [4]

Carbon dioxide capture can be retrofitted to an existing conventional combined cycle power plant. Due to the low $CO_2$ content of ~4 vol.%, the installation is quite large and cost intensive. Exhaust gas recirculation in a gas turbine is an interesting option to increase the $CO_2$ content in the exhaust gas so as to increase the efficiency of $CO_2$-separation and permit a smaller footprint and, hence, lower investment. A sketch of such an arrangement is shown in Fig. 1.2 for an ALSTOM KA26

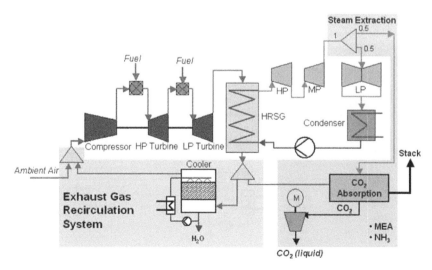

*1.2* The principle of a KA26 combined cycle power plant (CC) with exhaust gas recirculation (EGR)

(combined-cycle with a GT26 gas turbine) with sequential combustion. The air is compressed in the compressor and the fuel is added in the combustion chamber. The hot gas is expanded in the high-pressure turbine and then led to the second combustion chamber. Afterwards, the hot gas is expanded in the low-pressure turbine and fed into the heat recovery steam generator (HRSG), where the steam is heated for the steam turbines.

The exhaust gas recirculation system and the $CO_2$-separation device (Fig. 1.2) will be designed such that they could be added to an existing power plant. After the HRSG, ~50% of the exhaust gas is led through a cooler to reduce the temperature and $H_2O$ content. Then it is fed into the compressor right after the air intake. The remaining exhaust gas after the HRSG goes into the $CO_2$-separation device. There, $CO_2$ is separated from the exhaust gas with a chemical solvent (e.g. monoethanolamine (MEA), $NH_3$) and steam from the steam cycle.

The following components of the power plant will be affected by a different environment: the entire gas turbine with the compressor, the combustion chambers, both turbines and the HRSG. An enlarged view of these components is shown in Fig. 1.3.

The recirculation rate is limited by the oxygen content after the second combustion chamber. For combustion stability, the excess oxygen content is set to ~1% [4] and corresponds roughly to 50% of the recirculation rate. For natural-gas firing, the gas composition is calculated at three locations for an exhaust gas recirculation rate of 50% (EGR) and for a conventional combined cycle power plant (CC). The values for both conditions are provided in Table 1.1.

## Corrosion risk evaluation for the compressor and the cooling air side

At the compressor inlet, the gas composition has changed mainly in the carbon dioxide (5%) and water content (1.7%), whereas the oxygen content has dropped down to 11.8% (Table 1.1). This gas composition can be found in the whole compressor, in the cooling air piping and subsequently, on the cooling air sides of the hot

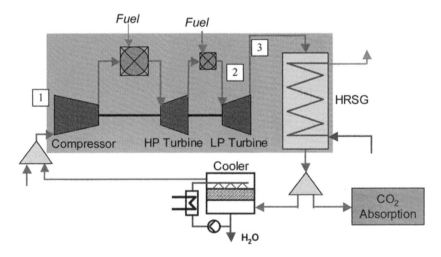

*1.3* Enlarged view of the components affected by the changed environment (shaded box). Locations for gas composition of Table 1.1: 1, compressor inlet; 2, 2nd combustor outlet; 3, LP turbine outlet

*Table 1.1*  Gas compositions at different locations for natural gas firing of power plant with exhaust gas recirculation (EGR) and a conventional combined cycle (CC). Locations for gas composition are highlighted in Fig. 1.2

|  | Compressor inlet, vol.% | | 2nd combustor outlet, vol.% | | LP turbine outlet, vol.% | |
|---|---|---|---|---|---|---|
|  | EGR | CC | EGR | CC | EGR | CC |
| $N_2$ | 80.6 | 78.1 | 76.5 | 73.5 | 77.3 | 74.2 |
| $O_2$ | 11.8 | 21.0 | 1.1 | 10.0 | 3.2 | 11.9 |
| $CO_2$ | 5.0 | 0.0 | 9.8 | 4.9 | 8.8 | 4.0 |
| $H_2O$ | 1.7 | 0.0 | 11.7 | 10.8 | 9.7 | 9.0 |
| Others | 0.9 | 0.9 | 0.9 | 0.8 | 0.9 | 0.9 |

gas path components, such as turbine and combustor components. In a conventional power plant, all of these parts encounter only ambient air (Table 1.1). The $H_2O$- and $CO_2$-contents in the EGR environment are assumed to have an influence on the oxidation/corrosion behaviour. The low-temperature and high-temperature regimes must each be evaluated for corrosion risk.

The low-temperature regime is defined here as the condition when water condensation is possible during operation (first rows of the compressor) and at standstill. In this regime, the main corrosion mechanism will be pitting corrosion. The increased $H_2O$ content will increase the number of locations where condensation is possible during operation. The $CO_2$ content can change the chemistry and could lead to an increased pitting corrosion tendency. The conditions during standstill are more difficult to evaluate, since the gas composition will change when the engine is not in operation. A more detailed evaluation of the corrosion risk requires assessment during shut-down conditions.

The high-temperature regime is characterised by the oxidation consumption of the materials, which are, in general, steels. In recent publications [17,18], the influence of carbon dioxide on oxidation behaviour has been investigated for different steels with chromium contents from 2.25% to 25%. The $CO_2$ content was varied between 0, 6 and 25 vol.% in testing for up to 360 h. In general, the mass gain increased with increasing amount of carbon dioxide. This effect was more pronounced for the ferritic steels. Water vapour generally increases the oxidation rate of steels compared to dry air [19–22]. The steepest increase is often between 0% and 10% of water. Thus, the increased $H_2O$ and $CO_2$ contents are apt to lead to accelerated oxidation rates. From the data available, it is not possible to quantify exactly the effects of the $CO_2$ and $H_2O$ contents in the case of exhaust gas recirculation.

## Corrosion risk evaluation for the combustion chambers and the HP and LP turbines

The gas composition after the second combustion chamber contains a relatively low oxygen content of 1.1% compared to a standard combined cycle with 10% (Table 1.1). Directly after the second combustion chamber, the oxygen content is at its lowest in the whole hot gas path. Afterwards, in the LP turbine, the oxygen content will increase to 3.2% due to the addition of cooling air at the outlet. The low oxygen content in the second combustion chamber is an average value. The combustion may be inhomogeneous and incomplete in some locations, which could lead to a locally reducing atmosphere. These changed conditions could increase the tendency for nitridation of combustor alloys such as Hastelloy X and Haynes 230, which are already prone to exhibit internal nitridation [23].

Exhaust gas recirculation will increase the water content only slightly from 10.8% to 11.7% (Table 1.1). This will probably not influence the oxidation behaviour of the combustor and turbine materials. The carbon dioxide content is doubled from 4.9% to 9.8%, when the exhaust gas is recycled to 50%. Although no systematic study has been published on the effect of $CO_2$ on the oxidation behaviour of turbine and combustor materials, only a small influence is assumed.

### Corrosion risk evaluation for the HRSG

The gas composition is quite different for the EGR compared to a standard combined cycle (Table 1.1), but is close to that for a coal-fired power plant (see section 1.3.2). The corrosion risk will be increased compared to a standard combined cycle plant. The experience for coal-fired boilers can be transferred, because the materials and gas compositions are comparable.

### Consequences for materials selection and testing

Combined cycle power plants with exhaust gas recirculation exhibit quite different gas compositions than a standard power plant. The changed environment raises the need to re-evaluate the oxidation/corrosion behaviour of the whole gas turbine. Two main corrosion risks have been identified. The increased $H_2O$ and $CO_2$ contents in the compressor and cooling air system could lead to higher oxidation rates and metal wastage. For the materials in the turbine and in the combustor, the low oxygen content could promote a higher tendency for internal nitridation or material consumption.

### 1.3.2   Pulverised fuel power plant with oxy-fuel combustion

#### Description of a pulverised coal boiler with oxy-fuel combustion [12]

Oxy-fuel firing is an interesting option for coal-fired power plants with $CO_2$ capture. Figure 1.4 shows a sketch of such a plant arrangement for a pulverised fuel (PF) boiler. Here, the coal is burned with $O_2$, coming from an air separation unit, instead of air. Following particle removal, a major part (~2/3) of the $CO_2$-rich exhaust gas is recycled back to the boiler to control the combustion temperature. Due to the high recirculation rate, the oxy-fuel fired boiler has approximately the same size as an air-fired boiler. The remaining part of the flue gas is cleaned, compressed and transported to storage or another suitable application, such as enhanced oil recovery (EOR).

The gas composition in the furnace is summarised in Table 1.2 for an oxy-fuel and an air-fired PF boiler. For a first evaluation, dried lignite from the Lausitz area (eastern Germany) was used to calculate the atmosphere in both cases. All of the components which are in contact with the flue gas are subjected to the changed environment: the furnace water walls, the heating surfaces, the flue gas cleaning and the flue gas recirculation piping. The steam-side of the power plant is not affected by the changes in the combustion.

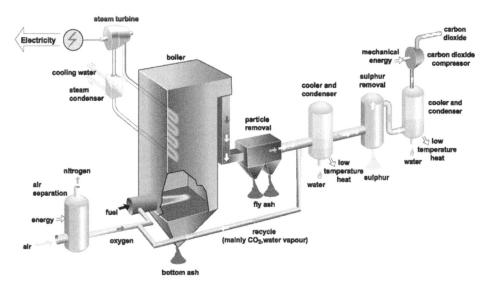

*1.4* The principle of oxy-fuel firing ($O_2$/$CO_2$ recycle) combustion in a pulverised fuel (PF) boiler [12]

## Corrosion risk evaluation for the components in the flue gas path

The gas composition is significantly changed when the combustion is changed from air-fired to an oxy-fuel fired condition (Table 1.2). The carbon dioxide content is raised from 15 to 59 vol.% and the water-content from 10 to ~32%. For the same coal, the $SO_2$-content in the flue gas increases by nearly a factor of 4 to ~0.5%. The oxygen content remains roughly the same.

So far, no data have been published on oxidation behaviour or corrosion under deposits in oxy-fuel fired atmospheres. The publications to date [17,18], which have already been discussed in the earlier subsection 'Corrosion risk evaluation for the compressor and the cooling air side', have only investigated $CO_2$ contents up to 25%. Both oxidation without deposits and corrosion under deposits exhibit higher rates for typical boiler steels with Cr contents from 2.25% to 25%. However, the atmospheres in the oxy-fuel case are very different from previously investigated atmospheres and extrapolation is not possible. Additionally, the deposit chemistry may be changed under the oxy-fuel gas composition, which could lead to higher metal wastage.

*Table 1.2*   Gas compositions (vol.%) in the furnace for dried lignite combustion for oxy-fuel fired and air-fired PF boilers

|  | Oxy-fuel fired PF, vol.% | Air-fired PF, vol.% |
|---|---|---|
| $N_2$ | 4.8 | 71.3 |
| $O_2$ | 1.9 | 2.5 |
| $CO_2$ | 58.9 | 15.3 |
| $H_2O$ | 31.8 | 10.0 |
| Others | 2.1 | 0.8 |
| $SO_2$ | 0.49 | 0.13 |

Low $NO_x$-combustion, which is typical for ultra-supercritical PF boilers, can lead to locally reducing environments. Such reducing environments could also occur for oxy-fuel combustion. Due to the high content of $CO_2$ in the flue gas, a change in deposit chemistry is assumed to occur, and the formation of carbonates in the deposit might be possible. As carbonate deposits are very corrosive, the metal wastage rate could increase.

### Consequences for materials selection and testing

The flue gas composition of oxy-fuel fired power plants is subject to an important change compared to air-firing. The corrosion behaviour needs to be re-evaluated for all of the components in contact with the flue gas for oxidising and reducing environments. The hottest zones of the convective heat-exchanging surfaces are assumed to have the highest corrosion risk. As no data from material testing have been published for oxy-fuel conditions so far, the increased corrosion risk cannot be quantified and new data must be generated.

The deposit chemistry and hence the corrosion rates are likely to be changed because of the different gas composition. By means of thermodynamic modelling, the change in deposit chemistry could be calculated for different fuels and operating conditions to determine a representative corrosion test for oxy-fuel firing.

## 1.4    Conclusions

New power plant concepts exhibit quite different gas compositions compared to current state-of-the-art plants. The changed environments will give many challenges for materials testing and development. Therefore, the testing conditions used for testing oxidation and corrosion behaviours must be adapted to the new plant concepts. Thermodynamic modelling could support the adaptation of the testing conditions, especially for the modelling of the deposit chemistries. The materials which have been used in standard plants so far need to be re-evaluated for oxidation and corrosion behaviour under the new conditions. If the standard materials are not adequate, development of improved materials and coatings will be necessary to resist the new corrosion conditions.

## References

1. World Energy Outlook 2004, IEA – International Energy Agency, Paris, France.
2. P. Dechamps and A. P. Sainz, *The EU Research Strategy for $CO_2$ Capture and Storage*, PowerGen Europe, 2004.
3. O. Bolland, H. M. Kvamsdal and J. C. Boden, 'A thermodynamic comparison of the oxy-fuel power cycles water–cycle, Graz-cycle and Matiant-cycle', in International Conference on Power Generation and Sustainable Development, Liège, Belgium, 8–9 October 2001.
4. M. Wolf, A. Brautsch, J. Gernert, G. Kaefer, A. Pfeffer and D. Winkler, 'Outlook on post combustion $CO_2$ capture in ALSTOM gas turbine plants', in Proceedings of PowerGen Europe, Cologne, Germany, 30 May–1 June 2006.
5. D. Fiaschi and F. Scatragli, 'Off-design analysis of the Semi-Closed Gas Turbine Cycle (SCGT) with low $CO_2$ emissions', in Proceedings of the ASME Turbo Expo 2006, Barcelona, Spain, 8–11 May 2006, Paper GT2006-91056.
6. F. Reiss, T. Griffin and K. Reyser, 'The ALSTOM GT13E2 medium BTU gas turbine', in Proceedings of the ASME Turbo Expo 2002, Amsterdam, The Netherlands, 3–6 June 2002, Paper GT-2002-30108.

7. G. Cau, D. Cocco and A. Montisci, 'Performance of zero emission integrated gasification hydrogen combustion (ZE-IGHC) power plants with $CO_2$ removal', in Proceedings of the ASME Turbo Expo 2001, New Orleans, Louisiana, 4–7 June 2001, Paper 2001-GT-0366.

8. T. Hasegawa, M. Sato, Y. Katsuki and T. Hisamatsu, 'Study of medium-BTU fuelled gas turbine combustion technology for reducing both fuel-NOx and thermal-NOx in oxygen-blown IGCC', in Proceedings of the ASME Turbo Expo 2002, Amsterdam, The Netherlands, 3–6 June 2002, Paper GT-2002-30666.

9. M. Vascellari, D. Cocco, and G. Cau, 'Comparative analysis of hydrogen combustion power plants integrated with coal gasification and $CO_2$ removal', in Proceedings of the ASME Turbo Expo 2006, Barcelona, Spain, 8–11 May 2006, Paper GT2006-90653.

10. H. M. Kvamsdal, O. Maurstad, K. Jordal and O. Bolland, 'Benchmarking of gas turbine cycles with $CO_2$ capture', in Proceedings of GHGT-7, 7th International Conference on Greenhouse Gas Control Technologies, Vancouver, Canada, 5–9 September 2004.

11. D. J. Dillon, R. S. Panesar, R. A. Wall, R. J. Allam, V. White, J. Gibbins and M. R. Haines, 'Oxy-combustion processes for $CO_2$ capture from advanced supercritical PF and NGCC power plant', in Proceedings of GHGT-7, 7th International Conference on Greenhouse Gas Control Technologies, Vancouver, Canada, 5–9 September 2004.

12. K. Jordal, M. Anheden, J. Yan and L. Strömberg, 'Oxyfuel combustion for coal-fired power generation with $CO_2$ capture – Opportunities and challenges', in Proceedings of GHGT-7, 7th International Conference on Greenhouse Gas Control Technologies, Vancouver, Canada, 5–9 September 2004.

13. R. J. Allam, R. S. Panesar, V. White and D. Dillon, 'Optimising the design of an oxyfuel-fired supercritical PF boiler', in Proceedings of the 30th International Technical Conference on Coal Utilization & Fuel Systems, Clearwater, Florida, USA, 17–21 April 2005.

14. C. Buzzuto and N. Mohn, 'Environmentally advanced clean coal plants', in 19th World Energy Congress, Sydney, Australia, 5–9 September 2004.

15. R. Naqvi, O. Bolland, O. Brandvoll and K. Helle, 'Chemical looping combustion – Analysis of natural gas fired power cycles with inherent $CO_2$ capture', in Proceedings of the ASME Turbo Expo 2004, Vienna, Austria, 14–17 June 2004, Paper GT2004-53359.

16. T. Mattisson, F. Garcia-Labiano, B. Kronberger, A. Lyngfelt, J. Adanez and H. Hofbauer, '$CO_2$ capture from coal using chemical-looping combustion', in GHGT-8, 8th International Conference on Greenhouse Gas Control Technologies, Trondheim, Norway, 19–22 June 2006.

17. S. Sroda, M. Mäkipää, S. Cha and M. Spiegel, *Mater. Corros.*, 57(2) (2006), 176–181.

18. M. Mäkipää and S. Sroda, 'The effect of $CO_2$ on the corrosion rate in simulated combustion atmospheres', in Proceedings of EUROCORR 2004, Long-term prediction and modelling of corrosion, Nice, France, 12–16 September 2004.

19. B. A. Pint and J. M. Rakowski, 'Effect of water vapour on the oxidation resistance of stainless steels', in NACE Corrosion 2000, Orlando, FL, March 2000. NACE Paper 00-259, Houston, TX.

20. R. J. Ehlers, 'Oxidation von ferritischen 9–12% Cr-Stählen in wasserdampfhaltigen Atmosphären bei 550 bis 650 °C', Dissertation, Rheinisch-Westfälische Technische Hochschule, Aachen, 2001.

21. H. Asteman, K. Segerdahl, J. E. Svensson, L.-G. Johansson, M. Halvarsson and J. E. Tang, 'Oxidation of stainless steel in $H_2O/O_2$ environments – Role of chromium evaporation', in Proceedings of 'High Temperature Corrosion and Protection of Materials 6', Les Embiez, France, 2004, *Mater. Sci. Forum*, 461–464 (2004), 775–782.

22. M. Schütze, M. Schorr, D. P. Renusch, A. Donchev and J. P. T. Vossen, *Mater. Res.*, 7(1) (2004), 111–123.

23. G. Lai, 'Nitridation attack in a simulated gas turbine combustion environment', in *Materials for Advanced Power Engineering*, Part II, 1263–1272, ed. D. Coutsouradis *et al.* Kluwer Academic Publishers, 1994.

# 2

# Corrosion behaviour of boiler steels, coatings and welds in flue gas environments*

## Jana Kalivodová and David Baxter

*European Commission, Institute for Energy JRC Petten,*
*1755 ZG Petten, The Netherlands*
*david.baxter@jrc.nl*


## Michael Schütze and Valentin Rohr

*Karl Winnacker-Institut der Dechema eV, Theodor-Heuss-Allee 25*
*Postfach 150104, D-60486 Frankfurt am Main, Germany*

## 2.1    Introduction

Over the last decade, steadily increasing attention has been paid to increasing energy efficiency as a means to reduce the negative impact of human activity on the climate, and in particular the effect that industrial emissions have on changes in the climate, specifically global warming. One of the key international agreements on climate came in the form of the Kyoto protocol that aims to reduce emissions of $CO_2$, and other compounds that similarly contribute to global warming. As far as power generation is concerned, emissions per unit of energy, heat or electricity produced must be reduced as much as possible. For a constant supply of energy, this would also mean that less primary fuel would be required meaning the import of less fuels thus improving security of energy supplies. The main way that energy efficiency can be achieved during combustion processes used for power generation is to increase the temperature and pressure of steam entering the steam turbine. Unfortunately, increasing the temperature of power plant components inevitably results in increased rates of mechanical degradation and corrosion. Corrosion is a constant concern for operators of power plant as it leads to lower plant reliability and reduced performance during operation due to the thermally insulating effects of corrosion product layers on component surfaces. A range of approaches are available to mitigate the higher risk of corrosion damage that will ultimately have an adverse impact on reliability of operation of power plants. With regard to materials, steps can be taken to reduce the rate of corrosion. One option is to use alloys that have higher inherent resistance to corrosive attack than traditional boiler steels. This approach includes use of alloys containing higher concentrations, particularly of Cr, Al or Si. The main drawback to this approach is cost, and mechanical properties may not necessarily be adequate for the conditions of use. Single or multiple element additions to existing

* Reprinted from J. Kalivodová et al.: Corrosion behaviour of boiler steels, coatings and welds in flue gas environments. *Materials and Corrosion*. 2008. Volume 59. pp. 367–73. Copyright Wiley-VCH Verlag GmbH & Co. KGaA.

boiler steels can also result in degradation of mechanical properties. As a consequence, composite tubes, tubes normally produced by extrusion with an inner boiler steel and outer corrosion-resistant annulus, have been used in some applications [1]. Co-extruded tubes are however very expensive. The main alternative is to use some form of surface treatment to improve corrosion resistance without negatively affecting the mechanical properties of the main cross-section of the component being protected. With regard to surface treatments, there are two favoured options, weld cladding [2] and thermal spray or pack diffusion coatings for large boiler components. Currently, the waste incinerator industry uses mainly weld cladding for both improving the performance of standard boiler steels and facilitating boiler repair, although this is expensive (approximately €2000 to €2500 per square metre). Thermal spray coatings are less expensive to apply (around half to two-thirds that of weld cladding), but their long-term performance has still to be proved. This paper considers the corrosion performance of pack diffusion coatings that have potentially lower cost than either of the two alternatives mentioned above. Corrosion test results are reported on a range of experimental coatings. In addition, account is taken of the need to weld coated components, either during installation in a plant or as a result of maintenance and/or repair. The performance of welded coated components must not be significantly poorer than that of unwelded coatings or weldments between uncoated components. The diffusion coatings used in this work were aluminium-rich applied to the conventional boiler steels P91 (9% Cr) and HCM12A (12% Cr) that are used for high efficiency power plants.

## 2.2    Experimental

The coated and uncoated boiler steels P91 and HCM12A used in this study contained 9% and 12% Cr, respectively, as a major alloying element. Table 2.1 shows the full elemental compositions of the two alloys. In order to enhance the corrosion resistance of the alloys to flue gases in high efficiency boilers, meaning higher than normal metal surface temperatures, the compositions of the surfaces of the alloys were modified via an in-situ chemical vapour deposition coating (pack cementation) process. Pack cementation is essentially an in-situ chemical vapour deposition coating process (CVD) [3,4]. The coating process is described below.

## 2.3    Pack cementation of 9–12% Cr boiler steels

The substrates to be coated were placed in a sealed or semi-sealed container together with a powder mixture consisting of metal elements to be deposited, halide activators and an inert filler (usually alumina). One-step aluminising of 9–12% Cr boiler steels was carried out by employing a pack cementation process. The substrate

*Table 2.1*   Composition of the substrate materials (wt-%)

| Alloy | Cr | Mn | Si | Ni | Mo | C | Other |
|---|---|---|---|---|---|---|---|
| P91 | 9.2 | 0.5 | 0.4 | 0.38 | 0.9 | 0.089 | V=0.22, Nb=0.062, P=0.013 |
| HCM12A | 12.5 | 0.54 | 0.25 | 0.34 | 0.36 | 0.071 | V=0.21, Nb=0.045, W=1.9, Cu=0.85, |
| Alloy 625 | 22 | | | Bal. | 9 | | Nb=3.5 |

samples were embedded in a powder mixture composed of the aluminium source (5% aluminium powder), activator (0.5% $NH_4Cl$) and an inert filler (94.5% $Al_2O_3$). The alumina-covered container, containing embedded samples, was heated under a protective atmosphere of Ar $+10\%$ $H_2$ to a temperature of 650 °C in a horizontal furnace and held at the desired temperature for 6 h. After the pack cementation treatment, the pack was cooled under argon to room temperature. After cooling, the samples were removed from the pack and ultrasonically cleaned in ethanol to remove any embedded pack material. Aluminising was also carried out using a two-step pack cementation coating process. The first step was chromium deposition at 1000 °C for 2 h followed by rapid cooling; (pack composition: 30% Cr, 5% $NH_4Cl$, 65% $Al_2O_3$). After simple cleaning of the samples (brushing loose powder remnants off the surfaces), the second aluminium deposition step was carried out at 650 °C for 1 h (pack composition: 20% Al, 1% $NH_4Cl$, 79% $Al_2O_3$).

## 2.4    Welding

In order to simulate necessary welding processes in boilers, some of the coated samples were also welded. Weld joints were made on uncoated and coated 9–12% Cr boiler steel coupons using the gas tungsten arc (GTA) welding process with cold wire feed and argon as a protective gas (see Table 2.2 for welding parameters). The weld joint was prepared by bevelling the joint edges forming a 'V' shaped joint geometry (Fig. 2.1). The welding parameters employed are presented in Table 2.2. In general, the weld was achieved using a single pass. The filler metal used was alloy 625, a nickel–chromium–molybdenum alloy with a significant amount of niobium (Table 2.1). Alloy 625 is very commonly used as a filler metal and a cladding alloy in applications where corrosion resistance is required. Alloy 625 also has a coefficient of thermal expansion similar to that of the substrate boiler alloys. The dimensions of the samples before welding were 20 mm × 10 mm × 3 mm.

*Table 2.2*  Welding parameters

| Welding current (A) | Voltage (V) | Gas flow (L min$^{-1}$) | Filler material |
|---|---|---|---|
| 75 | 10 | 7 | Alloy 625 |

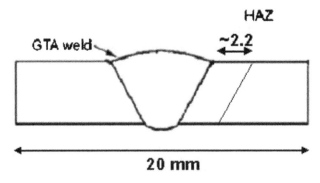

*2.1* Schematic diagram of two samples welded by GTA: Heat affected zone indicated

## 2.5    Corrosion testing and sample analysis

The corrosion behaviour of coated and welded materials was investigated under oxidising atmospheres consisting of 14% $H_2O$, 5% $O_2$, 1.2% $CO_2$, 0.2% HCl and balance $N_2$. Experiments were carried out in autoclaves at a pressure of 1.4 bar (absolute) and temperatures of 600 °C and 650 °C. The duration of exposure was 360 h, achieved in a single cycle. In all tests, uncoated samples of P91 and HCM12A were included for comparison.

The surface morphology and the cross-section microstructures were examined using an optical microscope and a scanning electron microscope (SEM) equipped with energy dispersive X-ray (EDX) analysis capability. The structure and chemical composition of the aluminised layer, welds and oxides formed were also analysed with energy dispersive spectroscopy (EDS). Hardness profiles were obtained across welded samples using a Vickers hardness tester with a load of 1000 g and loading time of 10 s.

## 2.6    Results and discussion

### 2.6.1    Weld microstructures and properties

The integrity of welded samples was investigated by means of hardness testing, microstructural examination and, finally, corrosion testing. Examples of the microstructure of the weld and P91 without and with aluminium coating are given in Fig. 2.2a and 2.2b, respectively. The micrograph of the interface between weld metal and coated P91 (Fig. 2.2b) shows a good smooth transition from coated P91 to weld metal. There is also clear evidence of AlN particles in the weld metal in the region close to the aluminide coating (Fig. 2.2b). These precipitates were mainly formed during the coating process and they have subsequently been transferred to the weld pool during the welding. The use of argon as a protective gas during welding should in principle prevent additional nitride formation, but this possibility cannot be excluded owing to the very high stability of AlN. The AlN precipitates were not dissolved and reformed during welding and cooling, due to their high melting point. No significant difference in microstructure was found between welded steels P91 and HCM12A. According to the literature [5] and the Shaeffler Diagram for alloy 625, the

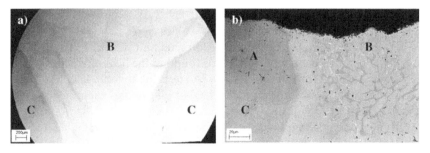

*2.2* (a) Cross-section micrograph of welded steel P91; (b) Cross-section micrograph of welded aluminised P91 (A, aluminium coating; B, weld material; C, steel P91)

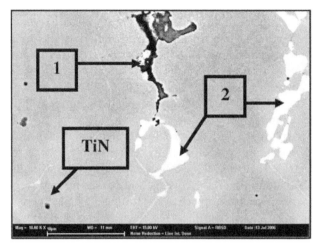

| Weight (%) | O | Al | Cr | Fe | Ni | Nb | Mo |
|---|---|---|---|---|---|---|---|
| 1 | 29 | - | 19 | - | 2 | 46 | - |
| 2 | - | - | 12 | 10 | 33 | 20 | 19 |

*2.3* Cross-section through an alloy 625 weld showing interdendritic phases (composition given in the table), porosity and occasional TiN particles

filler metal solidifies to austenite. The welds were found to show a distinct dendritic microstructure.

The interdendritic phase was found to be enriched in molybdenum and niobium that had segregated to the interdendrite regions during solidification of the weld. An SEM-EDX analysis of the interdendritic phases showed the concentration of molybdenum and niobium to be around 20% (Fig. 2.3). Molybdenum segregates as it is rejected from the growth of the dendrites to the liquid due to the low solubility of Mo in the austenitic phase during solidification of the weld. The literature [2] reports that weld metal zones, with dendritic arms depleted in Mo, can lead to localised poor corrosion resistance in the weld metal and this can lead to failure of weld overlay coatings in power plants. It is also evident in Fig. 2.3 that porosity is present in the interdendritic zones. Pores may contribute to crack formation and subsequent enhanced corrosion penetration into the metal.

Based on the measurement of hardness values across the weld joint, similar hardness values were observed for both weld metal and welded P91 and HCM12A steels. A significant increase in hardness was found in the heat affected zones (HAZ) (Fig. 2.4) that is typical for P91 and HCM12A steels [6]; this can lead to degradation of creep properties. The thickness of the HAZ of all samples examined in this work was about 2.2 mm.

The chemical composition linescan analysis of welded steel HCM12A in Fig. 2.5 shows a smooth interface between weld and parent material; there was little diffusion of Cr from the weld metal to the HCM12A, despite the difference in Cr concentration of the starting alloys. The micrograph in Fig. 2.6 of welded aluminised HCM12A shows good integrity of the interface between weld metal and both boiler steel

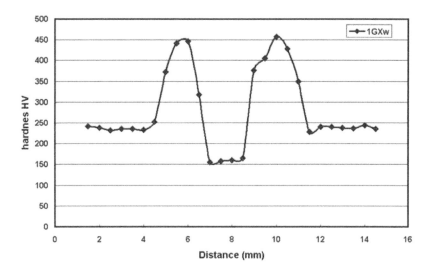

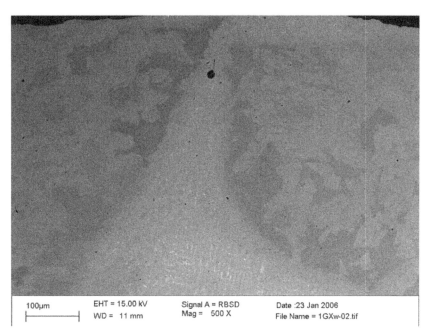

2.4 Cross-section micrograph showing weld close to the root and associated hardness profile taken horizontally across the middle of the field of view

substrate and coating, and no significant chromium diffusion from the weld into the HAZ was evident in the X-ray map. While hardening of the HAZ had occurred during welding, no cracks or other defects were observed. In welds where the welding heat input was maintained as low as possible, while still achieving good wetting and thus an acceptable weld, there was no significant nickel diffusion into the coating and very little diffusion of aluminium into the weld (Fig. 2.6).

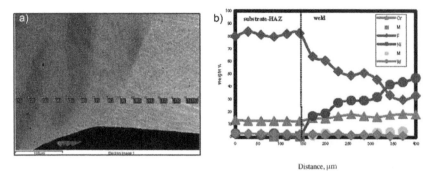

2.5  (a) Cross-section micrograph of welded steel HCM12A; (b) Corresponding chemical compositions (B, weld material; C, steel HCM12A)

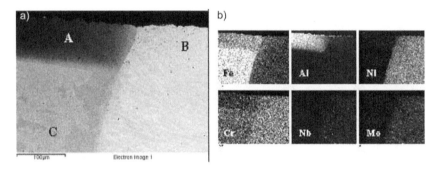

2.6  (a) Cross-section micrograph of welded aluminised steel HCM12A; (b) EDS X-ray map (A, aluminium coating; B, weld material; C, steel HCM12A)

### 2.6.2   Corrosion of welded coatings

The high-temperature corrosion test conditions were designed to simulate the aggressive environments commonly found in waste and biomass boilers. Exposures were carried out under oxidising atmospheres, with additions of HCl, at 600 °C and 650 °C. The corrosion behaviour of coatings on high-alloyed materials and 9–12% Cr steels was extensively studied in previous work [7]. Pack diffusion coatings containing Al+Cr exhibited relatively poor performance in simulated flue gas environments containing HCl. Tests on the unwelded coatings showed the aluminium content in a 30-µm-thick coating needs to be more than 50 wt-% in order to provide protective corrosion behaviour in oxidising atmospheres containing HCl at 600 °C for at least 1000 h. It was found that stable homogenous coatings, without defects, could substantially improve the corrosion protection afforded to common boiler materials. These results are compared with the same coatings, with welds, on P91 and HCM12A. Two types of coated (aluminised 9–12% Cr steel and Al + Cr coating) and subsequently welded material were studied under simulated combustion gas environments containing HCl and water vapour.

*2.7* Macrograph of welded Al coated samples after exposure in simulated gas (2000 ppm HCl, 360 h, 600 °C) (A, aluminised P91; B, weld metal; C, aluminised HCM12A)

### Aluminised 9–12%Cr steel

Significant formation of oxides on the coatings and welds was observed after 360 h of exposure; there was particularly aggressive corrosion in the regions of the HAZ where thick, apparently non-protective oxides had formed (Fig. 2.7). A corrosion-tested sample is shown in cross-section in Fig. 2.8. In this case, two different aluminised steels, P91 and HCM12A, were joined by welding. The micrograph (Fig. 2.8a) shows coating, weld metal and multilayer oxide formation. The oxidation clearly extends from the surface of the sample into the zone of metal between weld and coating,

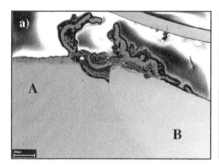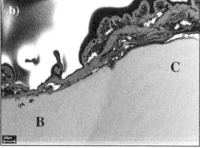

*2.8* (a) Cross-section micrograph of welded aluminised steel HCM12A-phase A; (b) Cross-section micrograph of welded aluminised steel P91-phase C (B-weld material) after exposure in simulated gas (2000 ppm HCl, 360 h, 600 °C)

possibly due to microstructural and compositional changes during welding. According to Singh Raman and Muddle [8], chromium forms secondary precipitates in the HAZ, mainly carbides, and consequently is less available for protective oxide formation. X-ray mapping of the non-protective oxide scale over the HAZ confirmed the presence of external chromium-rich oxide scale. Also, iron and aluminium oxides, with more iron in the oxide on the P91 compared to HCM12A, were identified. Enhancement of the concentration of tungsten was detected in the oxide layer on HCM12A. Tungsten is added to 9–12% Cr steels to enhance strength by stabilising the martensitic structure [9].

It was found that the heat input from welding has a significant influence on the corrosion performance of the coating and of the HAZ. The microstructures of the coatings immediately adjacent to the weld were investigated and compared with microstructures further away from the welded area and thus not affected by heating during the welding process. Coatings immediately adjacent to the weld exhibited a different distribution of Al and also significantly lower aluminium concentration. Also, more cracks were observed in coatings in the HAZ; this effect coincides with diffusion of aluminium from the coating to the weld during the welding process.

### Cr + Al coating onto 9–12% Cr steel

The corrosion behaviour of welded chromium–aluminium coatings on P91 and HCM12A was studied in a model waste combustion flue gas. Two-step pack cementation coatings are relatively thin, discontinuous coatings with a chromium-rich ($AlCr_4$) layer between the boiler steel substrate and an Al-rich aluminide outer layer. The X-ray mapping results in Fig. 2.9 show an oxidised surface of alloy 625 weld and Al + Cr coating on P91. Reaction of the Al-rich aluminide outer layer with gas resulted in the formation of a non-protective aluminium oxide scale that was mostly spalled off during cooling and collected in the sample crucibles. Internal oxidation of the weld was observed in the region adjacent to the boiler steel substrate. The amount of chromium in this region was lower than in the weld. Detailed analysis confirmed selective corrosion of chromium at the grain boundaries that is typical for welded or aged materials. No significant difference in corrosion behaviour was found between welded and Al + Cr coated steels P91 and HCM12A.

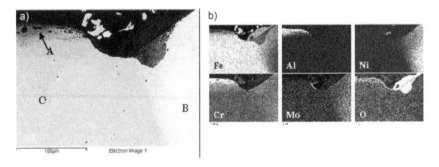

*2.9* (a) Cross-section micrograph of welded Al + Cr coating on P91; (b) EDS X-ray map (A, coating; B, weld metal; C, steel HCM12A) after exposure in simulated gas (2000 ppm HCl, 360 h, 600 °C)

### 2.6.3    Corrosion of alloy 625 welds

This alloy solidifies entirely to austenite with some possible lower melting point eutectic and/or carbide formation at the dendrite interstices. Evidence of localised corrosion and segregation of Mo and Nb was found in the weld metal of the samples exposed to simulated waste combustion flue gas. The segregation behaviour and three-step solidification of alloy 625 during welding was described by Dupont *et al.* [10]. A crack along the weld with length 2.1 mm was found in one of the welded samples. The crack followed grain boundaries enriched in a phase containing mainly niobium (NbC). Figure 2.10b shows the end of the deepest crack in the weld metal. The lighter shaded areas in the weld metal are enriched in molybdenum and niobium, and depleted in chromium. The surfaces of weldments were also oxidised during exposure in simulated flue gas, but the thickness of the oxide was generally less than on the aluminide coatings. The thickness of the oxide, containing mainly iron and chromium was 10 μm after 360 h. Localised corrosion and microcracks were observed in regions of depleted chromium content (Fig. 2.10a).

This investigation has shown that filler metal alloy 625 can be used to join aluminised 9–12% Cr steel, but that special precautions, particularly minimising heat input during the welding process and so minimising segregation of elements normally contributing to good corrosion resistance, may be necessary to avoid subsequent incidences of rapid corrosion in flue gas environments.

### 2.7    Conclusions

* Coatings enriched with Al provide improved corrosion resistance to P91 (9% Cr) and HCM12A (12% Cr) boiler steels in waste combustion flue gases containing HCl.
* Aluminised P91 and HCM12 can readily be welded using standard industrial welding techniques with alloy 625 as the filler metal. Welds and associated heat affected zones (HAZ) are crack-free if welded correctly, but diffusion of Al from the coating into the weld metal cannot be avoided. The loss of Al from the coating HAZ leads to reduced corrosion resistance compared to unwelded coatings. The amount of Al depletion can be reduced by maintaining the heat input during

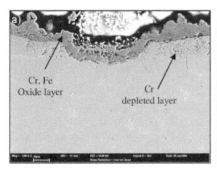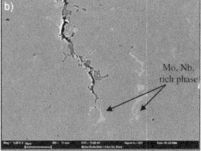

*2.10* (a) Cross-section micrograph of oxidised surface of weld metal alloy 625 after exposure in simulated gas (2000 ppm HCl, 360 h, 600 °C); b) detail of the crack in the weld metal alloy 625 after exposure

welding to the lowest possible level while still permitting sufficient wetting for a good weld to be produced.

- Welding leads to the formation of HAZ in the substrate boiler steel immediately adjacent to the weld metal. The HAZ has increased hardness, due mainly to carbide formation that is linked to depletion of elements providing corrosion resistance, particularly Cr, and also to degradation of creep properties in other work. Creep properties were not addressed in this study.
- Welding with alloy 625 was found to result in segregation in the weld metal, particularly of the elements Mo and Nb, which is associated with enhanced corrosion. The incidence of segregation and the severity of its effects can be reduced by minimising heat input during welding.

## Acknowledgements

The authors wish to acknowledge the financial contribution of the European Commission to the SUNASPO project (HPRN-CT-2001-00201) that supported the work reported in this paper.

## References

1. A. Wilson, U. Forsberg and J. Noble, in Proceedings of Corrosion 97, Volume 153, 1–15. NACE, Houston, TX, USA, 1997.
2. K. Luer, J. DuPont, A. Marder and C. Skelonis, *Mater. High Temp.*, 18(1) (2001).
3. C. Houngninou, S. Chevalier and J. P. Larpin, *Appl. Surf. Sci.*, (2004), 1–14.
4. Z. D. Xiang and P. K. Datta, *Mater. Sci. Eng., A356* (2003), 136–144.
5. E. J. Barnhouse and J. C. Lippold, *Welding J.*, 77(12) (1998), 477–487.
6. G. Scheffknecht and Q. Chen, Proc. 5th Int. Ch. Parsons Turbine Conference, 2000, 249.
7. J. Kalivodová, D. Baxter, M. Schütze and V. Rohr, *Mater. Corros.*, 56(12) (2005), 882–889.
8. R. K. Singh Raman and B. C. Muddle, *Int. J. Pressure Vessels Piping*, 79 (2002), 585–590.
9. J. Onoro, *Int. J. Pressure Vessels Piping*, 83 (2006), 540–545.
10. J. N. Dupont, S. W. Banovic and A. R. Marder, *Welding J.*, (2003), 125–135.

# 3

## Corrosion of alloys and their diffusion aluminide coatings by KCl:K$_2$SO$_4$ deposits at 650 °C in air*

### Vratko Vokál and Michael J. Pomeroy

*Materials and Surface Science Institute, University of Limerick,*
*Limerick, Ireland*

### Valentin Rohr and Michael Schütze

*Karl-Winnacker-Institut, DECHEMA, Frankfurt, Germany*

### 3.1    Introduction

The combustion of biomass in conjunction with coal gives rise to the formation of deposits of potassium chloride and potassium sulphate on the heat exchanger surfaces of coal–straw fired boilers. Other fuel impurities such as NaCl, Na$_2$SO$_4$ and CaSO$_4$ which cause significant hot corrosion [1,2] can also be deposited at the same time. The combustion of biomass, a so-called carbon neutral fuel because it consumes CO$_2$ whilst growing, is beneficial for environmental reasons. In addition, lower carbon dioxide emissions can be achieved by raising steam temperatures from the current 580 °C to 600 or 650 °C. For these higher temperatures, more creep-resistant steels and/or nickel-based alloys need to be used [3]. These higher temperatures can also result in significantly greater corrosion damage. Thus, for example, Mohanty and Shores [4] have shown that alkali sulphate–alkali chloride rich deposits cause catastrophic corrosion of a 25Cr–13Ni steel at temperatures in the range from 600 °C to 900 °C. Kalivodova *et al.* [5] have shown that aluminide coatings can be beneficial in substantially reducing corrosion of alloys by gaseous environments rich in HCl.

This paper addresses the corrosion of alloys and coatings typical of those which could be used for superheater tubes in coal–straw fired boilers operating with metal temperatures of 650 °C.

### 3.2    Experimental

The compositions of the alloys tested are given in Table 3.1. The faces of 15 mm × 10 mm × 3 mm samples of each of the alloys were surface ground to a 1200 grit finish. Sample thicknesses were then measured and then the specimens were cleaned using isopropanol. Alloy samples were then either corrosion tested, or aluminised then corrosion tested. Table 3.2 gives the aluminising conditions for each of the alloys and

* Reprinted from V. Vokál et al.: Corrosion of alloys and their diffusion aluminide coatings by KCl:K2SO4 deposits at 650 °C in air. *Materials and Corrosion.* 2008. Volume 59. pp. 374–9. Copyright Wiley-VCH Verlag GmbH & Co. KGaA.

*Table 3.1*    Composition of alloys tested (wt.%)

| Alloy | Fe | Co | Ni | Mo | Mn | Cr | Al | Si |
|---|---|---|---|---|---|---|---|---|
| P91 | Bal. | | 0.36 | 0.93 | 0.55 | 9.0 | 0.006 | 0.4 |
| 17Cr–13Ni | Bal. | | 12.36 | 2.21 | 1.32 | 16.5 | | 0.38 |
| Alloy 800 | Bal. | | 32.72 | | 0.95 | 21.1 | 0.42 | 0.29 |
| Inconel 617 | 1.14 | 11.35 | Bal. | 8.70 | 0.07 | 21.9 | 1.17 | 0.08 |

*Table 3.2*    Coating conditions for each alloy and phase assemblage of coating as determined by X-ray diffraction

| Alloy | Coating conditions | Phase assemblage in order of decreasing X-ray intensity |
|---|---|---|
| P91 | 650 °C for 6 h, powder mixture of 10% Al, 1% $NH_4Cl$, 89% $Al_2O_3$, in Ar + 10% $H_2$ | $Fe_2Al_5$ |
| 17Cr–13Ni alloy | 950 °C for 8 h, 5% Al, 0.5% $NH_4Cl$, 94.5% $Al_2O_3$, in Ar + 10% $H_2$ | (Ni, Fe)Al, $Fe_2Al_5$ |
| Alloy 800 | 950 °C for 8 h, 5% Al, 0.5% $NH_4Cl$, 94.5% $Al_2O_3$, in Ar + 10% $H_2$ | (Ni, Fe)Al, $Fe_2Al_5$ |
| Inconel 617 | 950 °C for 8 h, 5% Al, 0.5% $NH_4Cl$, 94.5% $Al_2O_3$, in Ar + 10% $H_2$ | $Ni_2Al_3$ |

All coatings cooled over 8 h in Ar + 10% $H_2$.

the phase assemblage of each coating. Aluminised and non-aluminised samples were then placed on a 10 mm deep bed of a mixture of 50 mol.% KCl + 50 mol.% $K_2SO_4$ within an alumina crucible and then covered with a further 10 mm thickness of the mixture.

The crucibles were then placed in a muffle furnace operating at 650 °C for 300 h. Following exposure, corroded samples were removed from the mixed salt powder and their surfaces lightly brushed to remove loosely adherent salt and corrosion product. Sample surfaces were then subjected to X-ray diffraction (XRD) using Cu Kα radiation. Specimens were then laid face down on two copper rods which supported the coupons above the base of a 32 mm inside diameter mounting cup. Epoxy resin was then poured into the cup such that the specimen was fully covered and allowed to cure. The mounted specimen was then sectioned through its midpoint, giving a sample cross-section of approximately 15 mm × 3 mm. This section was then mounted in epoxy resin in a standard 32 mm metallurgical mount. The cross-sectioned samples were then polished to a 1 μm finish. All cutting, grinding and polishing operations were conducted using non-aqueous lubricants. The thicknesses of unaffected alloy were then measured using a travelling optical microscope. Specimens were then carbon-coated and examined using scanning electron microscopy (SEM), and energy dispersive X-ray analysis (EDX) was used to map the spatial distribution of elements of interest. As will be shown later, high levels of porosity were evident after corrosion of the 17Cr–13Ni alloy and Alloy 800 materials. In order to determine whether this was a real effect or pull-out due to polishing, focused ion beam milling (FIBM) was used to mill beneath the polished surface of a corroded

Alloy 800 specimen to a depth of some 5 to 8 µm. The vertical milled face was then tilted at an angle of 45° and then imaged using secondary electrons. Samples for scanning transmission electron microscope (STEM) examination were prepared using focused ion beam milling. These were also subjected to EDX analysis.

## 3.3    Results

### 3.3.1    Analysis of coatings

Table 3.2 shows the phases present in the top 15 to 20 µm of the coating. It is seen that the P91 alloy formed an aluminium-rich $Fe_2Al_5$ coating corresponding to an Al content of about 71 at.%, as measured by EDX. The coatings formed on the 17Cr–13Ni alloy were predominantly (Fe, Ni)Al. However, they also contained small amounts of $Fe_2Al_5$, which is consistent with an aluminium content, as measured by EDX, of the order of 55 at.%. The coating formed on the Inconel 617 alloy comprised the aluminium-rich $\delta$-$Ni_2Al_3$ phase.

### 3.3.2    Corrosion of alloys

Figure 3.1 shows data for the loss of sound metal by the alloys due to corrosion. It is seen that the P91 alloy suffered the greatest metal loss (385 µm). Alloy 800 suffered the next greatest loss (255 µm) followed by the 17Cr–13Ni alloy (190 µm). The

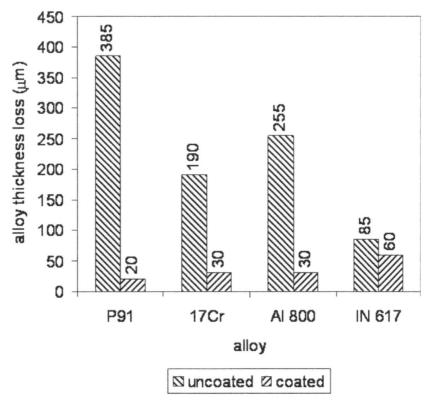

3.1 Sound metal loss data after corrosion at 650 °C for 300 h

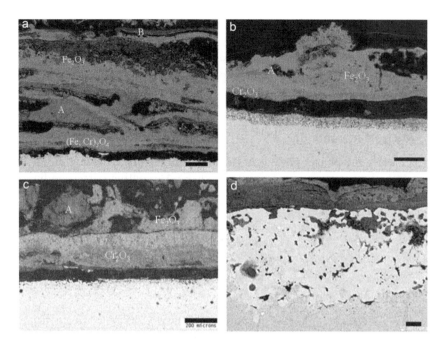

3.2 Corrosion morphology for alloys corroded under KCl + K₂SO₄ for 300 h at 650 °C; (a) P91 alloy (bar = 100 μm); (b) 17Cr–13Ni alloy (bar = 100 μm); (c) Alloy 800 (bar = 200 μm); (d) Inconel 617 (bar = 10 μm)

Inconel 617 alloy underwent the least corrosion losing some 85 μm of sound alloy. In contrast to the other alloys, the metal loss for the P91 alloy was due to surface corrosion only. The other alloys all suffered internal corrosion effects.

Figure 3.2 shows the corrosion morphologies of the alloys corroded at 650 °C for 300 h. For the P91 alloy (Fig. 3.2a), EDX showed that there was a thin sub-scale layer of chromium sulphide beneath a mixed iron–chromium oxide layer (typically 50 μm thick). The remainder of the scale comprised Fe₂O₃. Within this scale, particles rich in both K and Cl as well as Fe were observed (A in Fig. 3.2a) and a thin (10 μm) KCl covering was present on the top of the Fe₂O₃ (marked B in Fig. 3.2a). The thick external scale was confirmed as Fe₂O₃ by XRD of the surface layers before mounting and sectioning. XRD analysis of the surface of the corroded 17Cr–13Ni alloy showed the presence of Fe₂O₃ and KCl. This is consistent with the outer layers of the corrosion scale (Fig. 3.2b). Figure 3.2b also shows that the corrosion scale comprises two layers, the outer lighter Fe₂O₃ layer and an inner darker Cr₂O₃ layer (as determined using EDX mapping). What is not apparent in Fig. 3.2b is a thin CrS layer between the scale and the alloy. This sulphide forms above internal corrosion which extends some 30 μm into the alloy. This internal corrosion is typified by large voids which appear to result from the formation of volatile chlorides. A similar corrosion morphology is observed for Alloy 800 (Fig. 3.2c). In this case however, the internal corrosion typified by voiding (not readily visible in Fig. 3.2c) extends deeper into the alloy (~120 μm).

The corrosion morphology for the Inconel 617 alloy is shown in Fig. 3.2d. The bright contrast on the outside of the scale was shown by EDX to be cobalt oxide which forms above a thin (10 μm) chromia scale. Beneath the scale, significant Cr depletion effects were determined using EDX and these give rise to the very bright contrast. Within this Cr-depleted region, internal chlorides had formed (grey areas). There was appreciable sulphur in this region too, suggesting internal sulphide formation. XRD analysis of the surface before mounting the sample showed the presence of $Cr_2O_3$ and $CoO$, thus confirming the analysis inferred from EDX.

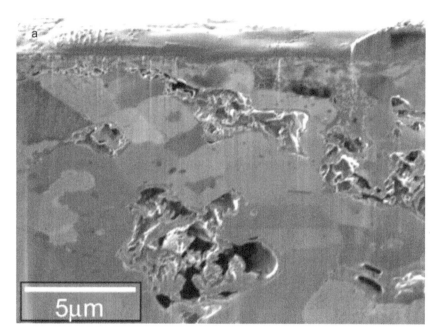

3.3 Secondary electron images of FIBM section beneath polished surface of Alloy 800 specimen corroded by $KCl + K_2SO_4$ for 300 h at 650 °C; (a) full section; (b) greater detail of a single corrosion induced void

As stated in the experimental section, FIBM was used to confirm whether the internal voids formed in the corroded Alloy 800 material were real effects or merely pull out during polishing. Figure 3.3 shows two electron images taken from sub-polished surface regions of the corroded Alloy 800 material and clearly shows that the internal voids are present beneath this surface, thus confirming that they are corrosion effects.

### 3.3.3   Corrosion of coated alloys

Figure 3.1 includes alloy thickness losses for the coated materials. All of these losses were due to the effects of alloy–coating interdiffusion leading to movement of the coating–alloy interface towards the centre of the specimen coupons. Clearly, with the exception of the Inconel 617 alloy, alloy thickness losses were significantly less than when the alloys were not protected by a coating.

Whilst alloy thickness losses after 300 h at 650 °C were slight, all coatings except the $Fe_2Al_5$ coating on the P91 alloy were significantly degraded, as Fig. 3.4 shows. For the P91 alloy, a thin (1 to 2 μm) alumina scale (identified as α-alumina by XRD) had formed on the surface of the coating. Above this oxide, a thin layer of KCl was observed, as indicated in Fig. 3.5a. This figure further shows that a thin sub-alumina scale containing chromium had formed. Beneath these oxides, bright regions richer in iron than the remainder of the coating can be seen by comparison of Fig. 3.4a and 3.5a. These were identified as FeAl by quantitative EDX and also by XRD analysis of the corroded surfaces. The remainder of the coating comprised $Fe_2Al_5$ which, as

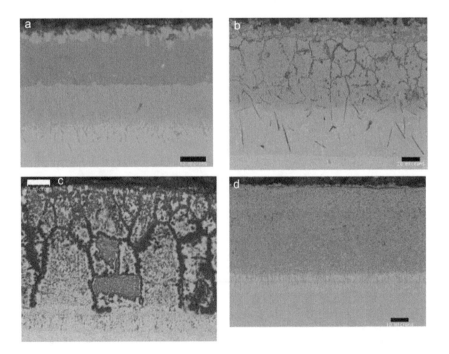

*3.4* Corrosion morphologies for aluminised alloys corroded under $KCl + K_2SO_4$ for 300 h at 650 °C; (a) P91 alloy (bar = 10 μm); (b) 17Cr–13Ni alloy (bar = 20 μm); (c) Alloy 800 (bar = 20 μm); (d) Inconel 617 (bar = 10 μm)

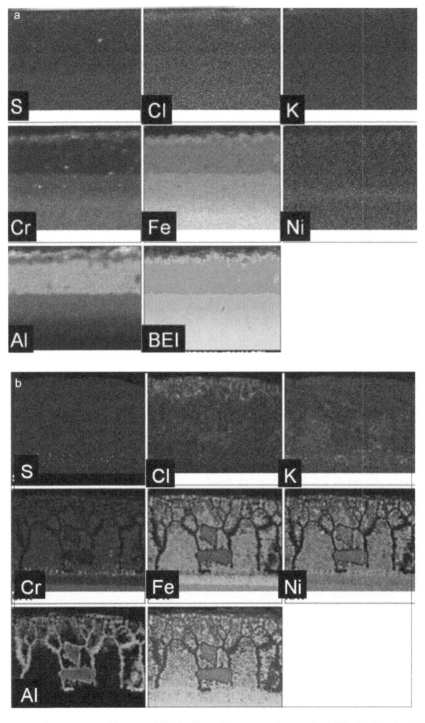

*3.5* X-ray maps; (a) coated P91 alloy after corrosion under KCl + K₂SO₄ for 300 h
at 650 °C; (b) Alloy 800 after corrosion under KCl + K₂SO₄ for 300 h at 650 °C

seen in Fig. 3.5a, contained small isolated regions of FeAl and regions of chromium enrichment.

In contrast to the $Fe_2Al_5$ coating on the P91 alloys, the (Fe,Ni)Al coating on the 17Cr–13Ni alloy underwent significant surface and intergranular corrosion, as shown in Fig. 3.4b. This was typified by a thicker scale (5 μm surface α-alumina scale) above which a thin layer of chromium sulphide had formed beneath the $KCl + K_2SO_4$ deposit. Figure 3.4b shows that beneath the alumina surface scale, grain boundaries were decorated with corrosion product. X-ray mapping using EDX showed that the corrosion product was alumina. In addition, in the 30 μm thickness of coating adjacent to the alloy, intergranular sulphides were observed. EDX analysis showed that significant aluminium depletion in the coating had occurred due to both corrosion and alloy–coating interdiffusion effects. The (Fe,Ni)Al coating on the Alloy 800 material underwent similar corrosion effects to that on the 17Cr–13Ni alloy, except that the corrosion effects were more significant (compare Fig. 3.4b and 3.4c). Thus, intergranular corrosion products were thicker and intragranular corrosion effects were substantially greater. This latter point is graphically illustrated in Fig. 3.5b where the X-ray maps show thick regions of aluminium and almost whole grains which are Al-rich implying the formation of much alumina. In contrast to the 17Cr–13Ni coating, chlorine is present within the coating, albeit more predominantly in the outer 20 μm (Fig. 3.5b). It also appears from Fig. 3.5b that potassium is present throughout the coating. In common with the situation for the 17Cr–13Ni alloy, intergranular sulphides formed at the base of the coating on the Alloy 800 material.

Figure 3.4d shows a backscattered electron image of the corrosion morphology typically shown by the coated Inconel 617 alloy. Beneath thin alumina and discontinuous chromia scales, aluminium depletion was observed to a depth of 10 μm. Above the scale, KCl was observed. Both chlorine and sulphur were observed within the coating, indicating internal chloridation/sulphidation effects. In contrast to the two austenitic steels, there was no intergranular alumina formation within the coating, indicating that less significant degradation of the $Ni_2Al_3$ coating had occurred after 300 h beneath $KCl + K_2SO_4$ at 650 °C. The $δ$-$Ni_2Al_3$ coating had however been converted to $β$-NiAl during the exposure. This probably explains the more significant alloy sound metal loss observed for the Inconel 617 alloy, with coating–alloy interdiffusion causing the $δ$ to $β$ transformation.

## 3.4    Discussion

### 3.4.1    Corrosion of alloys

The catastrophic corrosion of the iron-based alloys, in particular the P91 alloy and Alloy 800 materials, is typical of the chloride-induced active oxidation process described by Grabke et al. [6] and Zhas et al. [7]. In addition, the formation of internal voids during the corrosion of austenitic steels by chloride–sulphate mixtures has previously been observed by Mohanty and Shores [4]. For these mechanisms to operate, then either high partial pressures of HCl [6,7] must be present or molten salts must be present [4] on the surface of the alloy. In the work reported here however, the $KCl$–$K_2SO_4$ salt deposit did not melt during exposure in the muffle furnace, where of course no HCl would have been present. The only way in which molten Cl-rich phases could have formed at 650 °C was by reaction between KCl and the alloys in the same way as observed by Cha and Spiegel [8]. Thus, Fe and Cr would be expected to form molten phases, in air, at 300 and 500 °C, respectively. Once these liquids were formed, then they would be likely to dissolve potassium sulphate yielding K–Fe–Cr–S–Cl–O liquids facilitating active oxidation, internal chloridation leading to void formation and the formation of sulphides at the oxide–alloy interface. What is not known, is how extensive liquid formation is. It would seem that, whilst liquid formation is not excessive, it is sufficient to initiate and support the three degradation modes referred to above.

The lesser corrosion observed for the Inconel 617 alloy is consistent to some extent with the observation of Li *et al.* [9] for the corrosion of pure Ni by molten KCl–ZnCl$_2$. This consistency is, however, limited to the formation of residual NiCl$_2$ within the alloy. It was reported above that these chlorides are observed in the Cr depleted region beneath the scale. It thus appears that NiCl$_2$ will only form within the alloy once Cr levels decrease to a level where active Cr oxidation via CrCl$_2$ or CrCl$_3$ formation ceases. Liquid phase formation via a similar mechanism to that for the iron-based alloys would need to occur to explain the formation of internal chlorides and sulphides.

It is interesting to note that the greater nickel content of Alloy 800 does not make it more corrosion-resistant than the 17Cr–13Ni material. This may be because the presence of Mo (Table 3.1) confers better protection to the 17Cr–13Ni material, as Mo is known to be useful in protecting alloys from corrosion by chlorides and HCl-containing gases [10].

### 3.4.2   Protectiveness of coatings

It is clear from the data presented in Fig. 3.1 that sound alloy thickness losses for the coated materials are nearly an order of magnitude less than those for uncoated materials, except for the Inconel 617 alloy. It was also stated that these alloy thickness losses were due to alloy–coating interdiffusion. With respect to the alloys therefore, the coatings are highly protective under the corrosion conditions tested. However, as observed by Kalivodova *et al.* [5] alloy–coating interdiffusion effects are sufficiently great to reduce the effectiveness of the coatings after 1000 h at 600 °C in a gas representative of the flue gas of a coal-fired boiler.

### 3.4.3   Corrosion of coatings

In the work reported here, it is clear that only Fe$_2$Al$_5$ coating suffers the least degradation due to corrosion and interdiffusion. In contrast, the coatings on the austenitic stainless steels suffer extensive almost through coating thickness corrosion whilst the δ-aluminide coating on the Inconel 617 alloy induces significant interdiffusion as it transforms to β-NiAl. In order to try to identify the comparative reason for the more extensive intergranular corrosion of the coatings on the austenitic steels compared to the coating on the P91 alloy, FIBM was used to prepare TEM samples from areas near the surface of the coatings following their deposition. STEM examination of foils of the coatings on Alloy 800 and P91 alloys, in combination with EDX, indicated that, in the former, coating intergranular α-Cr regions were present (Fig. 3.6a). In contrast, Cr present in the coating formed on the P91 alloy appeared to be present as discrete particles of α-Cr or as chromium carbides. These different distributions of Cr would account for the differing corrosion morphologies observed because the presence of Cr on grain boundaries would result in grain boundaries

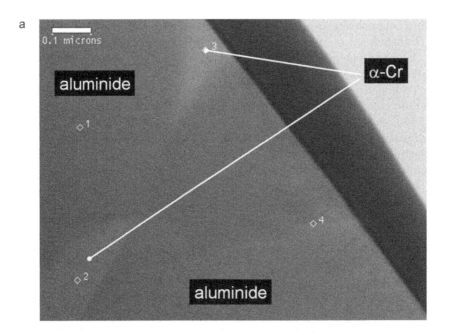

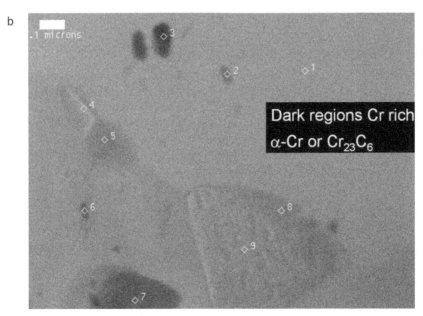

3.6 STEM images of outer regions of aluminide coatings; (a) Alloy 800 (bar = 0.1 µm); (b) P91 alloy (bar = 0.1 µm)

undergoing chloridation to form volatile $CrCl_x$ compounds. This would then allow alumina to form along grain boundaries as observed both for the coated Alloy 800 and 17Cr–13Ni materials. Whilst further work is required to fully substantiate these findings, it strongly appears that chromium enrichment of grain boundaries due to the coating process facilitates rapid coating degradation. At this time, it may be concluded that good corrosion protection from KCl + K$_2$SO$_4$ deposits at 650 °C can be afforded by coatings which are rich in aluminium and do not have grain boundary chromium enrichment.

## 3.5   Conclusions

- The iron-based alloys P91, 17Cr–13Ni and Alloy 800 undergo catastrophic corrosion via an active oxidation process. Whilst the corrosion is planar for the P91 alloy, it involves internal chloridation for the other two alloys leading to the formation of voids.
- The nickel-based alloy, Inconel 617, suffers much less corrosion that the iron-based alloys because nickel chlorides can form thus inhibiting the active oxidation process.
- The formation of chromium sulphide along the interface between the iron-based alloys and the oxides formed on them arises because of the fact that sulphides are less stable than chlorides and so any internal sulphides formed will be chloridised.
- For all coated materials tested for 300 h at 650 °C, metal losses are restricted to alloy–coating interdiffusion effects.
- The corrosion resistance of the coatings themselves depends on aluminium content and the distribution of chromium within the coatings. Coatings with grain boundary chromium enrichment undergo catastrophic intergranular corrosion.

## Acknowledgements

This work was funded under the Framework 5 Programme: Improving Human Potential – Training Networks via Contract HPRN-CT-2001-00201, 'European Network Surface Engineering of New Alloys for Super High Efficiency Power Generation' [SUNASPO]. The authors are indebted to the European Union for this funding.

## References

 1. K. Wieck-Hansen, P. Overgaard and O. H. Larsen, *Biomass Bioenerg.*, 19 (2000), 395.
 2. M. Montgomery and O. H. Larsen, *Mater. Corros.*, 53 (2002), 1185.
 3. R. Viswanathan and W. T. Bakker, *J. Mater. Eng. Perform.*, 10 (2001), 81.
 4. B. P. Mohanty and D. A. Shores, *Corros. Sci.*, 46 (2004), 2893.
 5. J. Kalivodova, D. Baxter, M. Schütze and V. Rohr, *Mater. Corros.*, 56 (2005), 882.
 6. H. J. Grabke, E. Reese and M. Spiegel, *Corros. Sci.*, 37 (1995), 1023.
 7. A. Zahs, M. Spiegel and H. J. Grabke, *Corros. Sci.*, 42 (2000), 1093.
 8. S. C. Cha and M. Spiegel, *Corros. Eng. Sci. Technol.*, 40 (2005), 249.
 9. Y. S. Li, Y. Niu and W. T. Wu, *Mater. Sci. Eng. A*, 345 (2003), 64.
10. Y. Kawahara, *Corros. Sci.*, 44 (2002), 223.

# 4

## Oxide scale formation on Al containing Ni–Cr-based high-temperature alloys during application as flame tube material in recirculation oil burners*

### H. Ackermann, G. Teneva-Kosseva, H. Köhne and K. Lucka

*Oel-Wärme-Institut gGmbH, Technologiepark Herzogenrath Kaiserstr. 100,*
*52134 Herzogenrath, Germany*
*h.ackermann@owi-aachen.de*

### S. Richter and J. Mayer

*Central Facility for Electron Microscopy GFE, RWTH Aachen,*
*Ahornstr. 55, 52074 Aachen, Germany*

### 4.1 Introduction

Flame stabilisation in modern recirculation oil burners for domestic heating is achieved in the flame tube of the burner. During operation, it is exposed to high temperatures (up to 920 °C, [1]) and rapid changes of temperature and gaseous atmosphere during the beginning and ending of burner operation. Furthermore, the inner side of the tube is in close contact with the combustion flame, while the outer surface is exposed to the exhaust gas, which is recirculated back to the burner. The material of the flame tube should possess very good creep strength and high-temperature corrosion resistance in order to provide a sufficient life time. Experience has shown that nickel–chromium-based alloys meet these requirements and they are widely used in this application.

The aim of the present work was to provide the burner industry with reliable data about the thermal limit for the application of high-temperature nickel-based alloys in the burners.

### 4.2 Experimental

In the present work, four nickel–chromium-based alloys with aluminium additions: Alloy 601; Alloy 602 CA; Alloy 617; and Alloy 693 were investigated in a recirculation oil burner for application as flame tube materials. The chemical compositions of the alloys as well as the exposure times are shown in Table 4.1. The good corrosion resistance of these alloys is achieved by their high chromium content. Additional

* Reprinted from H. Ackermann et al.: Oxide scale formation on Al containing Ni -Cr-based high-temperature alloys during application as flame tube material in recirculation oil burners. *Materials and Corrosion*. 2008. Volume 59. pp. 380–8. Copyright Wiley-VCH Verlag GmbH & Co. KGaA.

*Table 4.1*  Chemical composition and exposure time of the investigated alloys

| Material | Alloy 617 | Alloy 601 | Alloy 602 CA | Alloy 693 |
|---|---|---|---|---|
| DIN EN 10 027 | 2.4663 | 2.4851 | 2.4633 | |
| | Chemical composition in wt.-% | | | |
| Ni | Balance | Balance | Balance | Balance |
| Cr | 22.1 | 21.8 | 25.1 | 29.1 |
| Al | 0.96 | 1.2 | 2.29 | 3.36 |
| Fe | 0.95 | 13.9 | 9.25 | 4.56 |
| Mo | 8.5–10 | | | |
| Co | 11–14 | | | |
| Mn | 0.3 | 0.55 | 0.08 | 0.2 |
| Ti | 0–0.2 | 0.36 | 0.15 | 0.41 |
| Y | | | 0.08 | |
| Zr | | | 0.08 | |
| Exposure time at 1000 °C | 1170 h | 500 h, 1000 h, 2000 h | 50 h, 500 h, 1000 h, 2000 h, | 50 h, 500 h, 1000 h, 2000 h 3000 h |

alloying elements are aluminium and, in minor amounts, manganese and titanium. It is generally agreed that the addition of aluminium to iron–chromium and nickel–chromium alloys improves their oxidation resistance [2]. Manganese is added during the melting process for the purpose of desulphurising and deoxidising the melt [3,4]. On the other hand manganese may be added to enhance formability [3]. By alloying with titanium, an enhancement of the creep strength is achieved due to precipitation hardening [4].

Simultaneous investigations of all alloys were carried out using an assembled flame tube consisting of six exchangeable segments of 1 mm thickness (Fig. 4.1). The dimensions of the assembled tube were identical to those of the original burner flame tube. The segments of the alloys were manufactured as-received from the producer, without additional surface and thermal treatment. The experiments were carried out at intermittent burner operation with operation cycles consisting of 15 min burner operation followed by 5 min downtime simulating the switching cycles generally occurring in domestic heating systems. During operation, the flame tube temperature increased from approximately 760 °C at the tube inlet to a maximum temperature of 1000 °C at the tube outlet. The temperature variation range in the tangential direction approached 20 K at the tube outlet. The gaseous atmosphere near the flame tube also varied in both tangential and axial directions. During stationary operation, the following typical gaseous concentrations in dry gas were measured at the tube outlet: 13.3–13.87 vol.% carbon dioxide, 0.8–2.7 vol.% oxygen, 0.1–0.95 vol.% carbon monoxide, 2–39 ppm hydrocarbons as propane equivalents. For details concerning burner construction, temperature distribution and gaseous atmosphere near the inner flame tube wall measured in tangential and axial directions, see Refs 6 and 7. For the metallographic analysis, samples with dimensions 15 mm × 15 mm × 1 mm and 15 mm × 10 mm × 1 mm were cut from regions at the flame tube outlet (at 1000 °C) and the half flame tube length (at 950 °C). Metallographic investigations of the cross-section of the samples were carried out at the Central Facility for Electron

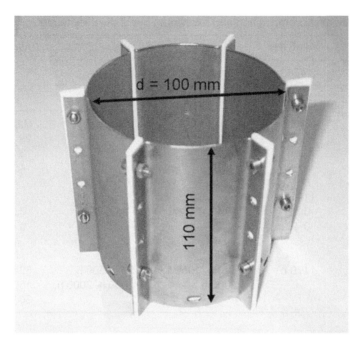

*4.1* Assembled burner flame tube

Microscopy of the University of Technology in Aachen. Electron probe microanaly-sis (EPMA) methods were used under the following conditions: 15 keV electron beam energy and 20 nA beam current, which result in an effective spot diameter of about 0.5 μm. Elemental maps showed the qualitative distribution of selected elements near the oxide–gas interface. The quantitative concentration of the elements was measured in line scans perpendicular to the alloy surface. Backscattered electron images of each analysed area were taken. The cross-sections of etched samples were observed by light microscopy.

## 4.3    Results

On all alloys, a similar microstructure of the scale was observed after an operation time of 500 h: a chromium oxide layer on top and aluminium oxides below. As can be seen from the elemental maps in Fig. 4.2 for Alloy 601 after 500 h at 950 °C, the chromia scale was enriched with titanium and manganese at the oxide–gas interface. Aluminium oxide formed as internal oxides predominantly along grain boundaries. By point microanalysis, the composition $Al_2O_3$ was proved for the aluminium oxides.

The quantitative distributions of the chemical elements within the chromia scale as measured in line scans of Alloy 601 after 500 h and Alloy 602 after 1000 h at 1000 °C are shown in Fig. 4.3. The concentrations of chromium, manganese and oxygen are consistent with a spinel-type oxide of $MnCr_2O_4$ on Alloy 601, while on Alloy 602, which originally contained less manganese than Alloy 601, a spinel of type $(Fe,Mn)Cr_2O_4$ was detected on top of the chromium oxide layer. In both alloys, titanium was distributed within the chromium oxide layer. For Alloy 602, an oxide

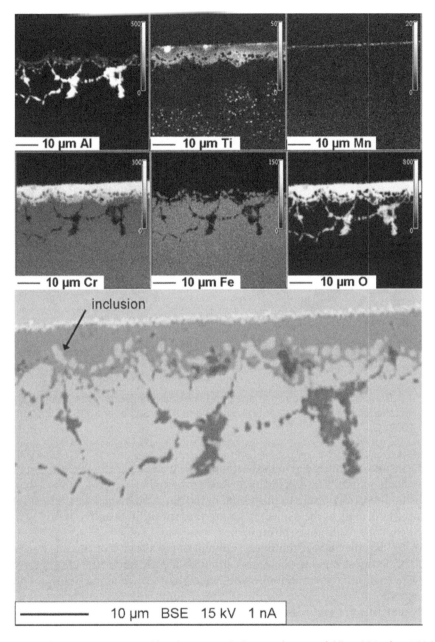

*4.2* Elemental maps and backscattered electron image of Alloy 601 after 500 h at 950 °C

particle of type $TiO_2$ was identified at the oxide–gas interface. Aluminium was also enriched in the chromium oxide layer with the maximum concentration at the metal–oxide interface.

Figure 4.4 shows the evolution of the microstructure with exposure time on Alloy 602 at 1000 °C. The thicknesses of the chromium oxidation and the internal aluminium oxide zones increase with exposure time. For all observed exposure

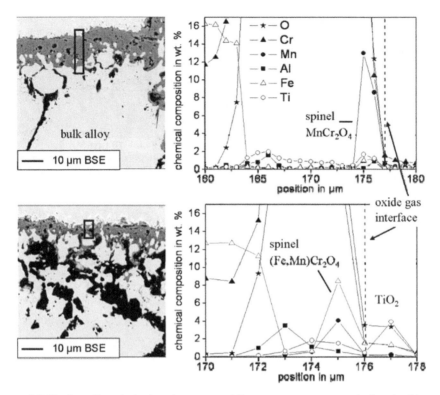

4.3 Backscattered electron images and line scan measurements (marked by rectangles) of Alloy 601 after 500 h (top) and Alloy 602 after 1000 h (bottom) at 1000 °C

periods, inclusions embedded in the chromia scale were detected. As can be seen from the elemental maps in Fig. 4.2, the inclusions are metallic. Figure 4.5 shows that their composition (corresponding to position **a** in the line scan image) was very similar to that of the bulk alloy at the oxide–metal interface (position **b**) and to the alloy matrix deeper below the oxide zone (position **c**).

Alloy 693 has the highest aluminium content of the alloys. For this alloy after 50 h at 1000 °C, in most regions, the grains situated directly under the chromium oxide scale were enclosed by aluminium oxide, which had grown along grain boundaries (Figs. 4.6 and 4.7a). At some locations, aluminium oxide was observed as a layer directly underneath the chromia scale. After 500 h at 1000 °C, a very thin aluminium oxide layer was detected at the alloy surface (Fig. 4.7b). Internal oxidation of chromium and aluminium occurred as well as formation of aluminium nitrides below the zone of internal oxidation. Figure 4.8a shows the inner corrosion products in the bulk of Alloy 693 after 1000 h at 1000 °C. The formation of aluminium nitrides was also observed for Alloy 601 after 1000 h at 1000 °C (Fig. 4.8b).

The chromia scale is prone to spallation due to growth stress and thermal stress during thermal cycling. For the duration of the presented investigations, on some alloys, cracks in the chromium oxide scale were observed indicating the occurrence of spalling. Quantification of the spallation was achieved in the following way: the thickness of a hypothetical chromia scale was calculated from the overall chromium loss derived from the line scans of the alloy samples. From backscattered electron

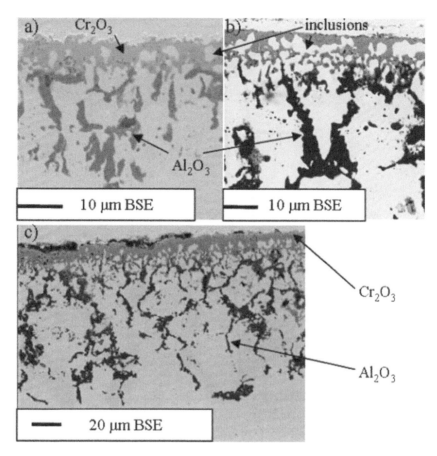

*4.4* Microstructure of Alloy 602 after 500 h (a), 2000 h (b) and 3000 h (c) at 1000 °C

images of the cross-section of the samples, the average thickness of the chromia scale was read. Figure 4.9 represents a comparison of these data for different exposure times at 1000 °C. The calculated values are higher than the scale thickness observed in the cross-sections for all alloys. This means that a part of the chromia scale spalled during the exposure. The spallation rate increases with exposure duration in the first 1000 h and seems to stay constant for Alloy 602 for times longer than 1000 h.

## 4.4    Discussion

Spinel-type oxides observed on top of the chromia scale are reported among others in Refs 8 and 9. On Alloy 601, Li [10] identified $MnCr_2O_4$ spinel and on Alloy 602, $FeCr_2O_4$ spinel by XRD analysis after 125 1-day oxidation cycles in air with a dwell temperature of 1000 °C. In the present investigation, the element maps of the samples of Alloy 602 after exposures of 50 h and 500 h [7,11] gave evidence that in the first hours of oxidation, $MnCr_2O_4$ forms on Alloy 602. But after manganese has been depleted in the bulk due to repeated spallation and oxidation, iron gradually substitutes manganese in this oxide phase. About 1 at.% Ti is dissolved in the spinel. On the sample of Alloy 602 after exposure for 1000 h, a particle with composition $TiO_2$

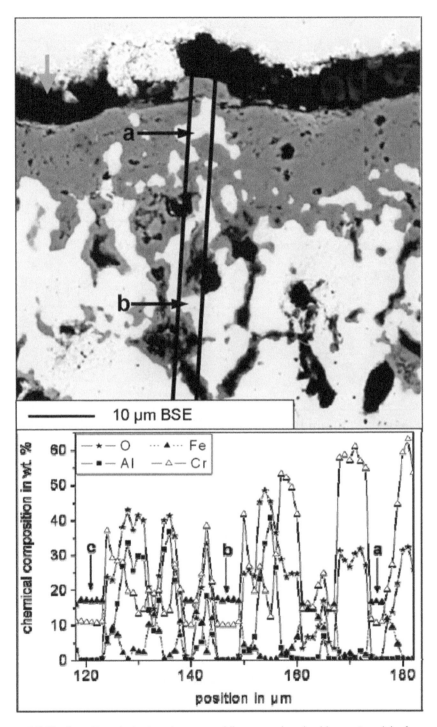

4.5 Backscattered electron image and line scan (marked by rectangle) of
Alloy 601 after 1000 h at 1000 °C. Positions a and b indicate metallic inclusion in
the chromia scale and the bulk alloy at the oxide–metal interface, respectively.
Position c corresponds to the alloy matrix below the zone of inner oxidation not
shown in the backscattered electron image

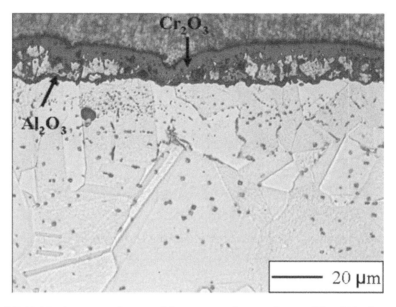

4.6 Light microscopy image of the near-surface region on Alloy 693 after 50 h at 1000 °C

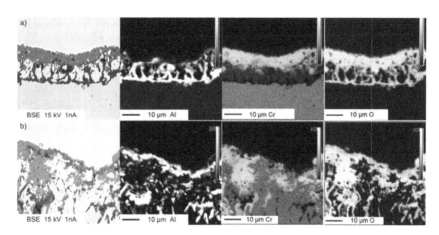

4.7 Backscattered electron images and elemental maps of Alloy 693 after 50 h (a) and 500 h (b) at 1000 °C

was observed. Titanium distributed in the chromia scale, as found in the present investigation, has also been reported in Ref. 12 for the model alloy Ni–10Cr–15W–2 Ti (wt.%) after exposure for 5000 h at 950 °C in simulated steam reforming gas. At the oxide–metal interface, the titanium concentration has a shallow peak in Fig. 4.3 for Alloy 601. This peak was frequently found in the line scans of the other samples and sometimes is also visible in the element maps. A corresponding concentration profile has been observed in Ref. 10 for Alloy 800 and Alloy 800 HT that contain about 0.4 mass% titanium. The phases and phase compositions in the oxide layer presented in this work and reported in the quoted publications correspond with the

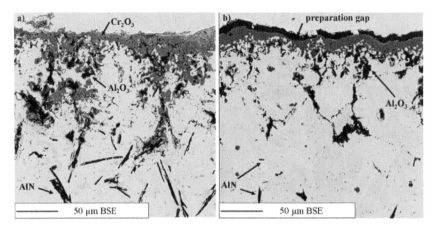

*4.8* Backscattered electron images of Alloy 693 (a) and Alloy 601 (b) after 1000 h at 1000 °C

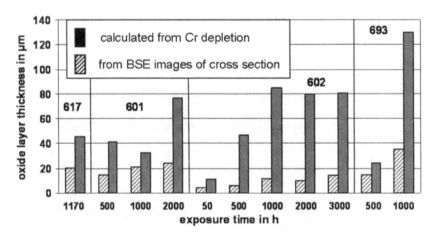

*4.9* Comparison between chromia scale thickness calculated from chromium depletion and observed in backscattered electron images after different exposure times at 1000 °C

phase equilibria of the chromium–manganese–titanium oxide system at 1000 °C obtained from experimental investigations by Naoumidis *et al.* [13]. At the oxide–gas interface where the oxygen partial pressure is between $10^{-2}$ bar and $10^{-1}$ bar, $Cr_2O_3$ does not dissolve manganese or titanium. Here, $Cr_2O_3$ is in equilibrium with the spinel $MnCr_2O_4$. $MnCr_2O_4$ forms a solid solution with the spinel $Mn_2TiO_4$ and dissolves up to 16% of it. The concentration of 1 at.% titanium observed in this work corresponds to a proportion of 7% $Mn_2TiO_4$. Considerable amounts of Fe are observed in the region of the mixed spinel. As the spinel $FeCr_2O_4$ exists and the compound $Fe_2TiO_4$ is stable in the system Fe–Ti–O at 1000 °C [14], Fe most probably substitutes for Mn in the lattice of the mixed spinel. The mixed spinel and $TiO_2$ form a two-phase equilibrium [13]. So $TiO_2$ is found on top of the oxide layer (Fig. 4.3). In the chromia scale, the oxygen partial pressure decreases to about $10^{-20}$ bar. At such low oxygen partial pressures, titanium has some solubility in the $Cr_2O_3$ lattice. In Ref. 13, by extrapolating the measured lattice parameter data, it is stated that 18% of the chromium atoms can be replaced by titanium atoms, corresponding to 11 mass% titanium in the solid solution. This seems quite a high value compared to 2 wt.% found in this work. It is known that $Al_2O_3$ and $Cr_2O_3$ form solid solutions [15]. Therefore, it is assumed that in addition to titanium, aluminium is dissolved in the $Cr_2O_3$ at the metal–oxide interface.

The formation of internal aluminium oxides below the chromia scale was observed for all investigated alloys. The same observation has been reported by other authors: for Alloy 602, after 6700 h of discontinuous exposure in flowing carbon monoxide–hydrogen–water vapour mixtures at 650 °C [16], for Alloy 601 and Alloy 602, after 1000 h cyclic oxidation in air at 1200 °C [17], and for Alloy 601, Alloy 602 and Alloy 617, oxidised in air after 18 1-day cycles at 1000 °C and after 40 1-day cycles of oxidation at 1100 °C [10]. In Ref. 18, the metallic inclusions observed in the oxide scale are attributed to the internal oxidation. The formation of internal corrosion products with lower density than that of the matrix generates compressive stresses. In order to balance the stress, the metal yields towards the surface. As a result, matrix inclusions are pushed forward into the oxide scale and the surface of the sample becomes undulated.

Alloy 617 contains a relatively high amount of molybdenum and cobalt. Generally, these elements are used in order to increase the high-temperature strength of the alloys. They enhance creep strength by solid-solution hardening [4]. Molybdenum helps to decrease the rate of diffusion, which controls the creep behaviour at high temperatures ($>820$ °C). Additionally, an enhancement of creep strength is achieved by precipitation of globular molybdenum carbides along grain boundaries [4]. In carburising environments, the addition of molybdenum may promote carbide formation and provide diffusional blocking of the carbon flux [19]. A beneficial effect of molybdenum on the sulphidation resistance is reported for binary chromium–molybdenum alloys and nickel-based alloys [20]. In contrast, concerning the oxidation resistance, detrimental effects associated with accelerated oxidation rates may occur [21].

The following explanation for the development of the corrosion layer on Alloy 693 is suggested, taking into account the high aluminium content of this alloy. Aluminium oxide which had grown along grain boundaries, enclosing the alloy grains, may act as a barrier constricting the diffusion of chromium from the bulk alloy towards the surface. Therefore, the chromium needed for oxide scale growth can be supplied only by the enclosed grains. This conclusion is approved by the observed chromium concentration profile. As can be seen from the line scan in Fig. 4.10, the chromium concentration of the enclosed grain is very low and almost no chromium depletion is observed in the bulk alloy. A chromia layer thickness of 4.2 μm was calculated from the chromium loss of the grains ($\Delta m_{Cr} = 19$ wt.%), which fits well with the thickness of 5 μm observed in the metallographic cross-section of the sample after 50 h at 1000 °C (Fig. 4.10). At prolonged exposure, the following process is suggested: oxide scale growth takes place until the chromium content of the grains is depleted below a critical value. After ongoing spallation, the oxide scale is not replaced by newly formed chromia and the other alloying elements Ni and Fe are oxidised. Eventually the surface grains flake off leaving a surface covered partially by a thin film of aluminium oxide, as observed after 500 h (Fig. 4.7b). This aluminium oxide layer cannot protect the bulk alloy from corrosive attack. Nitrogen from the gaseous atmosphere diffused into the alloy matrix leading to aluminium nitride formation.

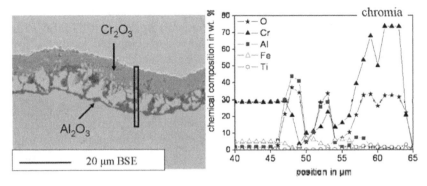

*4.10* Backscattered electron image and line scan (marked by rectangle) of Alloy 693 after 50 h at 1000 °C

The depth of the internal oxides and aluminium nitrides measured from backscattered electron images after different exposure times at 1000 °C is presented in Table 4.2. For Alloy 602 at 650 °C and 1000 °C, Grabke [16] and Li [10], respectively, observed aluminium oxide on grain boundaries, as was found in the present work. At temperatures of 1100 °C and 1200 °C the microstructure of the cross-sections presented in Refs 10 and 17 is quite similar to that of Alloy 693 in Fig. 4.6 but with wider regions where alumina forms a layer directly below the chromia scale. The depth of the internal aluminium oxides determined from cross-section images presented in the above publications is given in Table 4.3 for Alloy 602. Good agreement with the data in the present work (Table 4.2) results.

Assuming that the onset of aluminium nitride formation is related to the breakdown of the high-temperature resistance of the alloys, their lifetime can be estimated. For the investigated materials with a thickness of 1 mm at 1000 °C, the values for the lifetime are shown in Table 4.4. The long-term behaviour of the alloys correlates with the chromium consumption and the remaining chromium content in the bulk alloy. The time of breakdown of the chromia scale can be defined from the chromium depletion in the bulk alloy. If the chromium concentration drops below a critical value, a stable chromia scale cannot be established. For nickel chromium alloys, this value amounts to approximately 10 wt.% [22]. The chromium concentration of the

*Table 4.2*   Depth of internal corrosion products at 1000 °C

| Exposure time | Alloy 617 | Alloy 601 | | Alloy 602 | Alloy 693 | |
|---|---|---|---|---|---|---|
| | Depth of inner corrosion products in µm | | | | | |
| | Oxides (Cr + Al) | Oxides (Cr + Al) | Aluminium nitrides | Oxides (Cr + Al) | Oxides (Cr + Al) | Aluminium nitrides |
| 50 h | – | – | – | 43 | 27 | 0 |
| 500 h | – | 83 | 0 | 54 | 42 | 218 |
| 1000 h | – | 114 | 227 | 73 | 134 | 418 |
| 1170 h | 56 | – | – | – | – | – |
| 2000 h | – | 117 | 336 | 71 | 340 | 500 |
| 3000 h | – | – | – | 111 | – | – |

*Table 4.3*   Depth of internal aluminium oxides on Alloy 602 detected by other authors [10,16,17]

| Exposure conditions | Depth of internal aluminium oxides in µm on Alloy 602 |
| --- | --- |
| 6700 h discontinuous exposure in flowing $CO–H_2–H_2O$ mixtures at 650 °C [16] | 15 |
| 18 1-day cycles of oxidation in air at 1000 °C [10] | 44 |
| 40 1-day cycles of oxidation in air at 1100 °C [10] | 60 |
| 1000 h cyclic oxidation in air at 1200 °C [17] | 38 |

*Table 4.4*   Lifetime of the investigated alloys at 1000 °C

| 1 mm thick sheet of material | Life time at 1000 °C in oil burner |
| --- | --- |
| Alloy 617 | >1170 h |
| Alloy 601 | 1000 h |
| Alloy 602 | >3000 h |
| Alloy 693 | 500 h |

alloy matrix measured at the oxide–metal interface after different exposure times at 1000 °C is shown in Fig. 4.11. The corresponding line scan profiles after 500 h can be seen in Fig. 4.12. As can be seen from Fig. 4.10 for Alloy 601, the chromium concentration of 10 wt.% is reached after 1000 h. For Alloy 693, the chromium concentration is still higher than 10 wt.% after 500 h, but significantly below this value

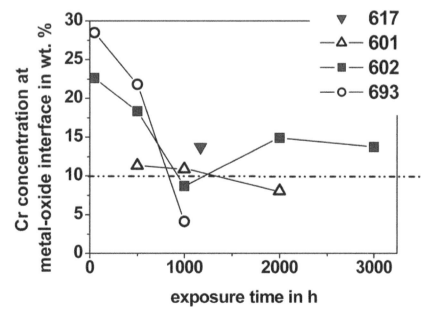

*4.11* Chromium concentration of the alloy matrix measured at the oxide–metal interface after different exposure times at 1000 °C

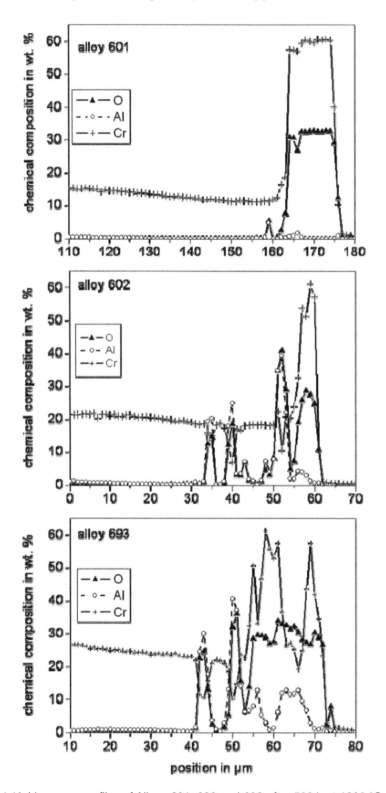

4.12 Line scan profiles of Alloys 601, 602 and 603 after 500 h at 1000 °C

after 1000 h. These observations are in good agreement with the determined lifetime from nitride formation for both alloys. In contrast, for Alloy 617 and Alloy 602, the chromium concentration at the end of the investigation time is still higher than the critical value, which also fits well with the lifetime defined from the time of onset of aluminium nitride formation. However, it should be mentioned that the chromium concentrations shown are derived from single line scan measurements representing local chromium distribution. Therefore, some data may deviate significantly from the average value, which is a possible explanation for the low value of the chromium concentration for Alloy 602 after 1000 h.

## 4.5    Conclusions

Under the described experimental conditions at 1000 °C, a probable adverse effect of aluminium was observed for the investigated alloys. Aluminium oxides grew along grain boundaries blocking the diffusion of chromium from the bulk towards the surface in the case of Alloy 693. Furthermore, if the oxide scale does not provide effective corrosion protection, aluminium nitride formation takes place as a result of nitrogen diffusion into the bulk. The high aluminium content of Alloy 693 had a detrimental effect on the lifetime of the alloy due to prohibition of chromia scale healing. A protective aluminium oxide layer did not form.

For application as a flame tube material for a recirculation burner, Alloy 602 showed the best performance. A lifetime of more than 3000 h can be expected. Alloy 617 had better corrosion resistance than Alloy 601. After an exposure time of about 1000 h, aluminium nitrides had formed in Alloy 601 whereas they were not detected in Alloy 617.

The lifetime of the materials can be determined from chromium depletion of the matrix at the metal–oxide interface. Lifetime ends at the breakdown of the chromia scale, indicated by internal nitride formation. It was shown that the breakdown of the chromia scale is correlated to a critical chromium concentration below the oxide scale.

It must be emphasised, that the presented results and conclusions are only valid for the above described service conditions and 1 mm thick metal sheets.

## Acknowledgements

The authors gratefully acknowledge the financial support of the Stiftung Industrieforschung within the research project S616.

## References

1. H. Ackermann, G. Teneva-Kosseva, K. Lucka and H. Köhne, *Heizungsjournal*, 3 (2006), 42.
2. P. Kofstad, *High Temperature Corrosion*. Elsevier Applied Science, London, New York, 1988.
3. R. Bürgel, *Handbuch Hochtemperaturwerkstofftechnik*. Vieweg Verlag, Braunschweig, Wiesbaden, 2001.
4. Krupp VDM, *Hochtemperaturwerkstoffe der Krupp VDM für den Anlagenbau*, VDM Report Nr. 25, 1999.
5. H. Pfeifer and H. Thomas, *Zunderfeste Legierungen*. Springer Verlag, Berlin, Göttingen, Heidelberg, 1963.

6. G. Teneva-Kosseva, H. Ackermann, H. Köhne, M. Spähn, S. Richter and J. Mayer, *Mater. Corros.*, 57 (2006), 122.
7. H. Ackermann, G. Teneva-Kosseva, K. Lucka, H. Köhne, S. Richter and J. Mayer, submitted for publication in *Corros. Sci.*
8. A. N. Hansson, M. Mogensen, S. Linderoth and M. A. J. Somers, *J. Corros. Sci. Eng.*, www.jcse.org, 6.
9. L. Mikkelsen and S. Linderoth, *Mater. Sci. Eng.*, A361 (2003), 198.
10. B. Li, PhD Thesis, Iowa State University, Ames, Iowa, 2003.
11. H. Ackermann, G. Teneva-Kosseva, K. Lucka, H. Koehne, S. Richter and J. Mayer, in *EMCR 2006*, 1–22. Dourdan, France, June 18–23.
12. P. J. Ennis and W. J. Quadakkers, in *High Temperature Alloys, Their Exploitable Potential*, 465, ed. J. B. Marriott, M. Merz, J. Nihoul and J. Ward. Elsevier Applied Science, London and New York, 1988.
13. A. Naoumidis, H. A. Schulze, W. Jungen and P. Lersch, *J. Eur. Ceram. Soc.*, 7 (1991), 55.
14. N. G. Schmahl, B. Frisch and E. Hargarter, *Z. Anorg. Allg. Chem.*, 305 (1960), 40.
15. S. T. Wlodek, *Trans. Metall. Soc. AIME*, 230 (1964), 1078.
16. H. J. Grabke, *Corrosion*, 56 (2000), 801.
17. U. Brill, *Met.*, 8, (1992), 778.
18. P. Huczkowski, S. Ertl, J. Piron-Abellan, N. Christiansen, T. Höfler, V. Shemet, L. Singheiser and W. J. Quadakkers, 'Microscopy of oxidation', in Proc. Mater. High Temp., Birmingham, UK, 2005.
19. B. A. Baker and G. D. Smith, *Metal Dusting in a Laboratory Environment – Alloying Addition Effects*, www.specialmetals.com/technical.htm
20. G. D. Smith, *Corrosion Resistance of Nickel-Containing Alloys in Petrochemical Environments*, www.specialmetals.com/technical.htm
21. G. D. Smith, *The Role of Protective Scales in Enhancing Oxidation Resistance*, www.specialmetals.com/technical.htm
22. J. E. Croll and G. R. Wallwork, *Oxid. Met.*, 1 (1969), 55.

# 5

# Development of thermally-sprayed layers for high-temperature areas in waste incineration plants*

D. Bendix, G. Tegeder, P. Crimmann,
J. Metschke and M. Faulstich

*ATZ Entwicklungszentrum, An der Maxhütte 1,
92237 Sulzbach-Rosenberg, Germany*

*bendix@atz.de*

## 5.1 Introduction

Waste incineration offers a favourable combination of safe waste disposal and energy generation. The balance between income from energy produced and cost of investment and maintenance often determines the means of energy production. Thus, about 90% of the waste incineration plants in Germany are characterised by energy produced by steam turbines with live steam parameters of 4 MPa and 400 °C. In practice, rather low energy efficiency is the result of this balance. Based on the increasing cost of energy, increasing energy efficiency is also desirable for economic reasons. Experience with plants which, for historical reasons, use higher steam parameters, has shown that higher steam parameters demand special corrosion protective layers to provide acceptable lifetimes for heat exchanger materials. Nowadays, the most usual coating process to protect pipes in waste incineration plants is cladding. Cladding has two essential disadvantages. The process is costly because of the high consumption of materials with rapidly increasing prices during recent years and because of the time-consuming cladding process. Also, it is known that clad pipes often fail in the high-temperature areas in waste incineration plants. That is why alternative processes to cladding have been under investigation in recent years. Thermal spraying is the most promising alternative technology. The specific costs of thermal spraying are much lower than the specific costs of cladding. Coating by thermal spraying reduces the risk of dilution of the substrate and coating materials. The variety of materials used for the protective layer is much higher than the materials which can be used for cladding. So different materials can be combined in one layer or in a multi-layer system (e.g. metal alloys and ceramics). The thickness of the layer needed for acceptable resistance to corrosion and wear can be drastically reduced. Thermal spraying has the potential to create cost-efficient coatings to protect components in the critical zones of incineration plants. So, a large number of layers have been applied successfully in several thermal plants. However, in waste incineration plants, there have been frequent failures of thermally-sprayed layers.

* Reprinted from D. Bendix et al.: Development of thermally-sprayed layers for high-temperature areas in waste incineration plants. *Materials and Corrosion*. 2008. Volume 59. pp. 389–92. Copyright Wiley-VCH Verlag GmbH & Co. KGaA.

The combination of the concentration of corrosive and erosive species in the gas, the temperatures of gas and pipes, and the gas velocity, is often the reason for the failure of the thermally sprayed layers. In particular, there is still no acceptable solution for the corrosion protection of the superheater pipes in incineration plants, where live steam temperatures higher than 400 °C are used. The use of a self-adopting multi-layer system is a new approach to solve these corrosion problems.

## 5.2    Thermally-sprayed layers under laboratory conditions

Thermal spraying represents a group of processes which employ heat and velocity to coat the surface of one material with another, using powder or wire feedstock. These processes are characterised by near-zero dilution of the substrate as a result of mechanical bonding, the ability to apply thin coatings, and a high rate of area coverage compared to arc welding processes. The low deposit temperatures (as compared to welding) mean no distortion or metallurgical degradation of the substrate. Thermal spray processes are all positional and can be operated in air, thus offering great flexibility for a wide range of applications (Fig. 5.1).

In a first step to develop thermally-sprayed layers for high-temperature areas in waste incineration plants, pieces of pipe material coated by thermal spraying were exposed to model corrosive conditions. The coated pieces were laid with the coated side exposed in synthetic salt (20% NaCl, 26% KCl, 24% $Na_2SO_4$, 30% $K_2SO_4$) and a corrosive gas atmosphere (72.7% $N_2$, 17.7% $O_2$, 10% $H_2O$, 0.08% HCl, 0.02% $SO_2$) was adjusted to simulate the conditions in a waste incineration plant (Fig. 5.2).

Uncoated pieces were included in this test for comparison purposes under waste incineration plant conditions. The expected symptoms of corrosion were to be seen

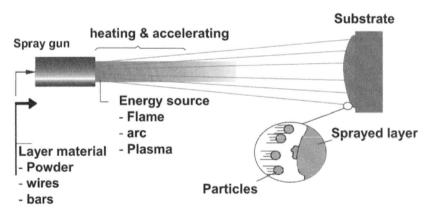

5.1 Thermal spraying – principle

5.2 Probes in synthetic salt

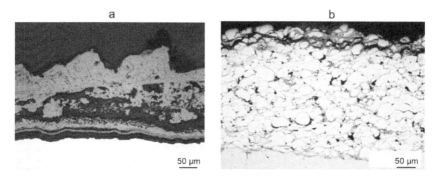

*5.3* (a) Layer-shaped corrosion, partial flaking of corrosion products (substrate 15Mo3, exposure temperature 430 °C, exposure time 470 h). (b) Local corrosion along the particle borders (layer material Alloy 690, exposure temperature 430 °C, exposure time 1115 h)

on the uncoated pieces after the exposure time. The porous layer-shaped structure of the corrosion products (as seen in Fig. 5.3a) is well known as characteristic of the corrosion in waste incineration plants.

Some of the thermally-sprayed coatings could not withstand the corrosive conditions without displaying evidence of corrosion. The reasons for this relate to the material composition and the parameters used for thermal spraying. Corrosion through the grains is evidence of the insufficient corrosion resistance of the coating material. Corrosion along the grain boundaries with layer-shaped flaking of grains was observed in many cases (Fig. 5.3b). It can be assumed that improved corrosion resistance can be achieved by changing the thermal spray parameters. Freedom from corrosion was displayed by an acceptable number of coating systems.

## 5.3    Results with material probes in incineration plants

Coating systems which demonstrated an acceptable corrosion resistance under laboratory conditions were investigated under the conditions of waste incineration plants. Material probes were used for these investigations so that the exposure of the coatings to the corrosive conditions could be started and stopped independent of the start and stop times of the incineration plant and to increase the useable parameter field of the corrosive conditions. The coated pipe on the outside of the material probe was air-cooled. The cooling air flowed through a central smaller pipe to the top of the material probe. In this way, a certain wall temperature (T0) was controllable (Fig. 5.4).

The material probes were placed in the second flue of the incineration plant. The off-gas temperature was in the range of 680–740 °C during the investigations. The wall temperature was varied between 330 °C and 600 °C. A certain number of coating systems, which had performed well when tested under laboratory conditions, failed during the tests on the material probes. The reasons were again associated with the corrosion resistance of the coating material (Fig. 5.5a and b) or problems, which seem to be surmountable by changing the parameters of thermal spraying. The porosity of the coating is one characteristic which can be influenced by the thermal spraying process. Coatings with high porosity lead to penetration by the corrosive species, resulting in corrosion of the substrate (Figure 5.5a). Too dense a coating leads to

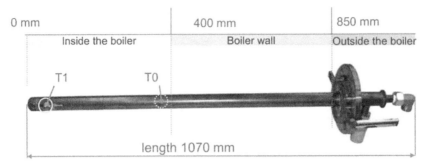

5.4 Material probe

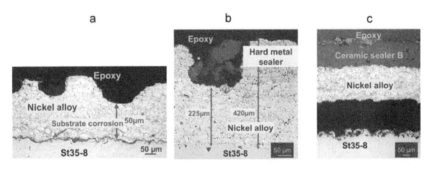

5.5 (a) Corrosion of the nickel alloy (670 h, 350 °C). (b) Corrosion of the sealer (700 h, 430 °C). (c) Flaking of the coating system (500 h, 430 °C)

thermal stresses which result in a flaking of the coating system (Figure 5.5c). In many experiments, no excessive porosity was found, which prevented substrate corrosion and flaking of the coating.

The idea for a successful coating system is self-adoption. Two layers should be combined in such a way as to create a thin dense film at the interface between them. The advantages of forming this dense film at the border between the two layers are that the coating can be formed independently of the substrate material; both outer layers can have a certain porosity to withstand thermal gradients, and the thin dense layer is protected against mechanical destruction via erosion by the ceramic sealer. This thin dense film can be created by diffusion and by solvo-thermal processes. The formation of the thin dense film was observable in the thermally sprayed layers of several material probes (Fig. 5.6). The coated layer of these probes always consisted of a nickel alloy and a zirconia-based ceramic sealer. The thickness of the layer was quasi-independent of the time of exposure in the waste incineration plant (the exposure time was changed from 500 h to 1500 h), but it strongly depended on the position. It can be assumed that the thin dense film is the result of a relatively fast reaction, which is strongly limited by temperature. There are many indications that a solvo-thermal process is the principal means of formation of this layer. For a solvo-thermal process a solvent is needed. This solvent can be formed by a reaction directly at the interface between the layers. One constituent of the reactants is included in the nickel alloy; another is included in the Zirconia-based sealer. All constituents travel by diffusion to the interface and there form the solvent. A strong temperature

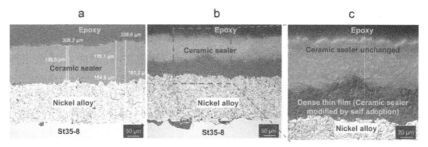

5.6 (a) Coating system as sprayed. (b) Coating system after use (500 h, 540 °C). (c) Detailed view of the dense thin film between ceramic and nickel alloy

dependency of the solubility of one component in the Zirconia will promote the dissolution of this component on the high-temperature side of the solvent and its release as dense material on the low temperature side of the solvent. So, only a very small amount of the solvent is needed. If the solvent reaches a temperature range where the temperature dependence of its solubility is not strong enough or where the solvent is no longer sufficiently stable, the process stops. Cracks in this film caused by local fast heating (sticking on of burning particles) or cooling (cleaning process) should be self-healed by restarting the film forming process as long as there is enough material for the formation of the solvent on the interface of the nickel alloy and the zirconia-based sealer.

The chemistry of the formation process is still not fully understood and will be the subject of further investigation.

## 5.4    Summary and conclusions

Commercially-available thermally-sprayed coating systems for corrosion and erosion protection often fail in the high-temperature areas of waste incineration plants. In the present study, a coating system has been found which withstood the corrosive conditions. The successful coating system is able to form a quasi-dense thin film between the two combined layers. Now certain coated pipes are in use in several waste-to-energy plants. In a next step, the film forming process will be investigated. In future, new self-adopting coating systems must be developed for other temperature ranges.

## Acknowledgments

This work was funded by the Bavarian Ministry of Economic Affairs, Infrastructure, Transport and Technology (BayStMWIVT).

## References

1.  C. Leyens, I. G. Wright and B. A. Pint, *Oxid. Met.*, 54 (2000), 401–424.
2.  Q. Wang and K. Luer, *Wear*, 174 (1994), 177–185.
3.  C.-D. Qin and B. Derby, *J. Mater. Sci.*, 28 (1993), 4366–4374.
4.  H. Chen, K. Zhou, Z. Jin and C. Liu, *J. Therm. Spray Technol.*, 13(4) (2004), 515–520.
5.  G. Dell'Agli and G. Masoco, *J. Eur. Ceram. Soc.*, 20 (2000), 139–S145.
6.  H. Cheng et al., *J. Mater. Sci. Lett.*, 15 (1996), 895–S897.

# 6

# High velocity oxy-fuel (HVOF) coatings for steam oxidation protection*

Alina Agüero, Raúl Muelas and Vanessa Gonzalez

*Instituto Nacional de Técnica Aeroespacial (INTA), Área de Materiales Metálicos, Ctra. Ajalvir Km. 4, 28850 Torrejón de Ardoz (Madrid), Spain*

aitgueroba@inta.es

## 6.1 Introduction

European COST Actions 522 (completed in 2003) and 536 (ongoing) have concentrated on designing and producing steels for steam power plants capable of operating at 600–625 °C in order to increase efficiency and reduce emissions [1–4]. However, efforts to produce ferritic steels for turbine components capable of operating at temperatures of 625 °C or higher have not yet been successful [5]. Materials such as P92 and COST-developed CB2 (9 wt.% Cr) with high creep strength up to 625 °C, have unacceptable oxidation resistance (Fig. 6.1), whereas materials with higher Cr content such as COST-developed FT4 (11 wt.% Cr) have better oxidation resistance but lower creep strength. When exposed to high-pressure steam at these temperatures, ferritic steels develop thick oxides which spall after relatively few hours of exposure [6]. Cross-section reduction, blockage and component damage due to erosion caused by the spalled oxides are some of the possible consequences.

A similar situation occurred with power generation and aeronautical gas turbines 45 years ago, when efforts to develop superalloys with the required mechanical properties as well as very low oxidation rates resulted in failure at higher operating temperatures. The solution was to employ coatings on superalloys with the required mechanical properties and presently, all new generation gas turbines require high-temperature oxidation resistance coatings as well as thermal barriers [7].

In 1998, efforts to examine the feasibility of applying coatings to steam turbine components were carried out within the context of COST 522 [8]. The results were very promising and the work continued within the framework of the European Commission project 'Coatings for Supercritical Steam Cycles' (SUPERCOAT) in which eight partners from different organisations across Europe participated [9,10]. A number of coatings, applied by means of slurry deposition, pack cementation and High Velocity Oxy-Fuel (HVOF) thermal spray were subjected to a variety of tests, including steam oxidation, creep strength, thermo-mechanical fatigue, etc. The project has recently been completed and some of the explored coatings were down-selected as candidates for industrial scale application on real components and for validation.

* Reprinted from A. Agüero et al.: HVOF coatings for steam oxidation protection. *Materials and Corrosion*. 2008. Volume 59. pp. 393–401. Copyright Wiley-VCH Verlag GmbH & Co. KGaA.

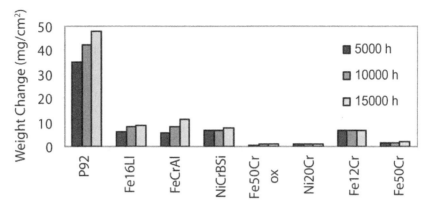

*6.1* Mass variation as a function of time for coated and uncoated P92 exposed to steam at 650 °C

Among the explored coating techniques, chosen on the basis of being potentially appropriate for coating large steam turbine components, HVOF thermal spray has emerged as one of the most successful. Abe and co-workers have also studied a number of HVOF deposited materials for this application [11]. This paper describes the steam oxidation behaviour of several alloyed materials deposited by this technique including the characterisation of the protective oxides formed on each material.

## 6.2    Experimental procedures

### 6.2.1    Materials

P92 (Fe: 87.69, Cr: 9.07, W: 1.79, Mn: 0.47, Mo: 0.46, V: 0.19, C: 0.12, Ni: 0.06, Nb: 0.063, N: 0.046, Si: 0.02, Al: 0.007, S: 0.006 wt.%) was obtained from Alstom Power, Switzerland. The powders employed for the deposition of coatings by HVOF are listed in Table 6.1, along with the corresponding commercial sources.

### 6.2.2    Coating deposition

The coatings were deposited by a Sulzer Metco Diamond Jet Hybrid HVOF unit (A-3120) mounted on a 6-axis robot (ABB) and fed by a twin rotation powder feeder.

*Table 6.1* Commercial powders employed for coating deposition by HVOF

| Powder (wt.%) | Source |
|---|---|
| Fe32Al | Osprey |
| Fe16Al | Osprey |
| Fe30Cr5Al1Mo1Si | Tafa-Castolín |
| Fe12Cr | Osprey |
| Fe50Cr | Osprey |
| Ni20Cr | Sulzer Metco |
| Ni17Cr4B4Si | Sulzer Metco |

### 6.2.3    Steam oxidation testing procedure

This test was carried out in a closed loop rig including a tubular furnace with pure, flowing, re-circulating steam (shown elsewhere) [12]. Before testing, air was displaced from the chamber by means of Ar which was kept flowing while heating up to the test temperature (at approximately 600 K h$^{-1}$). Once at temperature, the Ar flow was closed and the steam flow was switched on. To carry out weight measurements or to remove samples, the samples were furnace cooled to about 300 °C under argon and removed from the furnace. The reheating cycle (from 300 °C to the test temperature) was also carried out under Ar. Samples were removed at different intervals of exposure for metallographic analysis.

### 6.2.4    Characterisation

The coatings were characterised by optical and Field Emission Gun Scanning Electron Microscope (FESEM) (JEOL JSM-6500F equipped with an Oxford INCA 200 EDX/WDX micro-analyser) examination of metallographically polished cross-sections before and after exposure as well as by electron probe micro-analysis (EPMA) (JEOL JXA-8900 equipped with an Oxford EDX micro-analyser). Phase composition was examined by X-ray diffraction (Philips X'Pert) using the Cu K$\alpha$ line (15 418 nm).

### 6.3    Results

Commercially available powders of various compositions (Table 6.1) were sprayed by HVOF and optimised in order to reduce porosity and increase adhesion. The target thickness was 100–150 μm. The coatings were exposed to flowing steam at 650 °C and were characterised by FESEM and XRD before and after exposure. All coatings exhibited significantly lower mass gain than uncoated P92 when exposed to steam at 650 °C, as shown in Fig. 6.1.

### 6.3.1    Fe32Al

As-deposited Fe32Al (wt.%) exhibited through-thickness cracks and delamination cracks along the coating–substrate interface, causing partial coating separation (Fig. 6.2a). These cracks are similar to those observed on slurry deposited aluminide coatings [12] and probably originated due to stress caused by thermal expansion mismatch between brittle Al-rich FeAl intermetallics and the substrate. Delamination at the coating–substrate interface can be observed on exposure to steam for 1200 h; oxidation of the substrate through the above mentioned cracks occurs as shown in Fig. 6.2b. On its surface, the coating developed a mixed Al–Fe oxide which was not characterised since the coating was rejected as a candidate for protecting steam turbine components.

### 6.3.2    Fe16Al

In contrast to Fe32Al, Fe16Al does not develop cracks. EDX analysis and the XRD pattern indicate that β-FeAl is present (Fig. 6.3a and b). This coating exhibits excellent oxidation resistance up to at least 15 000 h despite presenting a relatively large degree of porosity obtained even after optimisation. The main phase present in a coating exposed for 10 000 h is still mostly β-FeAl, covered by a protective oxide

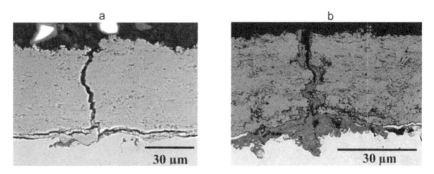

*6.2* FESEM image of the cross-section of Fe32Al deposited by HVOF on P92: (a) as-coated and (b) after 1200 h of exposure to steam at 650 °C

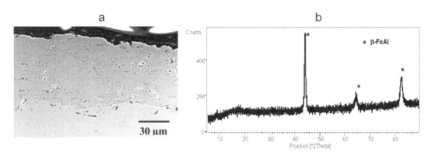

*6.3* HVOF deposited Fe16Al: (a) FESEM image and (b) XRD pattern

comprising a thin inner Al-rich oxide layer with a Fe-rich oxide on top (Fig. 6.4a). EDX analysis in combination with XRD (Fig. 6.4b) indicated that this outer oxide is $Fe_2O_3$ (probably with some $Al_2O_3$ which is soluble on $Fe_2O_3$ [13]), while some low intensity peaks can be attributed to $\gamma$-$Al_2O_3$ and $\alpha$-$Al_2O_3$, corresponding to the inner, thin, dark layer that can be observed in Fig. 6.4a. Although 650 °C is considered low for the formation of $\alpha$-$Al_2O_3$, it has already been observed on diffusion aluminide coatings also exposed to steam at 650 °C after 10 000 h of exposure [12]. A specimen removed after 4000 h already exhibited this relatively thick bi-layered oxide (~10 µm), which does not grow significantly in up to 10 000 h of steam exposure. In addition, no significant Al depletion could be measured below this layer indicating that this is a very protective oxide under steam at 650 °C. Moreover, no coating–substrate interdiffusion was observed and therefore Al does not seem to be lost by inward diffusion, which, in contradiction, is the principal degradation mechanism observed on diffusion slurry iron aluminide coatings [9].

### 6.3.3   Fe30Cr5Al1Mo1Si

Oxidation testing results for a similar coating (without Mo and Si) exposed for times of up to 10 000 h were published previously [12]. Although the commercial source was the same, the powder batch acquired for this set of new experiments contained 1 wt.% of both Si and Mo. The 'as-deposited' layer had the same composition measured by EDX as the starting powder and a uniform microstructure exhibiting only a single phase, as shown in Fig. 6.5a. This was confirmed by XRD as only peaks attributed to $\alpha$-Fe (containing Cr, Al, Mo and Si) could be observed (Fig. 6.5b). After

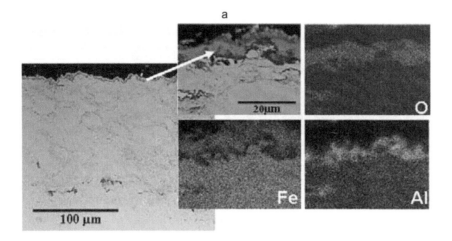

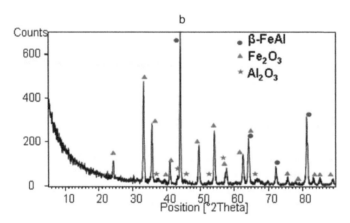

*6.4* HVOF deposited Fe16Al exposed to steam at 650 °C for 10 000 h: (a) FESEM image and (b) XRD pattern

testing under steam at 650 °C for 10 000 h, no substrate attack could be observed and at least three phases were present within the coating (Fig. 6.6a). The dark grey phase exhibits a composition similar to that of the as-deposited coating; the light grey phase is very rich in Cr, corresponding to the $\sigma$-FeCr phase, whereas the white precipitates are very rich in Mo with a composition close to that of the $Cr_6Fe_{18}Mo_5$ phase in agreement with the XRD pattern shown in Fig. 6.7. Remarkably, no oxide layer could be observed on the surface of the coating by FESEM (even at very high magnification) in contrast to the results obtained in the earlier experiments with 'pure' Fe30Cr5Al [12]. Mikkelsen and collaborators observed that adding small amounts of Si to Fe21Cr alloys significantly reduced the growth of the $Cr_2O_3$ scale [14]. In Fe30Cr5Al1Mo1Si, the growth of the protective oxide may be retarded by the presence of Si to such an extent that it is not possible to observe it by SEM even after 10 000 h of exposure to steam. However, on a specimen removed after 20 000 h, some sections of the coating surface were covered with a complex oxide (Fig. 6.6b) and EDX analysis indicated the presence of Cr, Fe and Al in this oxide. The XRD patterns of both samples taken out at 10 000 and 20 000 h are very similar and

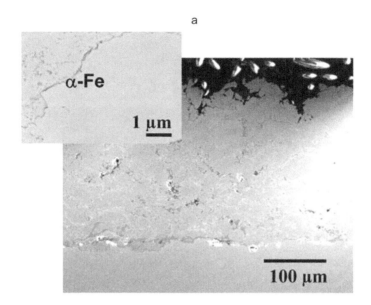

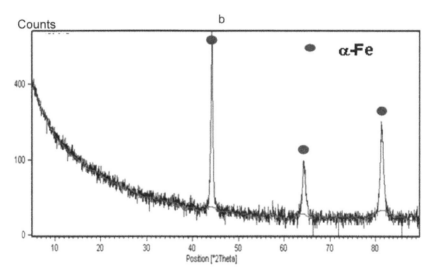

*6.5* HVOF deposited Fe30Cr5Al1Mo1Si: (a) FESEM image and (b) XRD pattern

consistent with the presence of $\gamma$-Al$_2$O$_3$ and $\alpha$-Al$_2$O$_3$ (Fig. 6.7). Transformation of $\gamma$-Al$_2$O$_3$ to $\alpha$-Al$_2$O$_3$ may occur slowly and it is also favoured by the presence of Cr [15,16]. Some interlamellar, as well as interfacial (coating–substrate) oxidation could be observed, but without any major impact on the protective nature of the coating.

### 6.3.4  Fe12Cr

This low-cost powder was deposited and optimised resulting in a coating with a Cr content of 16 wt.%. XRD exhibited peaks attributed to $\alpha$-Fe and some (Fe,Cr)$_3$O$_4$ probably resulting from partial interlamellar oxidation during spraying (Fig. 6.8a and b). After exposure to steam at 650 °C for 8100 h, the coating developed a protec-

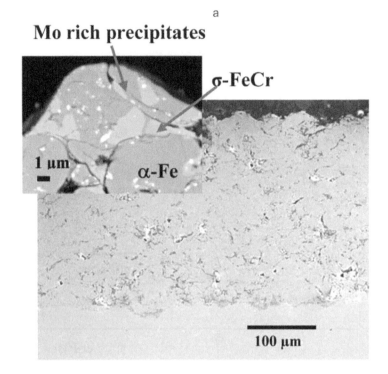

6.6 FESEM cross-section of HVOF deposited Fe30Cr5Al1Mo1Si after:
(a) 10 000 h and (b) 20 000 h of exposure to steam at 650 °C

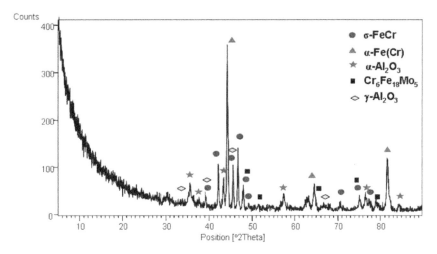

*6.7* XRD pattern of HVOF deposited Fe30Cr5Al1Mo1Si after 20 000 h of exposure to steam at 650 °C

tive mixed oxide scale consisting of outer Fe,Cr oxides (mostly $(Cr,Fe)_2O_3$) on top of a thinner Cr-rich oxide, as confirmed by EDX mapping and XRD (Fig. 6.9a and b), and in agreement with results obtained for 11 wt.% Cr steels by Segerdahl *et al.* [17]. This protective scale slowly grows from ~2–3 μm after 1000 h to ~8–10 μm after 8100 h. One of the proposed mechanisms for failure of the protective scale developed by Cr-containing steels when exposed to steam at temperatures of 600 °C and higher, is the evaporation of volatile chromium oxyhydroxides [17,18]. However, in this case, at 650 °C no significant Cr depletion was measured below the scale and no evidence of evaporation was found. Moreover, there are no Fe oxide nodules, which are typical of steels with a Cr content higher than 10 wt.%, when exposed to steam at temperatures higher than 600 °C [19,20].

### 6.3.5   Fe50Cr

As with Fe12Cr, the as-deposited Fe50Cr coating exhibited essentially one phase (Fig. 6.10a) which, according to XRD, is a solid solution of Cr (51 wt.% as measured by EDX) in α-Fe with some $(Fe,Cr)_3O_4$. The σ-FeCr phase did not form due to rapid cooling during the spray process. After 9000 h of exposure to steam at 650 °C, two new phases could be observed (Fig. 6.10b), the darker with a very high Cr content (~60 wt.% according to EDX analysis), while the lighter and more abundant phase exhibited a much lower Cr content (~27 wt.%). Moreover, some peaks attributed to σ-FeCr were present in the XRD pattern as shown in Fig. 6.11. It appears, therefore, as if the initial 50–50 wt.% solid solution alloy has transformed into two new solid solutions, of Fe in Cr and of Cr in Fe as well as into σ-FeCr. The coating has developed a protective, very slow growing $Cr_2O_3$ scale which, after 9000 h, has only reached a thickness of ~2–3 μm. Again, no evidence of evaporation of volatile chromium oxyhydroxides was found and for this higher Cr coating, no Fe,Cr oxides could be observed on the surface of the coating.

In a separate set of experiments, the HVOF deposition parameters were modified so as to maximise oxidation during spraying. Indeed, significant oxidation of the powder took place resulting in a cermet coating that has been designated $FeCrO_x$,

a

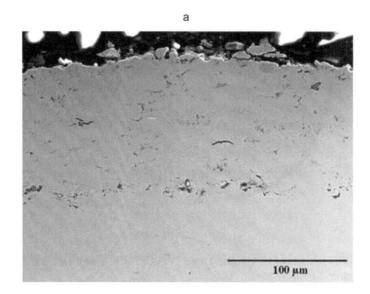

100 μm

b

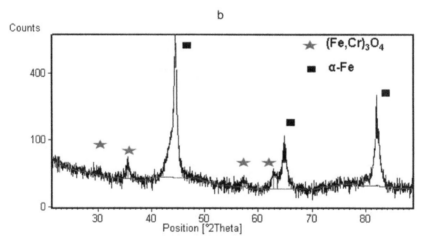

*6.8* HVOF deposited Fe12Cr: (a) FESEM image and (b) XRD pattern

composed of a $(Fe,Cr)_2O_4$ spinel on an FeCr matrix ($\alpha$-Fe), as confirmed by XRD, which also exhibits peaks attributed to $Cr_2O_3$ (Fig. 6.12a and b). After 20 000 h of exposure to steam at 650 °C, the coating developed a protective oxide layer composed of $Cr_2O_3$ and 'patches' of Cr and Fe mixed oxides and the content of the oxidised phases (dark) appeared to have increased when observed by FESEM, as seen in Fig. 6.13. EDX analysis showed that the Cr content in the metallic phase had decreased to 20–30 wt.% whereas that of the oxide species had become higher. On closer observation, some of the large metallic lamella within the coating showed different phases with different Cr contents, as was observed when Fe50Cr was exposed to steam for several hours at 650 °C (see above). The XRD pattern (Fig. 6.14) exhibits two new oxide phases while the intensity of the peaks attributed to $\alpha$-Fe present in the initial coating pattern has been significantly reduced. The peaks attributed to $Cr_2O_3$ have grown in intensity and the new peaks appear to

a

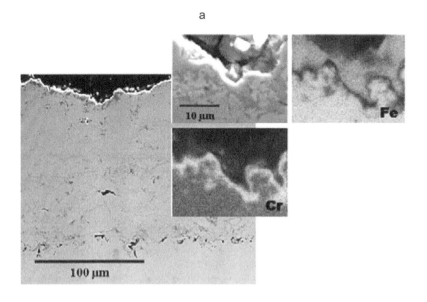

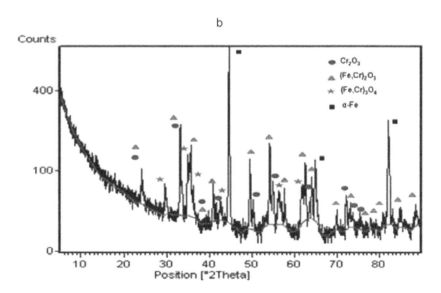

*6.9* HVOF deposited Fe12Cr exposed to steam at 650 °C for 8100 h: (a) FESEM image and (b) XRD pattern

correspond to $Fe_2O_3$ and $(Cr,Fe)_2O_3$. Although oxidation of the metallic matrix appeared to have increased on exposure to steam, no evidence of substrate attack could be observed.

### 6.3.6   Ni20Cr

This coating was also studied previously by this group [8,12] for exposures up to 10 000 h and by Abe and co-workers [11,20,21] for shorter exposures but also

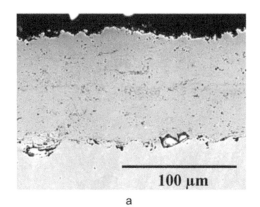

a

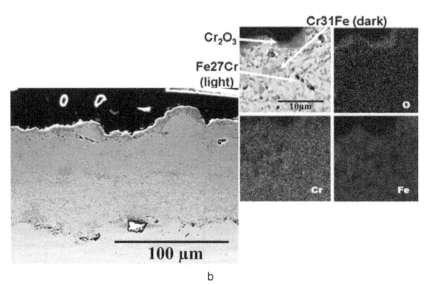

b

*6.10* FESEM image of the cross-section of Fe50Cr deposited by HVOF on P92: (a) as-coated and (b) after 9000 h of exposure to steam at 650 °C

at other temperatures. As-deposited Ni20Cr is a single phase layer, as shown in Fig. 6.15a, corresponding to a solid solution of Cr in Ni according to XRD. Excellent behaviour has been observed for exposures up to 30 000 h (test ongoing). FESEM of the cross-section of a sample taken out after 20 000 h, showed no coating degradation or substrate attack and a very thin protective $Cr_2O_3$ layer on the coating surface (Fig. 6.15b). However, a small degree of coating–substrate interdiffusion could be observed with mostly outwards diffusion of Fe. XRD of this sample still indicated mostly a solid solution of Cr in Ni as well as some peaks attributed to $Cr_2O_3$ (Fig. 6.16). Again, no evidence of evaporation of volatile chromium oxyhydroxides was found, as no significant Cr depletion within the coating and below the scale was observed.

### 6.3.7   Ni17Cr4B4Si

This coating exhibited an unexpected behaviour when exposed to steam at 650 °C. After 10 000 h of exposure, the coating developed a very complex thick (~30 μm)

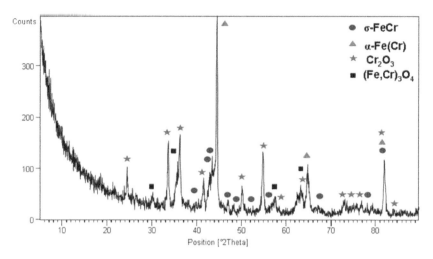

*6.11* XRD pattern of HVOF deposited Fe50Cr after 9000 h of exposure to steam at 650 °C

oxide containing all of its elements, as shown in Fig. 6.17. Significant B depletion under the scale was observed by EDX mapping whereas Cr and Si depletion also took place but to a minor extent. Unoxidised Ni was also present within the scale, as shown in a profile carried out by EPMA (Fig. 6.18). Despite its relatively high Cr content, this coating does not form protective $Cr_2O_3$ but it is nevertheless more resistant to steam at 650 °C than the uncoated substrate (Fig. 6.1).

## 6.4    Discussion

All of the deposited coatings showed better steam oxidation resistance than uncoated P92. Depending on their composition, pure or mixed protective oxides which were more or less stable were formed.

After several thousand hours, on Fe32Al and Fe16Al, Fe,Al mixed oxides were formed with, in the case of Fe16Al, a very thin inner layer of pure Al oxide. This layer appears to be a mixture of $\gamma$-$Al_2O_3$ and $\alpha$-$Al_2O_3$ according to XRD, despite the relatively low temperature. It has already been shown that slurry diffusion Fe aluminide coatings develop $\alpha$-$Al_2O_3$ after long time exposure to steam at 650 °C [12]. However, most results published in the literature indicate that $\alpha$-$Al_2O_3$ does form only at temperatures higher than 850 °C but the corresponding experiments were carried out for relatively short periods of time [15,16,22]. Moreover, Boggs reported the appearance of both $\gamma$-$Al_2O_3$ and $\alpha$-$Al_2O_3$ at temperatures higher than 600 °C, suggesting that $\alpha$-$Al_2O_3$ results from recrystallisation of $\gamma$-$Al_2O_3$ [23]. Under the present experimental conditions, $\alpha$-$Al_2O_3$ may also form slowly from less stable $\gamma$-$Al_2O_3$. On the other hand, the protective oxide formed on Fe30Cr5Al1Mo1Si exposed to steam for 10 000 h appears to be too thin to be observed by FESEM, probably due to the presence of Si. However, relatively intense peaks attributed to both $\gamma$-$Al_2O_3$ and $\alpha$-$Al_2O_3$ could be observed by XRD (they may be due in part to interlamellar oxidation, typical of thermally sprayed coatings). Despite the high Cr content of the latter coating, no $Cr_2O_3$ was observed.

As Fe12Cr forms a mixed Cr,Fe oxide whereas Fe50Cr forms pure $Cr_2O_3$, the critical content of Cr required to form pure protective $Cr_2O_3$ in FeCr alloys must be

a

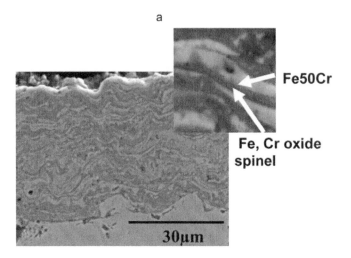

Fe50Cr

Fe, Cr oxide spinel

30μm

b

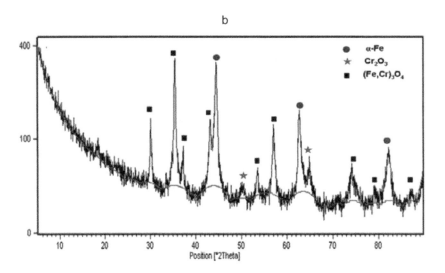

6.12 HVOF deposited FeCrOx: (a) FESEM image and (b) XRD pattern

somewhere within the range of 12–50 wt.%. The endurance of $Cr_2O_3$ under steam at 650 °C is remarkable for Fe50Cr and Ni20Cr. In order to form the volatile species $CrO_2(OH)_2$, responsible for the volatilisation of the oxide, $O_2$ is needed, as shown in the following equation:

$$Cr_2O_3 + 3/2O_2 + 2H_2O \rightarrow 2CrO_2(OH)_2$$

The present experiments were performed in the absence of $O_2$, under a pure flowing steam, employing thoroughly de-oxygenated water (some $O_2$ must be present in equilibrium). A plausible explanation for the observed stability of $Cr_2O_3$ may be that, under the experimental conditions used in this work, $CrO_2(OH)_2$ would only form in insignificantly low amounts, if any. Moreover, no mixed oxides were observed in Fe50Cr after several thousand hours, confirming the stability of the $Cr_2O_3$ layer. Surprisingly, in a relatively high Cr-containing Ni-base coating such as Ni17Cr4B4Si,

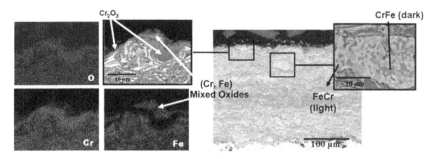

6.13 FESEM image of the cross-section of HVOF deposited FeCrOx exposed to steam at 650 °C for 20 000 h

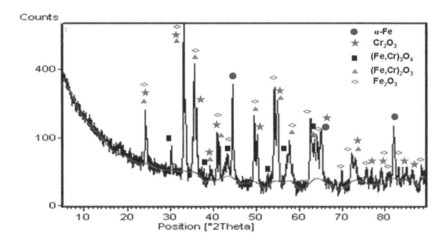

6.14 XRD pattern of HVOF deposited FeCrOx after 20 000 h of exposure to steam at 650 °C

no protective $Cr_2O_3$ formed. Either the critical content of Cr required to form $Cr_2O_3$ in NiCr alloys is higher than 17 wt.% (and lower than 20 wt.%) or Si and/or B have a significant effect in preventing its formation.

## 6.5    Conclusions

- Several coating compositions, including protective alumina, chromia and mixed oxide formers have been deposited by HVOF. The steam oxidation resistance at 650 °C of most of these coatings is very high and will probably exceed 100 000 h (steam power plant design criteria [24]).
- Despite the relatively low test temperature used in these experiments, $\alpha$-$Al_2O_3$ was observed on all of the studied Al-containing coatings.
- Protective $Cr_2O_3$ was formed on Fe50Cr and Ni20Cr and no evidence of oxide loss by Cr oxyhydroxide evaporation was found even after very long exposures to steam at 650 °C.
- A new composite Fe,Cr oxide coating has been deposited.

a

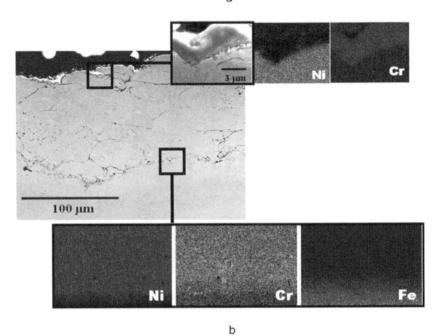

b

6.15 FESEM image of the cross-section of Ni20Cr deposited by HVOF on P92: (a) as-coated and (b) after 20 000 h of exposure to steam at 650 °C

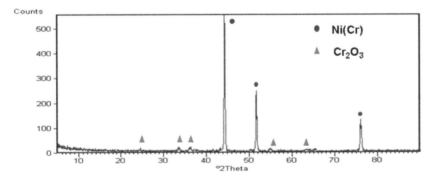

6.16 XRD pattern of HVOF deposited Ni20Cr after 20 000 h of exposure to steam at 650 °C

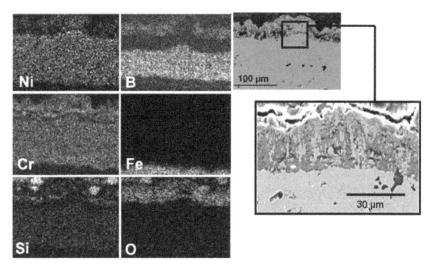

*6.17* FESEM image and EDX mapping of the cross-section of Ni17Cr4B4Si deposited by HVOF on P92 after 10 000 h of exposure to steam at 650 °C

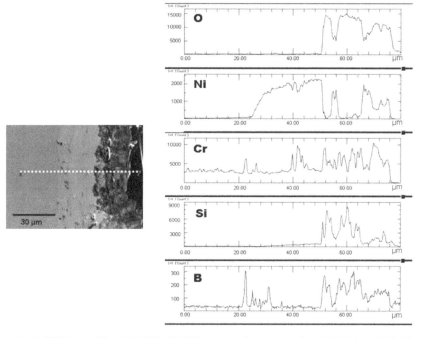

*6.18* EPMA profile of Ni17Cr4B4Si exposed to steam for 10 000 h at 650 °C

## Acknowledgements

The authors are grateful to the European Commission for financial support through the SUPERCOAT project, and to P. Vallés and M. C. García Poggio for their very valuable assistance in FESEM and XRD, respectively. All of the members of the Area of Metallic Materials at INTA are also acknowledged for their support during all of the stages of this work.

## References

1. T. U. Kern, M. Staubli, K. H. Meyer, K. Esher and G. Zeiler, in Proc. 7th Liege Conf., 30 Sept.–2 Oct. 2002, *Mater. Adv. Power Eng.* (2002), 1049.
2. M. Staubli, K. H. Meyer, W. Gieselbrecht, J. Stief, A. Di Gianfrancesco and T. U. Kern, in Proc. 7th Liege Conf., 30 Sept.–2 Oct. 2002, *Mater. Adv. Power Eng.* (2002), 1065.
3. T. U. Kern, M. Staubli, K. H. Meyer, B. Donth, G. Zeiler and A. Di Gianfrancesco, in Proc. 8th Liege Conf., 18–20 Sept. 2006, *Mater. Adv. Power Eng.* (2006), 843.
4. M. Staubli, R. Hanus, T. Weber, K. H. Meyer and T. U. Kern, in Proc. 8th Liege Conf., 18–20 Sept. 2006, *Mater. Adv. Power Eng.* (2006), 855.
5. J. Hald, in Proc. 8th Liege Conf., 18–20 Sept. 2006, *Mater. Adv. Power Eng.* (2006), 917.
6. W. J. Quadakkers, P. J. Ennis, J. Zurek and M. Michalik, *Mater. High Temp.*, 22 (2005), 47.
7. T. Narita, T. Izumi, T. Nishimoto, Y. Shibata, K. Zaini Thosin and S. Hayashi, *Mater. Sci. Forum*, 522-523 (2006), 1.
8. A. Agüero, F. J. García de Blas, R. Muelas, A. Sánchez and S. Tsipas, *Mater. Sci. Forum*, 369-372 (2001), 939.
9. A. Agüero, R. Muelas and M. Gutiérrez, *Mater. Sci. Forum*, 522-523 (2006), 205.
10. M. B. Henderson, M. Scheefer, A. Agüero, B. Allcock, B. Norton, D. N. Tsipas and R. Durham, 'Development and validation of advanced oxidation protective coatings for super critical steam power generation plants', in Proc. 8th Liege Conf., 18–20 Sept. 2006, *Mater. Adv. Power Eng.* (2006), 1553.
11. T. Sundararajan, S. Kuroda, K. Nishida, T. Itagaki and F. Abe, *ISIJ Int.*, 44 (2004), 139.
12. A. Agüero, R. Muelas, B. Scarlin and R. Knoedler, in Proc. 7th Liege Conf., 30 Sept.–2 Oct. 2002, *Mater. Adv. Power Eng.* (2002), 1143.
13. L. M. Atlas and W. K. Sumica, *J. Am. Ceram. Soc.*, 41 (1958), 50.
14. L. Mikkelsen, S. Linderoth and J. B. Bilde-Sørensen, *Mater. Sci. Forum*, 461-462 (2004), 117.
15. P. Tomaszewicz and G. R. Wallwork, *Rev. High Temp. Mater.*, 4 (1978), 75.
16. R. Prescott and M. J. Graham, *Oxid. Met.*, 38 (1992), 73.
17. K. Segerdahl, J. E. Svensson, M. Halvarsson, I. Panas and L. G. Johansson, *Mater. High Temp.*, 21 (2005), 69.
18. H. Asterman, K. Segerdahl, J. E. Svensson, L. G. Johansson, M. Halvarsson and J. E. Tang, *Mater. Sci. Forum*, 461-464 (2004), 775.
19. W. J. Quadakkers and P. J. Ennis, in Proc. 6th Liege Conf., *Mater. Adv. Power Eng.* (1998), 123.
20. T. Sundararajan, S. Kuroda, T. Itagaki and F. Abe, in Thermal Spray 2003: Advancing the Science and Applying the Technology, 495. ASM International, Materials Park, Ohio, USA, 2003.
21. T. Sundararajan, S. Kuroda, T. Itagaki and F. Abe, *ISIJ Int.*, 43 (2003), 104.
22. G. Berthomé, E. N'Dah, Y. Wouters and A. Galerie, *Mater. Corros.*, 56 (2005), 389.
23. W. E. Boggs, *J. Electrochem. Soc.*, 118 (1971), 906.
24. R. W. Vanstone, *Mater. Adv. Power Eng.* (2002), 1035.

# 7

# Identification of degradation mechanisms in coatings for supercritical steam applications*

R. N. Durham, L. Singheiser and W. J. Quadakkers

*IEF-2, Forschungszentrum Jülich GmbH, Jülich, Germany*

*j.quadakkers@fz-juelich.de*

## 7.1    Introduction and background

An increase in operating temperature in steam power plants up to 650 °C will lead to a significant reduction in fuel consumption and a subsequent reduction in $CO_2$ emissions and cheaper electricity [1,2]. The present materials utilised for steam power plants (9% Cr steels) are already being pushed to their physical limits and it is well known that in steam-containing environments, as seen in power plant service conditions, the oxidation rates of these materials are significantly faster than those in air [3].

The oxidation behaviour of the 9% Cr steels in steam has been the subject of exhaustive investigation [4–7]. These steels tend to oxidise forming very thick oxide scales consisting of an external layer of iron oxides, and an internal layer consisting mainly of Fe–Cr spinels. These scales spall during service causing enhanced metal cross-section loss and subsequent downstream damage of power plant components through erosion or blocking of gas flow, leading to extended downtime and plant maintenance programmes.

To enable the full potential of the 9% Cr steels to be achieved, a protective coating system has been proposed to further exploit the excellent mechanical properties of the materials, while at the same time offering a more oxidation-resistant barrier coating. As part of the EU funded project 'SUPERCOAT', the development and assessment of such coating systems on 9% Cr steels was undertaken and results concerning the oxidation and degradation mechanisms of these systems in steam bearing environments will be presented here.

## 7.2    Coating materials and processes

For use in supercritical steam conditions with a targeted operating temperature of 650 °C while maintaining high creep resistance, the present day materials (9% Cr steels) require protective coatings in order to be successfully utilised in such extreme environments. The substrates used for the coatings in the present study were either P91 or P92; details of the chemical compositions of these materials are given in

*Table 7.1*   Chemical composition of substrate materials

| Alloy | Fe | Ni | Cr | W | Mo | Nb | C | Mn | Si | V | B | Others |
|---|---|---|---|---|---|---|---|---|---|---|---|---|
| P91 | Bal. | 0.4 | 9 | – | 0.95 | 0.08 | 0.1 | 0.45 | 0.5* | 0.2 | – | 0.04 Al, 0.02% P, 0.01% S |
| P92 | Bal. | 0.4* | 9 | 1.75 | 0.45 | 0.0065 | 0.1 | 0.45 | 0.5* | 0.22 | 0.004 | 0.04 Al, 0.02% P, 0.01% S |

* denotes maximum.

Table 7.1. Numerous coating developments were devised as part of the EU Project 'SUPERCOAT' and several of these were successfully demonstrated in laboratory testing. Two main coating types will be concentrated on in this paper, namely diffusion coatings and overlay coatings, and their effects on the substrate material and subsequent degradation will be discussed. Details of the coatings used in this study are given in Table 7.2.

Seven different coatings were examined in this work, and a brief description of their manufacture is given below. The two high velocity oxy-fuel (HVOF) coatings were provided by Monitor Coatings (North Shields, UK) using commercially available Ni–20Cr and Ni–50Cr powders and were applied using a JP-5000 type gun. A coating thickness of 100 μm was desired.

The IPCOTE aluminium slurry coating was provided by Indestructible Paints (Birmingham, UK) and was dried between successive applications. After the final application, the samples were cured at 560 °C. In order to obtain a homogeneous coating, the samples were given a diffusion heat treatment in argon at 700 °C for 10 h.

*Table 7.2*   Details of investigated coatings

| SUPERCOAT name | Lab producer | Slurry/HVOF | Coating type | Coating description | Substrate |
|---|---|---|---|---|---|
| P91 | – | – | – | P91 substrate material | P91 |
| HVOF-1262F | Monitor | HVOF | Overlay | Ni-20Cr | P91 |
| HVOF-1260F | Monitor | HVOF | Overlay | Ni-50Cr | P91 |
| SCOAT/650/8-12 | UNN | Aluminised | Diffusion | Pack aluminised P92 | P92 |
| MX1 | Monitor | Slurry | Overlay | Silica-alumina-chromia slurry | P92 |
| IPCOTE 9183 R1 | IP | Slurry | Diffusion | Aluminium slurry, cured 560°C | P92 |
| Tungsten 6887 | IP | Slurry | Diffusion | Tungsten-magnetic sputtered, then topped with IPCOTE | P92 |
| Molybdenum 6883 | IP | Slurry | Diffusion | Molybdenum-magnetic sputtered, then topped with IPCOTE | P92 |

The two sputter-coated samples (tungsten and molybdenum) were supplied by Indestructible Paints and first sputter-coated with the respective element, then the IPCOTE slurry coating was applied in the same manner as that outlined above. The materials were then subsequently given a heat treatment at 700 °C in pure argon for 10 h.

The MX1 is an overlay coating consisting of silica particles embedded in a matrix of alumina and chromia, provided by Monitor Coatings. After each application of the slurry, the samples were subjected to a heat treatment at 520 °C. The slurry application process was repeated until a desired coating thickness of 50 μm was achieved.

The final coating to be examined was on a pack aluminised P92 sample, provided by the University of Northumbria (Newcastle upon Tyne, UK). Aluminium was coated onto the substrate using a pack aluminising process at 650 °C. Details of the aluminising procedure have been reported elsewhere [8]. The sample was given a subsequent diffusion heat-treatment to assist incorporation of aluminium into the substrate.

## 7.3    Experimental method

Before testing, a standard procedure was developed for all participating institutes in the project concerning steam testing. Details of the standard procedure are given in Table 7.3. This paper concentrates on the results of steam oxidation testing for 1000 h; results for longer testing times have been presented elsewhere [9–13].

For the oxidation studies, the specimens were exposed to an isobaric (1 bar) $Ar + 50\%\ H_2O$ atmosphere at a flow rate of 2.0 L h$^{-1}$ in a horizontal furnace equipped with a 100 mm diameter quartz tube. In all tests, the argon/water vapour mixture was generated by bubbling high-purity argon at atmospheric pressure through a saturator (humidifier), which was controlled at a fixed temperature. The exposures were carried out up to 1000 h at a temperature of 650 °C ($\pm 3$ °C), during which the specimens were cooled to room temperature under an argon atmosphere every 250 h for weight change measurements. After exposure, the oxidation products were

*Table 7.3*    Definition of laboratory steam testing method

| Parameter | Lower value | Upper value | Comments |
| --- | --- | --- | --- |
| Sample dimensions (mm) | $20 \times 10 \times 3$ | No value | Samples must be accessible to steam from all sides |
| Temperature (°C) | 648 | 653 | To be measured on dummy sample between test samples |
| Pressure (bar) | 1 | 300 | |
| Testing time (h) | 1000 (AUT, UCM, UNN, FZJ) | 10 000 (INTA, ALS) | |
| Cyclic times (weeks) | 2 | 4 | Samples to be removed and weighed periodically |
| Cooling time and regime | Report test conditions | Report test conditions | |
| Steam flow velocity (cm s$^{-1}$) | No value | 5 | |

characterised by optical metallography and scanning electron microscopy (SEM) with energy dispersive X-ray analysis (EDX). Before mounting the samples in resin for metallographic cross-section analyses, the specimens were sputter-coated with a thin gold layer and subsequently electroplated with nickel. This coating provided protection of the surface oxide layer during grinding and polishing and ensured better optical contrast between the oxide and mounting material.

## 7.4    Results and discussion

The 1000 h test results showed very low oxidation rates compared to the non-coated reference sample. The kinetics curves for the coatings used in Table 7.2 are shown in Fig. 7.1. The most important results from this figure can be summarised as follows:

A broad range of oxidation rates was seen for the coated samples, however, all coated samples gained less weight than the non-coated substrate material. While most of the samples continually gained weight with oxidation time, the MX1 sample initially lost weight and after around 500 h, no further alteration in weight change was noted (end weight change $-1.0$ mg cm$^{-2}$). The pack aluminised P92 sample gained almost no weight for the duration of the experiment (end weight change $+0.05$ mg cm$^{-2}$). For comparison, after 1000 h exposure time, the non-coated P91 substrate material had an end weight change of $+8.4$ mg cm$^{-2}$. To understand the differences in oxidation behaviour, it is necessary to examine the exposed samples in cross-section. The next section explains and discusses the observations found.

Beginning with the non-coated P91 substrate sample, it can be seen in Fig. 7.2 that the sample developed a very thick, dense oxide scale (up to 100 μm) that consisted of two layers. The outer layer consisted mainly of iron oxide, and the inner layer consisted of a mixture of iron and iron–chromium oxides. For the most part, the two-layer oxide was uniform in thickness, however, it can be seen in Fig. 7.2 that in certain locations, the scale was very thin, consisting solely of the external iron oxide layer. No iron–chromium mixed oxide layer was found beneath the iron oxide in this instance. Within the external iron oxide layer, another form of iron oxide was found, which when examined closely had a whisker-like morphology. It was assumed that this oxide grew first and was later consumed by a much faster growing iron oxide. From EDX analyses, no difference in chemical compositions of the two iron oxides was noted.

The HVOF-applied Ni–20Cr coating consisted of a single-phase γ-Ni microstructure. A thin oxide layer was also found on the splat boundaries within the coating.

After 1000 h of oxidation, the external surface of the Ni–20Cr coating formed a dense, adherent chromia scale with a thickness of around 1 μm (Fig. 7.3). At the metal/coating interface, a thin non-continuous mixed iron–chromium oxide was also found that tended to coincide with open pores exposed to the oxidising environment. No cracks were seen in the coating after oxidation and the coating was very adherent to the substrate.

The HVOF-applied Ni–50Cr coating consisted of a two-phase microstructure (inset in Fig. 7.4), which was identified as γ-Ni (light phase) and α-chromium (dark phase). A thin oxide layer was also found on the splat boundaries within the coating.

After 1000 h oxidation, the external surface of the Ni–50Cr coating formed a dense, adherent chromia scale with a thickness of around 2 μm (Fig. 7.4). At the metal/coating interface, a thin non-continuous mixed iron–chromium oxide was also found that

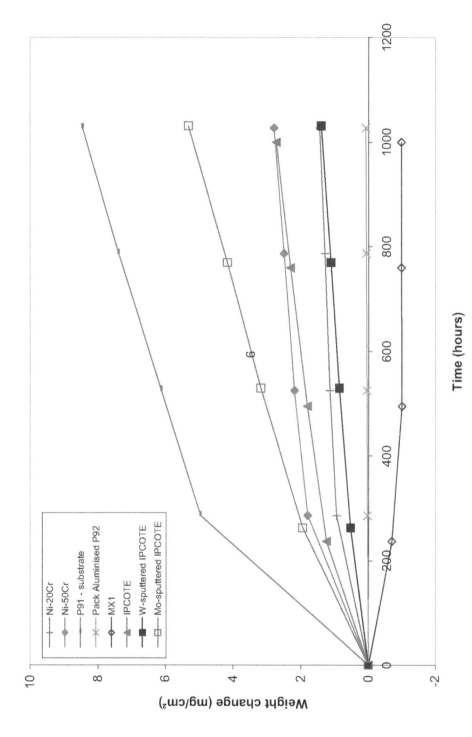

7.1 Kinetics curves for selected coatings (coating details given in Table 7.2)

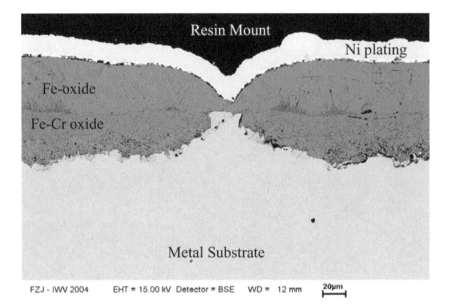

*7.2* P91 substrate material after 1000 h exposure to Ar + 50% $H_2O$ at 650 °C

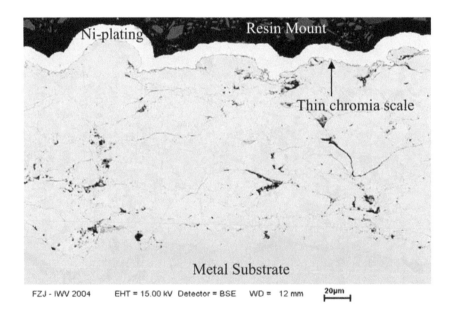

*7.3* Ni–20Cr after 1000 h exposure to Ar + 50% $H_2O$ at 650 °C

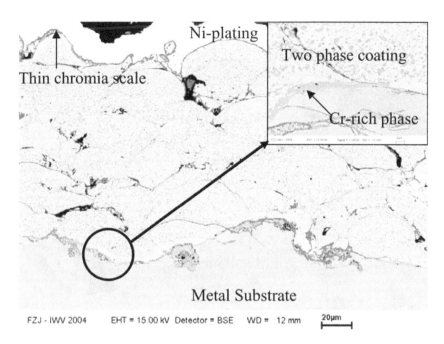

Thin chromia scale

Ni-plating

Two phase coating

Cr-rich phase

Metal Substrate

FZJ - IWV 2004     EHT = 15.00 kV  Detector = BSE    WD =  12 mm     20µm

*7.4* Ni–50Cr after 1000 h exposure to Ar + 50% $H_2O$ at 650 °C

tended to coincide with open pores exposed to the oxidising environment. No cracks were seen in the coating after oxidation and the coating was very adherent to the substrate.

It is important to note that at the metal/coating interface, a chromium-rich phase was observed as a thin line along the interface boundary (inset in Fig. 7.4). This was later determined to be chromium carbide, and had attained a thickness of around 2 µm. The presence of the carbide suggests chromium diffusion into the substrate, and combination with the dissolved carbon in the alloy. Prolonged high-temperature exposure would eventually lead to further thickening of the brittle carbide layer and subsequent spallation and destruction of the coating from the substrate would be expected.

A further point when comparing the two Ni–Cr HVOF-coated samples is the apparent reversal of the kinetics results. From Fig. 7.1, it can clearly be seen that the weight change of the Ni–50Cr-coated alloy was almost twice that of the Ni–20Cr coating. This at first seems questionable due to the higher chromium content of the former coating. However, inspection of the coating microstructures (Figs. 7.3 and 7.4) indicates that the differences in measured weight changes are related to differences in the porosities of the two coatings with the result that, in the case of the Ni–50Cr coating, more oxygen access occurred into the coating up to the coating/ steel interface (Fig. 7.4).

The IPCOTE slurry coating exhibited aluminium diffusion into the metal substrate to a depth of around 30 µm (Fig. 7.5). Coating coverage appeared to be somewhat inconsistent due to the irregular mode of oxidation attack. An external iron oxide scale formed uniformly over the substrate surface, with a thickness of around 30 µm. Internal attack was also observed and consisted of a mixed iron–chromium oxide,

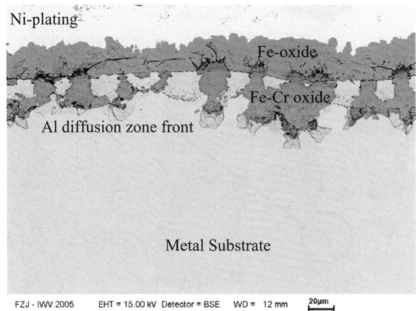

FZJ - IWV 2005    EHT = 15.00 kV  Detector = BSE    WD =  12 mm    20µm

*7.5  IPCOTE after 1000 h exposure to Ar + 50% H₂O at 650 °C*

although it appeared to have formed in regions where aluminium diffusion had not taken place. In the regions where aluminium had diffused into the substrate, very fine acicular aluminium nitride particles were observed. This result appears to contradict those of other authors who have examined similar aluminium slurry coatings in steam-bearing environments, and observed protective oxidation behaviour for times up to 40 000 h [10,13]. It was found that improper surface preparation of the substrate in the present study led to non-uniformity in the coating and hence the results observed.

In order to assist in inhibiting the inward diffusion of aluminium into the substrate, two variations of the IPCOTE coating were produced. The first consisted of sputter coating a molybdenum interlayer on the surface of the substrate and then applying the IPCOTE slurry coating. It was hoped that the Mo-layer would inhibit aluminium diffusion into the substrate.

After 1000 h of exposure to the steam-bearing environment, it can be seen that the molybdenum interlayer was not successful in inhibiting aluminium diffusion into the substrate (Fig. 7.6). A clear white line indicates the molybdenum interlayer and either side of it extensive oxidation took place. The oxidation behaviour appeared very similar to the normal IPCOTE (see Fig. 7.5) where an external iron oxide scale formed above the interlayer surface, with a thickness of around 30 µm. Internal attack was also observed and consisted of a mixed iron–chromium oxide, although it appeared to have formed in regions where aluminium diffusion had not taken place. Again, in the regions where aluminium had diffused into the substrate, very fine aluminium nitride needles were observed.

The second sample was treated with a sputter-coated interlayer that used tungsten. After 1000 h exposure to the steam-bearing environment, it can be seen that the

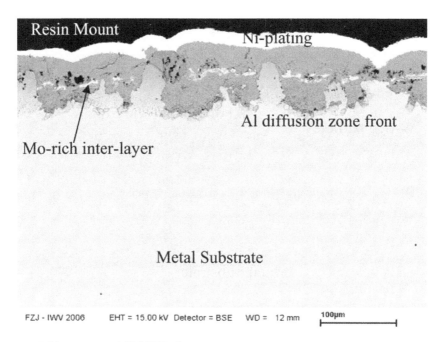

FZJ - IWV 2006     EHT = 15.00 kV  Detector = BSE    WD =   12 mm    100µm

7.6  Mo-sputtered IPCOTE after 1000 h exposure to Ar + 50% $H_2O$ at 650 °C

tungsten interlayer was not totally successful in inhibiting aluminium diffusion into the substrate (Fig. 7.7). However, the attack is not as extensive as with the molybdenum interlayer. This too can be seen from the lower weight gain of the tungsten-coated sample from the kinetics curves in Fig. 7.1. The tungsten can be seen as dispersed white flecks on the surface of the sample, and in comparison to the molybdenum-coated sample, no external iron oxides formed with the tungsten-coated sample. Isolated regions of internal attack in the substrate are noted and these regions formed a mixed iron–chromium oxide scale. Aluminium diffusion is seen into the coating to a depth of around 50 µm and where the aluminium has inwardly diffused, acicular aluminium nitride particles were also observed. This nitridation zone had a depth of 20 µm into the substrate.

The pack aluminised P92 sample formed an intermetallic coating of thickness about 20 µm (Fig. 7.8). The phase was identified as FeAl. After 1000 h of oxidation, the coating formed an extremely thin alumina scale, less than 1 µm in thickness. Beneath the coating, a series of fine pores formed, assumed to be Kirkendall porosity. Immediately beneath these pores, fine acicular precipitates were observed and these were determined to be aluminium nitride. In some instances, through-cracks in the coating were also observed. An increase in volume fraction of the tungsten-containing particles in the P92 substrate was found immediately beneath the coating. It was assumed that the FeAl phase has very limited solubility for tungsten and as the coating grew into the substrate, tungsten was continually being rejected from the intermetallic coating and reprecipitated at the coating/substrate interface.

The final sample to be examined was the MX1 coating. This coating consisted of silica particles embedded in a matrix of alumina and chromia. The coating had a thickness of 50 µm. After 1000 h exposure, very little oxidation damage was observed

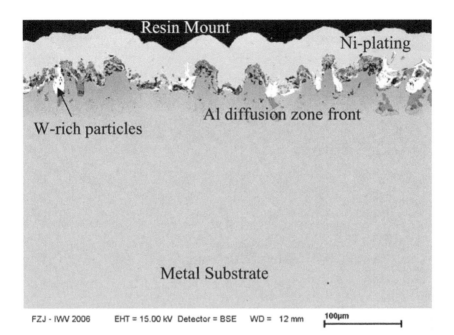

7.7  W-sputtered IPCOTE after 1000 h exposure to Ar + 50% H$_2$O at 650 °C

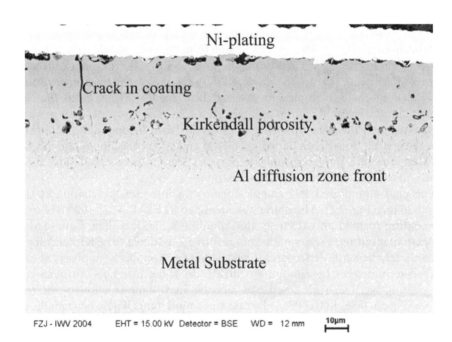

7.8  Pack aluminised P92 after 1000 h exposure to Ar + 50% H$_2$O at 650 °C

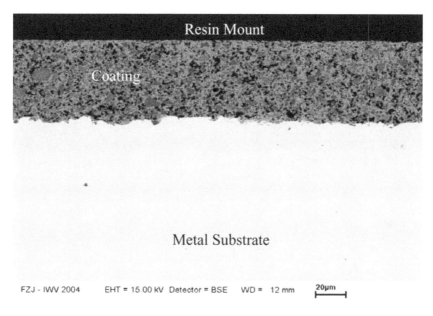

FZJ - IWV 2004      EHT = 15.00 kV   Detector = BSE     WD =   12 mm      20µm

*7.9* MX1 after 1000 h exposure to Ar + 50% $H_2O$ at 650 °C

(Fig. 7.9). The coating remained adherent and the only oxidation attack observed was at the coating/metal interface where a very thin, non-continuous iron–chromium oxide was found.

A batch of samples was subsequently tested in a high-pressure rig with operating conditions of 650 °C and 300 bar steam pressure. It was shown by Morey [16] that supercritical steam can be used as a solvent for many minerals, and that silica is highly soluble in this fluid. Therefore, it was considered essential to assess critically the behaviour of coating MX1 under such conditions to determine its suitability as a candidate for steam turbine usage.

The high-pressure test was conducted in a specially constructed rig for a period of 1500 h, with intermittent sample weight change measurements. Results of the experiment for coating MX1 appear in Fig. 7.10 in the form of a weight change versus time curve.

From the kinetics curves, it can be seen that after the first weight change measure-ment (24 h), a dramatic weight loss was recorded ($-5.0$ mg cm$^{-2}$), indicating that severe damage to the coating had already taken place. On further testing, it was seen that a continual weight increase was recorded; however, the slope of the curve was much greater than that recorded at 1 bar steam pressure. Surface examination of the exposed sample after 733 h of exposure revealed a thinning of the coating in the inner regions of the sample (Fig. 7.11: dark areas, compared to non-damaged coating, light coloured areas) and the coating had spalled off at the sample edges.

Examination of the solubility data presented by Morey [16] shows that at 600 °C and at 333 bars, quartz has a solubility of around 0.04 wt.% in the superheated steam. The author also performed experiments on silica glass at temperatures up to 500 °C, and said that silica had a higher solubility in steam than quartz. It is therefore assumed that the solubility of silica at 650 °C and 300 bar is higher than 0.04 wt.%. As the steam was not in a stagnant environment, but continually being moved through

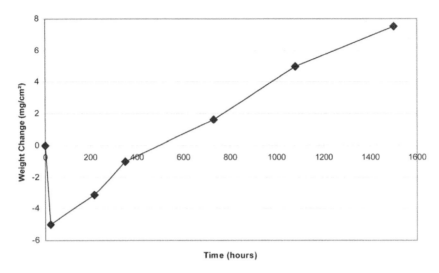

*7.10* Kinetics curves for MX1 coating at 650 °C and 300 bar steam pressure

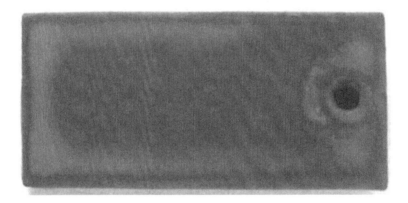

*7.11* Surface macro-photograph of MX1 coating after 733 h exposure to steam at 650 °C and 300 bar pressure

the test rig, it was continually able to dissolve silica and hence the constant thinning of the coating with increasing exposure time.

For conciseness, the main results of degradation for the different coatings are presented in Table 7.4. Degradation mechanisms have been categorised into four different fields, and each coating was allotted the main degradation mechanism in the table. It can be seen that those coatings containing aluminium all tended to degrade by an interdiffusion mechanism. Attempts to reduce the inward diffusion of aluminium through an interlayer of either molybdenum or tungsten failed to do so. In one case with aluminium-bearing coatings, the associated Kirkendall porosity was also seen as a further degradation mechanism.

The high chromium containing Ni–Cr coating formed a brittle interlayer at the coating/substrate interface, which on growing thicker, would eventually lead to

*Table 7.4* Summary table of degradation mechanisms for coated samples

| SUPERCOAT name | Slurry/HVOF | Coating type | Coating description | Degradation mechanism | | | |
| --- | --- | --- | --- | --- | --- | --- | --- |
| | | | | Al diffusion | Kirdendall porosity | Cr-carbide | Mechanical failure |
| HVOF-1262F | HVOF | Overlay | Ni-20Cr | | | | |
| HVOF-1260F | HVOF | Overlay | Ni-50Cr | | | ✓ | |
| SCOAT/650/8-12 | Aluminised | Diffusion | Pack aluminised P92 | ✓ | ✓ | | |
| MX1 | Slurry | Overlay | Silica-alumina-chromia slurry | | | | ✓ |
| IPCOTE 9183 R1 | Slurry | Diffusion | Aluminium slurry, cured 560°C | ✓ | | | |
| Tungsten 6887 | Slurry | Diffusion | Tungsten-magnetic sputtered, then topped with IPCOTE | ✓ | | | |
| Molybdenum 6883 | Slurry | Diffusion | Molybdenum-magnetic sputtered, then topped with IPCOTE | ✓ | | | |

coating failure through spallation. The presence of the carbide phase in the Ni–20Cr coating was not observed and hence this coating could be utilised as a candidate coating for steam turbines, subject to further mechanical testing assessments.

Finally, at 1 bar steam pressure, the MX1 coating behaved in a rather stable manner, however testing at 300 bar revealed severe degradation through solubility of the silica coating in supercritical steam. The continual removal of the coating in the steam environment led to the eventual breakdown of the coating and subsequent substrate oxidation.

## 7.5    Conclusions

It has been successfully demonstrated here that different coatings on 9% chromium steels can reduce the oxidation rates in steam-bearing environments remarkably. Coatings containing aluminium, chromium and silicon were applied using a number of different coating methods and tested in atmospheric and high-pressure steam at 650 °C. An in-depth examination of the degradation mechanisms of selected coatings was made and the results were summarised.

Coatings containing silicon performed poorly at high pressure due to dissolution of the silica in the supercritical steam, despite showing adequate protection at atmospheric pressure. The two HVOF Ni–Cr coatings showed excellent behaviour; however, the Ni–20Cr coating performed much better than the Ni–50Cr candidate. Finally, coatings containing aluminium formed very thin external alumina scales but also encountered continual inward diffusion of aluminium into the substrate. Means to try to arrest this behaviour, including the production of metallic interlayers before application of the aluminium containing coating, proved unsuccessful.

## Acknowledgments

The authors are grateful to the SUPERCOAT project partners for their invaluable efforts during the course of the programme and the kind permission of ALSTOM Power to publish this work. SUPERCOAT is an RTD Project, funded by the European Community under the 'ENERGY' (EESD) Programme (1998–2002). We express our sincere gratitude to the E.U. for financing this work through Contract No: ENK5-CT-2002-00608.

Sincere thanks must also be made to Prof. Dr H. Vogel from the TU-Darmstadt, Germany for conducting the high-pressure testing.

## References

1. K. Weizierl, *VGB Kraftwerkstechnik*, 74 (1994), 2.
2. R. Blum, J. Hald, W. Bendick, A. Rosselet and J. C. Vaillant, *VGB Kraftwerkstechnik*, 74 (1994), 8.
3. S. Jianian, Z, Longjiang and L. Tiefan, *Oxid. Met.*, 48(3/4) (1997), 347.
4. P. Mayer and A. V. Manolescu, in *High Temperature Corrosion*, 368–379, ed. R. A. Rapp. National Association of Corrosion Engineers, Houston, TX, 1983.
5. A. Rahmel and J. Tobolski, *Corros. Sci.*, 5 (1965), 333.
6. Y. Ikeda and K. Nii, *Trans. JIM*, 26 (1984), 52.
7. J. Ehlers, D. J. Young, E. J. Smaardijk, A. K. Tyagi, H. J. Penkalla, L. Singheiser and W. J. Quadakkers, *Corr. Sci.*, 48 (2006), 3248.

8. Z. D. Xiang, S. R. Rose and P. K. Datta, *J. Mater. Sci.*, 41 (2006), 7353.

9. A. Agüero, R. Muelas, A. Pastor and S. Osgerby, *Surf. Coatings Technol.*, 200 (2005), 1219.

10. A. Agüero and R. Muelas, *Mater. Sci. Forum*, 461–464 (2004), 957.

11. M. Scheefer, M. B. Henderson, A. Agüero, B. Allcock, B. Norton, D. N. Tsipas, and R. Durham, *Mater. Adv. Power Eng. Energy Technol.*, 53 Pt III (2006), 1553.

12. M. Scheefer, R. Knödler, B. Scarlin, A. A. (Agüero) Bruna and D. N. Tsipas, *Mater. Corros.*, 56 (2005), 907.

13. A. Agüero, *Mater. Adv. Power Eng. Energy Technol.*, 53 Pt II (2006), 949.

14. B. Sundman, B. Jansson and J. O. Andersson, *CALPHAD*, 9 (1985), 153.

15. S. M. Dubiel, *Hyperfine Interact.*, 111 (1998), 211.

16. G. W. Morey, *Geol. Soc. Am. Bull*, 61 (1950), 1488.

# 8

# Thermogravimetry-mass spectrometry studies in coating design for supercritical steam turbines*

## F. J. Pérez and S. I. Castañeda

*Grupo de Investigación de Ingeniería de Superficies,*
*Universidad Complutense de Madrid, Departamento de Ciencias de Materiales,*
*Facultad de Ciencias Químicas, Madrid 28040, Spain*

## 8.1    Introduction

Many industrial processes involve oxidation, i.e. a metal reacts in air to form and sustain a protective oxide. There can be several oxide products, some of which are less desirable. For example, wüstite is a defective oxide of iron that forms rapidly at about 843 K on steel. The influence of steam on the high-temperature oxidation of Cr, Fe, and FeCr and FeCrNi alloys has received considerable attention [1–5]. A number of authors [6,7] have studied the oxidation kinetics of ferritic/martensitic steels such as P91 and P92 under different conditions, including temperatures of 600–700 °C and various atmospheres ($O_2$, $N_2$, Ar, water, steam, etc.). These studies investigated the initial stages of oxidation but no attempt was made to verify experimentally the volatile species formed. Pérez-Trujillo and Castañeda [8] have conducted experimental work on P91 and P92 steels oxidised at 650 °C and a pressure of 1 atmosphere in Ar + 10% steam. They combined thermogravimetric measurements (TG) and mass spectrometry (MS) during the oxidation process, determining volatile chromium hydroxides and oxyhydroxide species such as: $CrOOH(g)$, $Cr(OH)_3(g)$, $Cr(OH)_6(g)$ and $CrO_2(OH)_2(g)$ for P91 steel and $Cr(OH)_2(g)$, $Cr(OH)_3(g)$, $CrO(OH)_2(g)$, $Cr(OH)_4(g)$ and $CrO(OH)_4(g)$ for P92 steel. These results indicate that the evaporation of chromium oxides and oxyhydroxides occurs when steam is present. Under these oxidation conditions, Pérez-Trujillo and Castañeda did not find evidence for the presence of volatile species $CrO_2OH(g)$ in the P91 and P92 ferritic steels studied, but they did detect $CrO_2(OH)_2(g)$ species in the P91. The authors also reported the formation of $FeO(g)$ and $FeO_{1.5}(g)$ volatile species during the oxidation of P92 steel. They concluded that breakaway oxidation begins earlier for P92 steel than for P91 steel.

The oxyhydroxide vapour species, $CrO_2OH(g)$ [9,10] and $CrO_2(OH)_2(g)$ [11,12] have been identified as volatiles of chromium of the oxidised solid samples of $Cr(s)$ and $Cr_2O_3(s)$ in the presence of air and steam atmospheres at high temperatures. Also, the species $Cr_3O_9(g)$, $Cr_4O_{12}(g)$ and $Cr_5O_{15}(g)$ have been identified as volatiles of chromium in air at temperatures below 500 K [13,14]. Several researchers [15–20] have attributed the formation of volatile species such as $CrO_3(g)$, $CrO_2OH(g)$ or $CrO_2(OH)_2(g)$ to the breakdown of chromia scales at temperatures near 1000 °C, and the subsequent depletion of Cr in the scale formed.

* Reprinted from F. J. Pérez et al.: Thermogravimetry-mass spectrometry studies in coating design for supercritical steam turbines. *Materials and Corrosion*. 2008. Volume 59. pp. 409–13. Copyright Wiley-VCH Verlag GmbH & Co. KGaA.

Iron aluminides are well known to have good resistance to oxidation and sulphidation due the formation of an external, protective alumina scale [21–29], which is more stable under these environmental conditions and has already been shown to be very resistant to steam oxidation at 650 °C [6,30,31].

Slurry coatings are suitable for internal and external application to small parts (e.g. blades) and large components (e.g., steam pipes, casings, and valve internals) and can be relatively easily applied to objects of complex shapes. Slurries of aluminium can be considered to be diffusion coatings since the adhesion between the deposited layer and the substrate is of a chemical nature due to the heat treatment of the coating. Migration of the elements between the coating and the substrate takes place. The main mechanism of degradation during oxidation of these coatings is interdiffusion with the substrate where Al spreads towards the interior of the substrate and Fe from the base material moves out, leading finally to a minimum critical aluminium concentration in the surface of the coating below which the capacity to form a protective oxide layer is lost [30].

In the present work, the oxidation of iron, aluminium and P91 steel with and without an FeAl-slurry coating at 600–650 °C in $Ar + 80\%$ $H_2O$ for different times has been investigated. The main objective was to find a direct relationship between the results of TG-MS experiments and the formation of volatile inorganic oxyhydroxide species that appear for these samples during the initial period of oxidation under these conditions. Comparisons with thermodynamic predictions for the system and previously published work have also been made. The findings of the study provide important inputs for coating design based on experimental findings and valid thermodynamic calculations.

## 8.2    Experimental

Oxidation tests have been carried out on Al, Fe (references supplied by Goodfellow), ferritic/martensitic steel (P91), and FeAl slurry-coated P91. In Table 8.1, the chemical composition of the P91 steel is given. Specimens with dimensions of 10 mm × 20 mm × 4 mm were cut for each test and polished to 600-grit with SiC paper. Before oxidation, all samples were cleaned with acetone in an ultrasonic bath for 10 min.

The paint used for the FeAl slurry coating of FeAl was IPCOTE 9183 (supplied by 'Indestructible Paint'): $Cr_2O_3$ (2.5–10%), $H_3PO_4$ (10–25%) and Al powder (25–50%). Water and diluted chromic acid were used as solvents for the paint. The samples were subjected to several thermal treatments and were then lightly polished with SiC paper (15 μm) to remove excess paint. The coatings were applied at INTA [6].

The oxidation tests of Al, Fe and P91 steel involved exposure at 650 °C to an atmosphere of Ar + 80% $H_2O$ for an exposure time of 100 h. The oxidation test of FeAl-slurry/P91 was of 150 h duration under the same conditions. This test was undertaken in two stages without removing the sample from the system. The first stage lasted for 100 h then, after a small interval, was continued for another 50 h in order to provide more data.

*Table 8.1*    Chemical composition of the stainless steel studied

| Material | Wt.% | | | | | | | | | | | | | |
|---|---|---|---|---|---|---|---|---|---|---|---|---|---|---|
| | Fe | Cr | Mo | Mn | Si | C | Ni | P | V | Al | Nb | N | S | W B |
| P91 | 89.35 | 8.10 | 0.92 | 0.46 | 0.38 | 0.100 | 0.33 | 0.020 | 0.18 | 0.034 | 0.073 | 0.049 | 0.002 | – – |

Routinely, for the oxidation tests, the thermobalance was first calibrated (verifying level zero) and the sensitivity of the mass spectrometer and background of the system were measured. This was done using two quartz crucibles as calibration samples placed in each furnace and maintained under the same conditions of oxidation. In all of the results of mass spectrometry, the respective background measurements were subtracted. In this way, the data for the sample and not the system (the apparatus or any gas contaminants) were obtained. In order to evaluate the results of mass spectrometry of the oxidised samples, a database based on the composition of each material and all possible chemical reactions with the environment was needed. This was obtained by means of theoretical thermodynamic simulations using two computer programs. The Thermo-calc program (Software AB: Foundation of Computational Thermodynamics 1995–2003, Stockholm, Sweden) and the HSC Chemistry for Windows program (Software Outokon: version 3.02, Finland, 1994) were used. The simulated conditions of oxidation were the same as those used in the experiments. All of the theoretical results obtained were compared with those in the literature.

Oxidation experiments and thermogravimetric (TG) measurements were made using a SETARAM–TAG 16 thermobalance (with a sensitivity of $10^{-7}$ g) and a symmetrical thermoanalyser. Initially, the furnaces were evacuated by a membrane pump to a pressure of $10^{-2}$ mbar. A constant flow of 500 mL min$^{-1}$ of high purity argon gas (99.999% purity) was then used to fill both furnaces (that containing the sample being studied and that containing the reference $SiO_2$ crucible) until atmospheric pressure was reached. The furnace system was equipped with a humidifier producing the reaction gas consisting of Ar $+ 80\%$ $H_2O$. A flow rate of argon of 16 mL min$^{-1}$ through a water flask (humidifier) was used.

Microstructural analysis by scanning electron microscopy (SEM) and X-ray energy dispersive spectrometry (EDS) measurements were performed in a JEOL-JSM 6400 scanning electron microscope operated at 20 kV.

## 8.3    Results and discussion

Figure 8.1 shows the volatile species found by mass spectrometry of an oxidised Fe sample at 650 °C in the Ar $+ 80\%$ $H_2O$ atmosphere for 100 h. These species have been

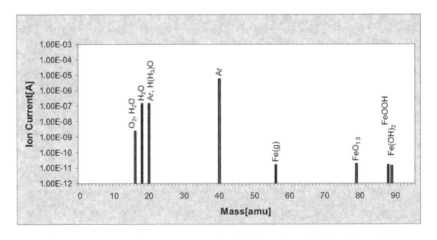

*8.1* Mass spectrum of iron oxidised at 650 °C in Ar $+80\%$ $H_2O$ for 100 h

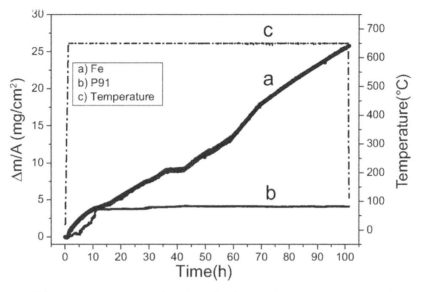

*8.2* TG measurements as a function of oxidation time and temperature for samples at 650 °C in Ar + 80% $H_2O$ for 100 h: (a) Fe and (b) P91 steel

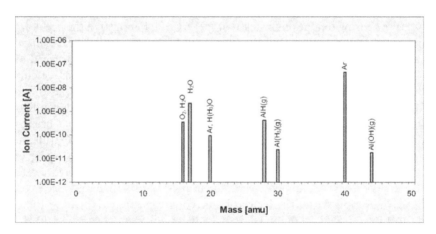

*8.3* Mass spectrum of aluminium oxidised at 650 °C in Ar + 80% $H_2O$ for 100 h

identified as Fe(g), $FeO_{1.5}$, FeOOH and $Fe(OH)_2$. The results indicate that, during the initial stages of oxidation of iron under these conditions, iron evaporates at the sample surface together with oxide and oxyhydroxide species. These volatile species are present during breakaway corrosion. The mass change as a function of test time (curve a) during oxidation of the iron and the temperature profile (curve c) are shown in Fig. 8.2. As can be seen from this figure, in the first 100 h of oxidation, the sample was oxidised constantly with an almost linear behaviour and a mass gain of around 25 mg $cm^{-2}$.

The results of the mass spectrometry for aluminium are shown in Fig. 8.3. The spectrum was obtained at 600 °C in an Ar + 80% $H_2O$ atmosphere for 100 h. AlH(g), $Al(H_3)(g)$, and Al(OH)(g) were identified as the volatile species. The presence of these

*Table 8.2*  Volatile species detected by the mass spectrometry in samples oxidised at 650°C in Ar+80% $H_2O$ for different times

| Sample | Volatile Species | | |
|---|---|---|---|
| | 15h | 100h | 150h |
| Fe | – | Fe(g), $FeO_{1.5}$, FeOOH, $Fe(OH)_2$ | – |
| Al* | – | AlH(g), $Al(H_3)$(g), Al(OH) | – |
| P91 | – | Cr(g), HNi(g), $Fe_{0.947}O$, $FeO_{1.5}$, $CrO_2$(g), CrOOH, FeOOH, Mo(g), $CrO_2(OH)_2$, $Cr(OH)_6$, $Fe_2O_3$(g), CO and $CO_2$ | – |
| FeAl/P91 | Al(g), AlO(g) | Al(g), AlO(g) and FeOOH | Al(g), AlO(g) and FeOOH |

* Sample oxidised at 600°C.

volatile species confirms that, under these oxidation conditions, aluminium reacts with steam, evaporating in the form of aluminium hydrides and hydroxide.

Mass spectrometry results for oxidised P91 steel under the same conditions are shown in Table 8.2. These species have been identified as Cr(g), HNi(g), $Fe_{0.947}O$, $FeO_{1.5}$, $CrO_2$(g), CrOOH, FeOOH, Mo(g), $CrO_2(OH)_2$, $Cr(OH)_6$, $Fe_2O_3$(g), CO and $CO_2$. Previously, some of these species were identified by theoretical calculations using the HSC and Thermocalc programs. The presence of the volatile species $CrO_2(OH)_2$(g) during oxidation in steam of P91 for 100 h indicated the breakdown of chromia scale on the sample. It is known that chromium depletion has important implications for the long-term stability of the protective oxide scale formed on P91 steel, species that some authors have proposed by thermodynamic calculations for other oxidised ferritic/martensitic steels in steam [11–20]. The timescale for the breakdown of the protective oxide scale that formed on P91 steel depends on the evaporation rate of $CrO_2(OH)_2$(g), a higher rate giving a more rapid breakdown. This oxidation may result in the formation of $Fe_2O_3$ (haematite)/$Fe_3O_4$ (magnetite)/ FeO (wüstite), and 'breakaway corrosion' as confirmed by mass spectrometer measurements with the appearance of other species such as: oxides, hydroxides and oxyhydroxide of iron. In addition, the fragmentation of other volatile species of $H_2O$, CO, $CO_2$, $O_2$, $N_2$ and Ar molecules was detected in this study. However, the details relating to these species in the mass spectra have not been the subject of detailed analysis because they are related to gases that belong to part of the background measurements.

Table 8.2 includes the results of mass spectrometry for FeAl-slurry/P91 samples oxidised at 650 °C in Ar + 80% $H_2O$ for 15–150 h. The species were identified as Al(g) and AlO(g) after 15 h of exposure. Al(g), AlO(g), and FeOOH(g) species were identified after exposure for 100 and 150 h.

The presence of these volatile species confirms the initial loss of aluminium from the coating in gas and oxide form. This loss and gradual depletion of aluminium from the sample would also relate to the loss of the protective alumina coating and, therefore, being unprotected, to oxidation of the sample. This oxidation would be related to the presence of the volatile species of oxyhydroxide of iron. In addition, it may be supposed that during oxidation, some of the aluminium from the coating diffused inwards or reacted with the substrate (P91 steel) and then formed a protective intermetallic coating.

The thermogravimetric results for the P91 steel oxidised at 650 °C in the Ar + 80% $H_2O$ atmosphere for 100 h are shown in Fig. 8.2 (curve b). The results demonstrate that this sample oxidised six times less than iron, and a mass gain of approximately 4 mg cm$^{-2}$ was observed. In accordance with the results, the good resistance to steam oxidation of the P91 steel compared to pure iron is attributed to the presence of the additional alloying elements (Cr, Mo, Ni, C, etc.). These elements contribute to the formation of protective films that help to avoid catastrophic corrosion. The active presence or loss of some of them was verified by means of the results of the mass spectrometry.

Figure 8.4a and 8.4b show the results of the TG measurements for the oxidised samples of Al and FeAl-slurry/P91. Both were oxidised under the same conditions as the P91 steel. The oxidised aluminium (Fig. 8.4a) displays parabolic behaviour and an increase in mass of around 0.06 mg cm$^{-2}$. For the FeAl-slurry/P91 sample (Fig. 8.4b), a constant loss of mass is observed, amounting to approximately 3.6 mg cm$^{-2}$. The thermogravimetric results for these samples have helped to explain the results of the mass spectrometry. The great difference during oxidation was the detection of an intense mass line for aluminium vapour in the FeAl/P91 sample with respect to the Al sample, as previously reported by the authors. This is due to the evaporation of Al from the FeAl coating in the form of gas and oxide. However, in the pure aluminium sample, mass lines of aluminium hydrides and hydroxides were detected. The results of thermogravimetric measurements and temperature measurements during the additional time of 50 h until completing 150 h of oxidation of FeAl-slurry/P91 under the same conditions as used previously are shown in Fig. 8.5. In this stage of oxidation, the sample gained a small additional mass of ~0.1 mg cm$^{-2}$. The small mass gain for this sample during oxidation is related directly

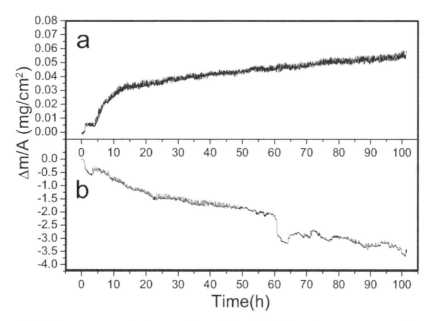

*8.4* TG measurements for samples oxidised at different temperatures and under the same conditions as the previous samples: (a) Aluminium at 600 °C and (b) FeAl-slurry/P91 at 650 °C

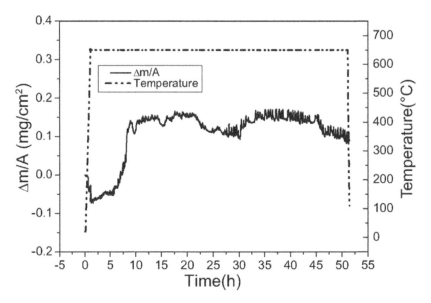

8.5 TG measurements of an FeAl-slurry/P91 sample oxidised at 650 °C in Ar + 80% $H_2O$ for an additional time of 50 h until completing a total time of exposure of 150 h

to the presence of FeOOH volatile species and the iron oxide layers that would have formed on the surface. These results indicate that, during the initial stages of oxidation under these conditions, the FeAl slurry coating on P91 steel shows higher oxidation resistance than the P91 steel substrate without a coating. The P91 steel with the slurry FeAl coating formed a protective alumina scale on its surface during oxidation.

The SEM images of the cross-section of the P91 steel without and with the FeAl coating after oxidation at 650 °C in Ar + 80% $H_2O$ for 100 and 150 h are shown in Fig. 8.6a and 8.6b, respectively. On the surface of the P91 sample without the coating (Fig. 8.6a), an oxide thickness of ~50 μm is observed. The EDS analysis of a small area of the cross-section of the oxide confirmed that a compound of three layers is present: a very superficial layer of haematite ($Fe_2O_3$), followed by a magnetite ($Fe_3O_4$) layer and, at the interface with the P91 base material, a mixed spinel of iron and chromium ($Cr_2FeO_4$).

Figure 8.6b shows the micrograph at a different magnification of the cross-section of FeAl-slurry/P91 after being oxidised under the same conditions as the P91 sample, but for 150 h. In the image, an 'oxide node' is observed. This had a diameter of around 40 μm. This node was probably formed by a crack, pore and/or some other imperfection in the FeAl protective coating. The EDS analysis and elemental mapping of the oxide node indicated that it contained aluminium in the outer part, forming a protective coating of alumina ($Al_2O_3$) and in other cases, it was absent. Superficial layers of haematite ($Fe_2O_3$) and magnetite ($Fe_3O_4$) were also observed. The inner part of the node displayed other layers: an intermetallic of mainly $Fe_2Al_5$ and a $Cr_2FeO_4$ mixed spinel, as was also observed for P91 without an FeAl coating. These layers that had formed in oxide on the P91 sample with an FeAl coating were also verified by means of the XRD technique in previous work [32].

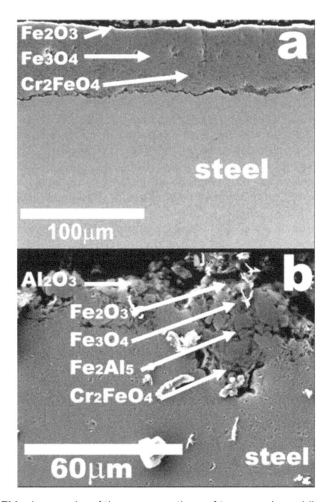

*8.6* SEM micrographs of the cross-sections of two samples oxidised at 650 °C in Ar + 80% $H_2O$ for different times: (a) P91 steel without coating for 100 h and (b) P91 steel with FeAl-slurry coating for 150 h

## 8.4    Conclusions

The oxidation behaviour of Fe, Al, and of P91 steel with and without an FeAl-Slurry coating was investigated. The samples were oxidised at 600–650 °C in an Ar + 80% $H_2O$ atmosphere for 100 to 150 h.

The volatile species Fe(g), $FeO_{1.5}$(g), FeOOH(g) and $Fe(OH)_2$(g) were detected for the Fe sample. Under the same oxidation conditions, AlH(g), $Al(H_3)$(g) and Al(OH)(g) volatile species were detected for the Al sample.

The presence of the volatile species $CrO_2(OH)_2$(g) during 100 h of steam oxidation indicated that the P91 steel loses its protective chromia ($Cr_2O_3$) scale due to the strong deterioration caused by the presence of steam. This result confirms observations by other authors on the oxidation of these steels and the results of theoretical calculations under the same conditions [11–19].

The loss of Al in the sample during the oxidation of the FeAl-slurry/P91 sample would be related later to the subsequent loss of alumina ($Al_2O_3$) and for that reason it would no longer be protected against oxidation. This loss of Al during oxidation is related thermally to its diffusion towards the interior of the substrate (P91 steel) followed by its reaction with iron to form $Fe_2Al_5$ intermetallic. In oxidation of the FeAl coating sample for up to 150 h, no evidence was found of another volatile species of chromium that could lead to catastrophic corrosion of the sample.

On the basis of these results, it is concluded that the samples of iron, aluminium, and P91 steel with and without the FeAl coating give rise to many volatile species by reaction of the material with the environment under investigation from the initial stages of oxidation and these soon form oxide layers on the surface.

According to the results of the TG measurements for the P91 steel with and without an FeAl coating, it has been verified that the FeAl-slurry increased the oxidation resistance of the P91 steel by up to 40 times, and SEM observation of the cross-sections of the oxidised samples further confirmed that the benefits of P91 with an FeAl coating over uncoated P91 steel are due to the smaller amount of oxide formed.

### Acknowledgments

The authors acknowledge the financial support of the European Community under project N° ENK5-CT-2002-00608-SUPERCOAT.

### References

1.  S. Jianian, Z. Longjiang and L. Tiefan, *Oxid. Met.*, 48 (1997), 347.
2.  H. Asteman, J.-E. Svensson, J.-G. Johansson and M. Norell, *Oxid. Met.*, 54 (2000), 11.
3.  A. Holt and P. Kofstad, *Solid State Ionics*, 69 (1994), 137.
4.  D. Caplan and M. Cohen, *J. Electrochem. Soc.*, 108 (1961), 438.
5.  C. S. Tedmond, Jr, *J. Electrochem. Soc.*, 113 (1966), 1966.
6.  A. Agüero, R. Muelas, R.B. Scarlin and R. Knödler, in *Materials for Advanced Power Engineering*, 1143. Université de Liége, European Commission, Ed. Forschungszentrum-Jülich, September 2002.
7.  W. J. Quaddakers and P. J. Ennis, 'The oxidation behaviour of ferritic and austenitic steels in simulated power plant service environments', in Proc. 6th Liége Conf. on Materials for Advanced Power Engineering, 123. Université de Liége, European Commission Ed. Forschungszentrum-Jülich, September 1998.
8.  F. J. Pérez-Trujillo and S. I. Castañeda, *Oxid. Met.*, 66(5/6) (2006), 231.
9.  Y. W. Kim and G. R Belton., *Met. Trans.*, 5 (1974), 1811.
10. E. M. Bulewicz, and P. J. Padley, *Proc. R. Soc. Lond.* A, 323 (1971), 377.
11. O. Glemser and A. Müller, *Z. Anorg. Allg. Chem.*, 334 (1964), 151.
12. M. Farber and R. D. Srivastava, *Combust. Flame*, 20 (1973), 43.
13. O. Glemser, A. Müller and U. Stocker, *Z. Anorg. Allg. Chem.*, 344 (1964), 25.
14. J. D. McDonald and J. L Margrave, *J. Inorg. Nucl. Chem.*, 30 (1968), 665.
15. H. Asteman, J. E. Svensson, L. G. Johansson and M. Norell, *Oxid. Met.*, 52(1/2) (1999), 95.
16. C. S. Tedmond, *J. Electrochem. Soc.*, 113 (1966), 766.
17. H. C. Graham and H. H. Davies, *J. Am. Ceram. Soc.*, 54 (1971), 89.
18. C. A. Stearns, F. J. Kohl and G. C. Fryburg, *J. Electrochem. Soc.*, 121 (1974), 945.
19. B. B. Ebbinghaus, *Combust. Flame*, 93 (1993), 119.
20. A. S. Khanna and P. Kofstad, in Int. Conf. Microsc. Oxid., 113. Institute of Materials, London, 1990.
21. Y. Zhang, B. A. Pint, G. W. Garner, K. M. Cooley and J. A. Haynes, *Surf. Coatings Technol.*, 188 (2004), 35.

22. P. F. Tortorelli and J. H. DeVan, *Mater. Sci. Eng., A. Struct. Mater. Prop. Microstruct. Process.*, 153 (1992), 573.
23. S. W. Banovic, J. N. DuPont and A. R. Marder, *Oxid. Met.*, 54 (2000), 339.
24. P. F. Tortorelli and K. Natesan, *Mater. Sci. Eng., A. Struct. Mater.: Prop. Microstruct. Process.*, 258 (1998), 115.
25. K. Natesan, *Mater. Sci. Eng., A. Struct. Mater. Prop. Microstruct. Process.*, 258 (1998), 126.
26. B. A. Pint, P. F. Tortorelli and I. G. Wright, *Mater. High Temp.*, 16 (1999), 1.
27. J. Stringer, *Mater. Sci. Eng.*, 87 (1987), 1.
28. R. Sivakumar and E. J. Rao, *Oxid. Met.*, 17 (1982), 391.
29. T. H. Wang and L. L. Seigle, *Mater. Sci. Eng., A. Struct. Mater. Prop. Microstruct. Process.*, 108 (1989), 253.
30. A. Aguero, R. Muelas, M. Gutiérrez, R. Van Vulpen, S. Osgerby and J. Banks, *Surf. Coatings Technol.*, 201 (2007), 6253.
31. A. Agüero, 'Coatings for protection of high temperature new generation steam plant components: a review', in Proc. 8th Liege Conf. on Materials for Advanced Power Engineering, ed. J. Lecomte-Beckers, M. Carton, F. Schubert, P. J. Ennis. European Commission, Schriften des Forschungszentrum – Jülich, Germany, September 2006, *Energy Technol.*, 53 (2006), 949, Part III.
32. F. J. Pérez and S. I. Castañeda, *Surf. Coatings Technol.*, 201 (2007), 6239.

# 9

# Void formation and filling under alumina scales formed on Fe–20Cr–5Al-based alloys and coatings, oxidised at temperatures up to 1200 °C*

## D. J. Potter, H. Al-Badairy and G. J. Tatlock

*Materials Science and Engineering, Department of Engineering,*
*The University of Liverpool, Brownlow Hill, Liverpool, L69 3GH, UK*
*D.J.Potter@liv.ac.uk*

## M. J. Bennett

*Materials Consultant, Oxford, UK*

## 9.1  Introduction

FeCrAl alloys possess excellent oxidation and corrosion resistance to high-temperature aggressive environments $> 1000$ °C [1–4]. Protection from degradation is provided by the formation of a slow growing, adherent alumina scale. The attractive properties of FeCrAl alloys at temperatures $> 1000$ °C have increased the demand for their usage in industrial applications, such as automotive catalytic converters, for example. However, one life limiting factor of these foils is when the aluminium reservoir within the alloy reaches a critical level [5,6] and ceases to heal the protective alumina scale. The component may then fail through breakaway oxidation of iron and chromium.

Many authors have studied the development, growth rate and adherence of alumina scales upon FeCrAl alloys together with the effects of impurities and elements such as S, P, C. [7,8]. Other papers commonly report upon the favourable minor additions of reactive elements, such as Y, Hf, Zr, La [9–12] as these are crucial to the alumina scale adherence.

Recent commercial interest, and the life limitations of similar alloys at high temperatures $> 1000$ °C has led to the deposition of these alloys as coatings. This is a novel application and can provide substantial extensions in service lifetimes if used with compatible substrates. This allows for degradation of the coating before that of the component, since the coating has its own aluminium reservoir. MCrAl alloys can also be used as bond coats for Ni and Co superalloys below deposited thermal barrier coatings (TBCs). Many reports demonstrate the advantage of TBC usage to create a favourable temperature gradient across the coating and therefore, increase the substrate oxidation resistance and prolong life [13,14].

* Reprinted from D. J. Potter et al.: Void formation and filling under alumina scales formed on Fe -20Cr -5Al-based alloys and coatings, oxidised at temperatures up to 1200 °C. *Materials and Corrosion.* 2008. Volume 59. pp. 414–22. Copyright Wiley-VCH Verlag GmbH & Co. KGaA.

FeCrAl bond coats are susceptible to the influence of diffusive elements from the substrate [15]. This also works in reverse when, for example, Al from the coating diffuses into the substrate and depletes the Al reservoir eventually to below the critical level needed to sustain alumina growth. Therefore, it is necessary to match both coatings and substrates in order to avoid any significantly deleterious interaction between the two.

Alumina scale adherence is of great importance to the survival of a FeCrAl coating or an alloy. The formation of voids at the metal/oxide interface at high temperatures $> 1000\,^\circ\mathrm{C}$ has rarely been discussed [16–18], but is considered detrimental to the lifetime of a component. The experiments described in this paper are designed to investigate the formation of voids beneath alumina scales formed on FeCrAl alloys and give an indication of the influence of void formation on the behaviour of FeCrAl coatings on selected substrates. In particular, a silicon-rich braze has been used in conjunction with FeCrAl foils to give insight into the possible behaviour of FeCrAl coatings on silicon-rich substrates. Three different void formation and filling mechanisms are described within this work. They vary under the influence of temperature and silicon content of the coating/alloy combination.

## 9.2    Experimental

Five Fe–20Cr–5Al-based alloys with dimensions of 10 mm × 20 mm and thicknesses ranging from 30 µm to 100 µm have been investigated. These consisted of two commercial alloys, Aluchrom YHf and Kanthal AF, and three model alloys designated A1–A3. The compositions of the alloys are shown in Table 9.1. Test samples, prepared from the supplied alloys were ultrasonically cleaned in acetone and degreased using isopropanol before oxidation. The samples were isothermally oxidised in laboratory air at temperatures ranging from $900\,^\circ\mathrm{C}$ to $1200\,^\circ\mathrm{C}$ for times of up to 700 h using a horizontal furnace. The samples were air-cooled after oxidation.

Another series of samples was produced to simulate the behaviour of FeCrAl coatings on a silicon-rich substrate. This was achieved by brazing two pieces of 70 µm thick alloy foil in a sandwich-type structure at high temperature. Alloy A3 and a commercial brazing material BNi-5 (analysis Table 9.1) in tape form were used in the construction. The average thickness of the complete sandwich was measured to be approximately 0.2 mm. It was thought that the amount of silicon in the braze alloy (10.2 wt.%) would be sufficiently high to influence the oxidation mechanism of the alloy.

Plan view and cross-sectional samples for metallographic examination were prepared from all of the oxidised samples. They were all progressively ground and polished to a 1 µm mirror finish.

## 9.3    Results

### 9.3.1    Initial observations

Initial, visual observations indicated substantial and continuing deformation of the thin alloy foils which were oxidised at 1200 °C. This is expected to be a consequence of stress introduced by oxide scale growth and relief by creep of the substrate alloy [19]. Initial curvature of the foils occurred in the rolling direction. Thin foils that were oxidised for short time periods (up to 300 h) exhibited a grey colour. This was the

Table 9.1 Alloy compositions

| Alloy | Fe, wt.% | Cr, wt.% | Al, wt.% | Ni, wt.% | Si, wt.% | Y, wt.% | Hf, wt.% | S, ppm | C, ppm | N, ppm | P, ppm | V, ppm | Ca, ppm | Ti, ppm |
|---|---|---|---|---|---|---|---|---|---|---|---|---|---|---|
| Aluchrom YHf | Bal. | 20 | 5.5 | – | 0.29 | 0.05 | 0.04 | 11 | 220 | 40 | 130 | – | – | 99 |
| Kanthal AF | Bal. | 21 | 5.2 | – | 0.19 | 0.03 | * | 1.5 | 280 | 150 | 140 | – | – | 940 |
| A1 | Bal. | 20 | 5 | – | – | 0.03–0.05 | – | 5–10 | 100 | <10 | 80 or 83 | 80 | – | – |
| A2 | Bal. | 20 | 5 | – | – | 0.03–0.05 | – | 5–10 | 100 | <10 | 80 or 64 | – | 10 | – |
| A3 | Bal. | 20 | 5.5 | * | ~0.3 | 0.05 | 0.04 | – | * | * | – | – | – | – |
| Braze Alloy BNi-5 | – | 19 | – | Bal. | 10.2 | – | – | – | 600 | – | – | – | – | – |

*Trace elements.

underlying metallic substrate that was visible through a thin, transparent, and protective alumina layer. Increased oxidation times produced green coloured patches of oxide, especially in areas near corners and edges.

It was calculated that after 600 h oxidation in air at 1200°C, most of the aluminium in the 70 μm thick foil should have been consumed by oxide formation. A continuous layer of chromia formed beneath the outer alumina scale, together with a large number of voids (Fig. 9.1). At longer times, a large number of these voids filled with chromia. The number of voids appeared to differ slightly from one side of the sample to the other.

### 9.3.2   Low temperature oxidation (900 °C)

The metallographic examination of mounted and polished cross-sectional samples from alloy A1 which was oxidised at 900 °C for 100 h showed discontinuous, scallop-shaped, pits containing chromia beneath the outer duplex-alumina scale, as established by the respective EDX analysis (Fig. 9.2). This contrasts sharply with the morphology of chromia scales formed on a similar alloy at higher temperatures

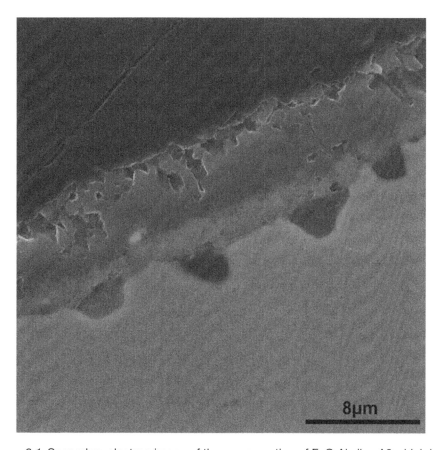

8μm

*9.1* Secondary electron image of the cross-section of FeCrAl alloy A3 which has been oxidised for 600 h at 1200 °C. Image shows the formation of a two-layered oxide containing alumina and chromia, and the presence of voids beneath this dual layer

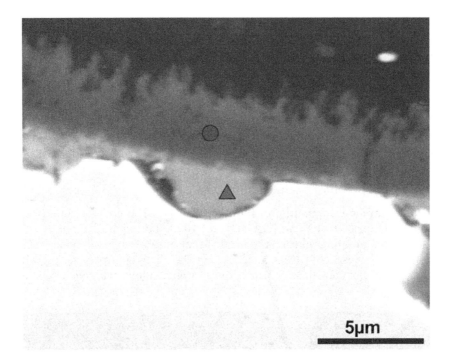

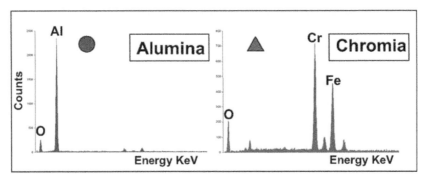

*9.2* Back-scattered scanning electron image of a metallographically mounted and polished cross-section of alloy A1 showing a void partially filled with chromia after 100 h at 900 °C. EDX spectra taken from the regions indicated on the micrograph

(1200 °C), where a continuous double-layered scale was observed (Fig. 9.1). In this case, voids formed below the chromia layer.

Other interesting observations, best illustrated in Fig. 9.3, are that the surface of the metal round the edges of the scallop-shaped areas is always smooth, the edges of the adjacent chromia layers are usually serrated and small gaps or a series of pores are usually present between the two regions. In contrast, the interface between the chromia and alumina is usually smooth and pore free.

Further SEM and EDS examinations revealed that some voids were partially filled with chromia, as shown, for example, in Fig. 9.2. These areas also often contained small bright particles, which were usually associated with the base of the chromia

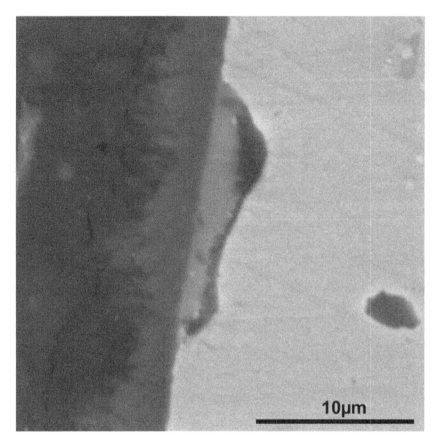

*9.3* Back-scattered scanning electron image of a metallographically mounted and polished cross-section of alloy A1 showing a void partially filled with chromia after 100 h at 900 °C

regions, between the oxide and the metal substrate. EDX analysis showed these bright regions to be silicon-rich.

### 9.3.3    High-temperature oxidation (1200 °C)

The examination of cross-sectional samples from alloy A3 foil after 600 h and 700 h at 1200 °C showed evidence of a further mechanism of void formation and filling (e.g. Fig. 9.4). Below the outer layers of alumina and chromia, a large void had formed between the substrate and the oxide. SEM examination of these voids showed large cracks through the pseudo-protective oxide film and the growth of an oxide upon the substrate side of the void. EDX analysis confirmed this oxide to be chromium-rich. On the same micrograph, a band of dark particles can be seen set back into the substrate at a constant distance around the void which chemical analysis revealed were silicon oxide.

Further examination of the same oxidised foils revealed voids of a similar size without either cracks or the presence of silica within the same region. The only chromia present is that which was growing continuously beneath the outer alumina

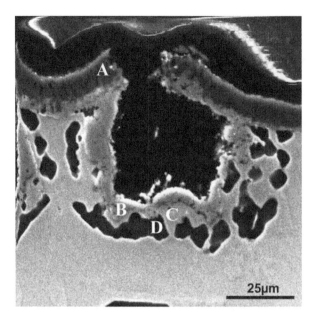

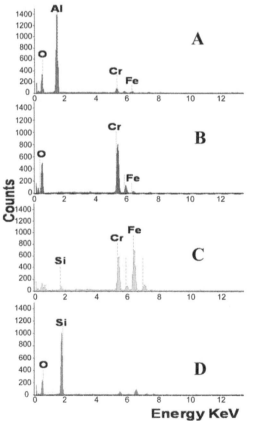

9.4 Void formation on FeCrAl alloy A3 when oxidised for 600 h at 1200 °C in air, together with the EDX spectra taken from four regions A–D. Region A is the alumina, B chromia, C substrate alloy, and D silica

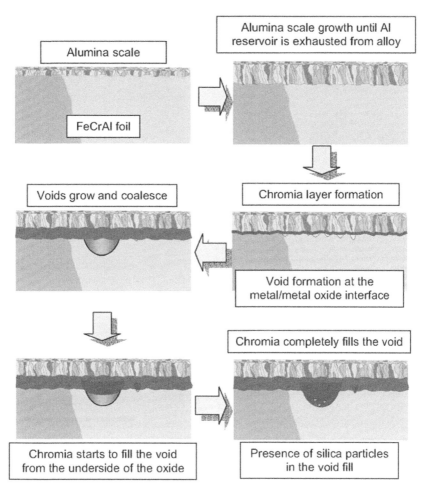

*9.5* Schematic diagram of the mechanism of void formation and filling upon thin foils of alloy A3 at high temperatures (1200 °C) and long exposure times

scale. Figure 9.5 gives a schematic representation of this void formation and filling. This internal growth of silicon oxide around the voids appeared to be associated with cracks that had propagated through the dual-layered oxide above the void. It was thought initially that the concentration of silicon within this alloy was insufficient for internal oxide growth.

Alloy A3 was tested isothermally, for times up to 700 h at 1200 °C in air. This alloy showed excellent oxide scale adherence whilst alumina was the protecting oxide. After 500 h and the exhaustion of aluminium from the substrate, a continuous chromia layer started to build up under the alumina scale. It was at this point that void formation was clearly seen beneath the chromia layer. These voids may then start to fill with chromia (Fig. 9.6). Analysis of numerous voids on the A3 alloy, oxidised at 1200 °C for 700 h revealed that the voids may not only fill with chromia. A small number have been recorded to have filled with silica (Fig. 9.7). However, it is important to note that the chromia filling mechanism of the voids seems to be the most dominant.

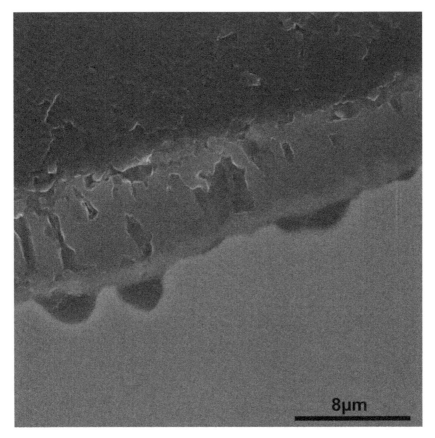

*9.6* Secondary electron image of the cross-section of FeCrAl alloy A3 which has been oxidised for 500 h at 1200 °C. This shows a continuous layer of chromia formed beneath the alumina scale after exhaustion of aluminium from the substrate. Three unfilled voids can be seen in the micrograph, below the dual-layered oxide

### 9.3.4    High silicon samples

In contrast, the silicon-rich brazed sandwich structure with an A3 FeCrAl alloy displays a different void filling mechanism. Silicon-rich foils oxidised at 1200 °C for 500 h display void formation directly beneath the protective alumina scale. Figure 9.8 shows an example of a void from one of these test specimens accompanied by the relevant EDX data for chemical analysis of the oxides formed. This example shows a crack in the outer alumina scale with a thin layer of chromia within the void beneath the alumina. This chromia layer is restricted to the void as a discontinuous layer. There is also evidence of chromia oxide growth within the crack in the alumina scale. Further analysis showed that the rest of the void was filled with silica. In tests of a shorter duration, only silica was observed to be growing on the substrate side of the void.

Interstitial voids have been found to form along the metal/oxide boundary of the silicon-rich sandwich structures at times as short as 300 h at 1200 °C. These voids also filled completely with silica but without the presence of a thin discontinuous chromia

*9.7* Back-scattered scanning electron image of a metallographically mounted and polished cross-section of a void formed upon alloy A3 after 700 h at 1200 °C. This void has completely filled with silica

layer (Fig. 9.9). The same alloy, when oxidised without the presence of a silicon-rich substrate, showed good oxide adherence beyond 500 h, without the appearance of voids.

## 9.4    Discussion

The reason why voids form beneath oxide scales is not yet completely understood. Void formation between a metal substrate and an oxide interface has been discussed previously [16–18,20]. All present a variety of mechanisms for void formation at the metal/oxide interface, including creep relaxation, influx of vacancies, Kirkendall porosity, etc. In the present case, thin foils tested at 1200 °C showed little or no void formation until the aluminium level had been exhausted from the substrate and a thin continuous layer of chromia had started to grow on the underside of the alumina. It is therefore considered that void nucleation is associated with a change in the oxide growth mechanism. Once a layer of chromia starts to grow beneath the alumina scale, outward diffusion of chromium ions to the base of the oxide could influence the movement of vacancies and cause this void formation.

Other observations of the oxidised thin foils indicate substantial growth of voids along one side of the specimen. This pointed to the source being a creep mechanism relieving stresses due to scale growth and causing void formation. The positioning of the voids, however, suggested that this may not be the complete story, since voids were also found within the middle of alloy grains as well as on grain boundaries. Voids would have been associated solely with grain boundaries if grain boundary

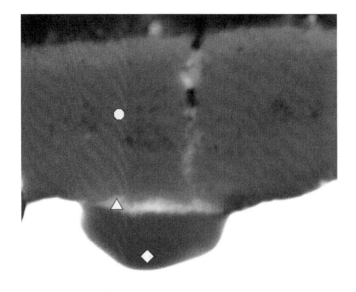

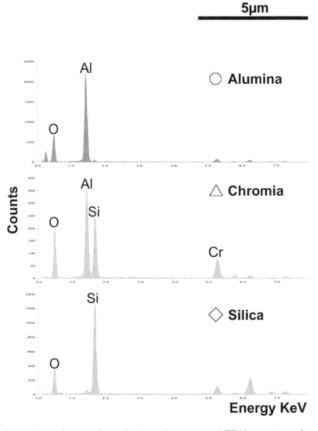

*9.8* Back-scattered scanning electron image and EDX spectra of a silican filled void formed after 500 h oxidation in air at 1200 °C. The micrograph shows the presence of a crack through the alumina scale and the presence of chromia between the silica and the alumina components of the void

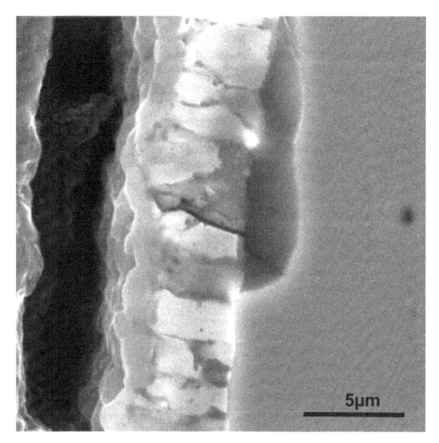

*9.9* Secondary electron micrograph of a silican-filled void found after 300 h at 1200 °C on alloy A3

alloy creep was responsible. This random positioning suggests otherwise, although some form of stress relaxation is thought to be responsible for the greater number of voids along one side.

In comparison, the alloys tested at the lower temperature of 900 °C show void formation after only 50 h. After this short time, there still remained sufficient aluminium within the metal substrate to sustain the growth of alumina. Tatlock *et al.* [20] suggested that the longer growth time for transition alumina at 900 °C before the formation of α-alumina, could cause the movement of vacancies to the metal/oxide interface resulting in void formation after short times.

Void formation was soon followed by void filling at the higher temperature of 1200 °C. After a continuous layer of chromia is formed beneath the alumina, a void may form. Many voids have been studied; few are found to be vacant but most are completely filled or partially filled with an oxide. For alloys low in silicon, once the aluminium activity in the alloy drops below the critical level [5], the formation of alumina will stop. Then the formation of a layer of chromia will occur beneath the outer oxide. If a void forms behind the layer of chromia, chromia may start to fill the void. This void filling commences from the base of the oxide by what is proposed as a vapour transport mechanism of chromium [21]. For this vapour transport mechanism to exist, the void region must have a substantially low partial pressure of oxygen. Otherwise the growth of oxide would be expected on the substrate/void interface. At 1200 °C, it is not unrealistic to propose a vapour cloud of chromium within the void as long as the pseudo-protective oxide scale remains intact. This proposal suggests that, as time increases, oxygen will diffuse through the dual-layered oxide and react with the chromium. New chromium-rich oxide is then deposited onto the existing chromia layer and grows into the hemispherical void. As long as the dual-layered scale remains intact, the process may continue until the void is completely filled.

Another possible mechanism for void filling is through surface diffusion of chromium round the void. However, it seems more likely that a vapour transport mechanism may lead to the smooth finishing of the substrate surface, observed in partially-filled voids.

Another type of void formation is seen within FeCrAl alloy A3, which contains ~0.3 wt.% Si. Similar to the previous mechanism, a void forms beneath the dual-layered oxide of alumina and chromia. However, this type of void grows much larger than those shown previously. This excessive growth may be responsible for cracking in the external scale. This crack provides a hot gas path and rapidly increases the partial pressure of oxygen within the now open void, and allows the nucleation of the less stable chromia on the substrate wall of the void. Interestingly, the growth of silica is also found near these voids, usually as small particles or as a continuous band (this silica nucleation is observed as an internal oxide in the substrate offset by a constant distance around the void). When first observed, these silicon-rich particles were thought to be debris from the polishing suspension used for sample preparation. However, reproducibility of this result and the lack of other silicon-rich particles embedded within other regions of the metal substrate make this unlikely.

The presence of silican internal oxide is possibly linked to a localised depletion of chromia round the void together with oxygen diffusion into the substrate. As long as the partial pressure of oxygen in the substrate is sufficiently high at a limited distance away from the void, silica may nucleate and grow. It is noted that further internal oxidation of silica does not appear to occur once a continuous band is formed around the void. The formation of silica is often observed in conjunction with the presence of chromia growth on the alumina on the outer wall of the void and with large cracks which penetrate the outer dual-layered oxide. It is suggested that these are linked to the degradation mechanism of the alloy.

Thermodynamically it is not unexpected to get the growth of alumina, chromia and silica at 1200 °C upon a FeCrAl alloy containing silicon. In fact, silica formation would be expected before the formation of chromia. The limited amount of silicon ($\leq 0.3$ wt.%) in the alloys may restrict the nucleation and growth of the silica, and chromia void filling is observed instead. However, local variations in the silicon content of the alloy may occur and lead to some voids being filled with silica while others are filled with chromia, as has been observed on a limited number of occasions.

In some cases the high-silicon substrate, when oxidised at 1200 °C, shows interstitial void formation directly beneath the α-alumina scale after only 300 h. These voids may then fill with silica, even though the aluminium level in the alloy is still at about 2 wt.%. However, in most cases, these voids are also associated with cracks through the outer alumina scale, and it may be that the remnant aluminium concentration is insufficient to re-heal the cracked scale effectively. However, one key feature of all of the voids which fill with silica is the growth mechanism within the void. The silica appears to nucleate and grow from the substrate side of the void unlike chromia which appears to always nucleate on the alumina side of the void.

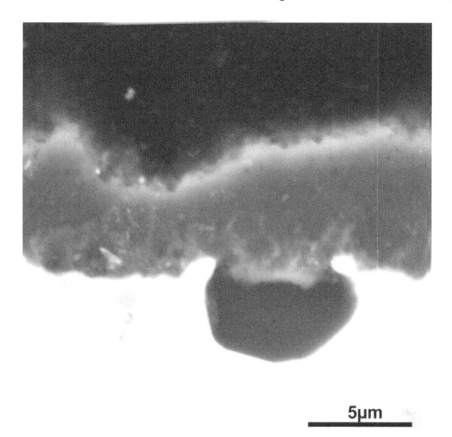

5μm

*9.10* Back-scattered scanning electron image showing a silica-filled void on the cross-section of alloy A3, oxidised for 560 h at 1200 °C in air

Silicon-rich substrates oxidised for over 560 h at 1200 °C exhibit an additional effect in the void filling mechanism. Figures 9.8 and 9.10 are back-scattered electron micrographs and clearly show that two different species have filled the voids in these cases. Both voids are filled largely with silica. However, a thin discontinuous layer of chromia has formed at the underside of the alumina which separates it from the silica. The chromia formation may have occurred after a crack has grown through the alumina scale, as shown in Fig. 9.8, although no crack is evident in the image in Fig. 9.10 and further work is under way to investigate this effect in more detail.

When FeCrAl materials are deposited as high-temperature protective coatings upon compliant substrates, these coatings can increase the lifetime of a component and improve the oxidation resistance by producing an alumina scale at temperatures >950 °C. However, protective coatings can also suffer from interstitial void formation, along with voids formed within the coating. The silicon-rich sandwich construction was developed to simulate the possible diffusion of silicon from a compliant substrate, through an FeCrAl coating, to the oxide/coating interface. This experiment demonstrates the possible behaviour of interstitial void formation within these coatings at temperatures between 900 °C and 1200 °C and shows that silica may grow within the voids and limit the lifetime of the coating by disrupting the protective alumina scale. Internal stresses, coating quality and thickness and the overall adherence of the coating may also play an important role, but these factors have not been included in the present study.

## 9.5    Conclusions

The difference in behaviour of the samples during oxidation may be related to different oxidation mechanisms in the two temperature regimes. At 900 °C, the formation of interstitial voids is soon followed by the growth of a discontinuous layer of chromia beneath the alumina scale. This contrasts with the behaviour at the higher temperature of 1200 °C where a continuous layer of chromia forms beneath the entire scale and voids are found beneath the chromia layer. The nucleation of chromia is thought to be linked to the depletion of aluminium from the alloy reservoir.

Alloy A3, when tested at the higher temperature (1200 °C), exhibited large voids and cracking of the alumina scale. Chromia growth can be observed upon all free surfaces of these voids. Evidence of silica internal oxide growth also indicates an additional degradation mechanism in FeCrAl alloys/coatings with moderate silicon content.

This work suggests that chromia vapour transport may occur across voids at both the higher (1200 °C) and lower (900 °C) temperatures. This could explain the uniform chromia layer thickness across the whole sample at the higher temperature and would explain the nucleation and growth of chromia at the underside of the alumina at the lower temperature (900 °C). This process is thought to occur once the level of aluminium in the alloy/coating drops below the critical level needed to sustain alumina formation.

The simulated silicon-rich substrate also demonstrated high levels of interstitial void formation. However, these voids filled with silica rather than chromia. This would be expected, thermodynamically, since silica is more stable than chromia at 1200 °C. The void filling mechanism is also different, since the silica starts to form at the metal interface of the void, rather than beneath the alumina scale, although some chromia-rich regions between the silica and alumina are also observed to form in the voids occasionally.

## Acknowledgements

The authors would like to thank Professor Jean Le Coze, Saint Etienne, France for the supply of the model alloys and Krupp VDM, Germany and Kanthal AB, Sweden for the supply of the commercial alloys, and to Neomet, UK for the supply of the braze alloy.

## References

1. G. C. Wood and F. H. Stott, Proc. Int. Conf. on High Temperature Corrosion NACE-6, 227. San Diego, CA, USA, 1983.
2. H. Bode (ed.), *Metal-Supported Automotive Catalytic Converters*. Werkstoff-Informationsgesellshaft, Frankfurt, 1997.
3. H. Bode (ed.), *Metal-Supported Automotive Catalytic Converters (MACC 2001)*. Werkstoff-Informationsgesellshaft, Frankfurt, 2001.
4. P. Kofstad, *High Temperature Corrosion*, 342–385 and 408–412. Elsevier Applied Science, 1988.

5. G. Strehl, P. Beaven, B. Lesage and G. Borchardt, *Mater. Corros.*, 56 (2005), 778.
6. I. G. Wright, R. Peraldi and B. A. Pint, *High Temp. Corros. Protect. Mater.*, 6 (2004), 579.
7. V. Kochubey, H. Al-Badairy, G. J. Tatlock, J. Le-Coze, D. Naumenko and W. J. Quadakkers, *Mater. Corros.*, (2005), 848.
8. J. S. Schaeffer, W. H. Murphy and J. L. Smialek, *Oxid. Met.*, (1995), 427.
9. S. Chevalier, C. Nivot and J. P. Larpin, *Oxid. Met.*, (2004), 195.
10. B. A. Pint, *Oxid. Met.*, 45 (1996), 1.
11. T. Amano, S. Yajima and Y. Saito, *Trans. Jpn. Inst. Metals Suppl.*, (1983), 247.
12. D. R. Sigler, *Oxid. Met.*, 32 (1989), 337.
13. J. Nunn, S. Saunders and J. Banks, *Microsc. Oxid.*, (2005), 219.
14. W. Braue, U. Schulz, K. Fritscher, C. Leyens and R. Wirth, *Microsc. Oxid.*, (2005), 227.
15. A. Littner, F. Pedraza, A. D. Kennedy, P. Moretto, L. Peichl, T. Weber and M. Schütze, *Microsc. Oxid.*, (2005), 245.
16. P. Y. Hou, Y. Niu and C. Van Lienden, *Oxid. Met.*, 59 (2003), 41.
17. P. Y. Hou and K. Priimak, *Oxid. Met.*, 63 (2005), 113.
18. M. W. Brumm and H. J. Grabke, *Corros. Sci.*, 34 (1993), 547.
19. M. J. Bennett, J. R. Nicholls, N. J. Simms, D. Naumenko, W. J. Quadakkers, V. Kochubey, R. Fordham, R. Bachorczyk, D. Goossens, H. Hattendorf, A. B. Smith and D. Britton, *Mater. Corros.*, 56 (2005), 854.
20. G. J. Tatlock, H. Al-Badairy, M. J. Bennett and J. R. Nicholls, *Microsc. Oxid.*, (2005), 303.
21. F. H. Stott, G. C. Wood and F. A. Golightly, *Corros. Sci.*, 19 (1979), 869.

# 10

## Influence of a TiO$_2$ surface treatment on the growth and adhesion of alumina scales on FeCrAl alloys*

### A. Galerie, E. N'Dah and Y. Wouters

*SIMAP, INP-Grenoble, BP 75, 38402 Saint Martin d'Hères, France*
*Alain.Galerie@ltpcm.inpg.fr; Ndah_Eugene@yahoo.fr; Yves.Wouters@ltpcm.inpg.fr*

### F. Roussel-Dherbey

*CMTC, INP-Grenoble, BP 75, 38402 Saint Martin d'Hères, France*
*Francine.Roussel-Dherbey@cmtc.inpg.fr*

### 10.1 Introduction

FeCrAl alloys are largely used for their good resistance to high-temperature oxidation above 1000 °C due to the growth, in service, of a protective α-alumina layer. This oxide is an excellent barrier to $Al^{3+}$ ion diffusion and slowly grows by inward oxygen transport [1,2]. Nowadays, such alloys are used for various metallic parts working at high temperatures, including reactor walls and tubes, burners, high-temperature (HT) filters, etc. However, for temperatures typically between 850 and 950 °C, metastable aluminas (γ, δ and θ) grow rapidly [3,4] and may consume the entire Al reservoir of the alloy, particularly for thin foil parts such as car catalyst supports, and lead to pseudo-protection by a chromia scale before the onset of catastrophic oxidation where iron oxides form [5,6].

In previous studies [7,8], it was observed that slurry TiO$_2$ surface treatment of FeCrAl alloys reduced the time where transition aluminas predominate in the temperature range 850–925 °C, as determined by scanning electron microscope (SEM) observations, X-ray diffraction and fluorescence experiments. When treated, the alloys exhibited a rapid decrease in the parabolic rate constant, determined by thermogravimetry, before stabilising to a value corresponding to transport through α-Al$_2$O$_3$. These results are in excellent agreement with the results of an earlier study [9] on the same treatment applied to FeCrAl fibres. In the present work, the TiO$_2$ slurry effect is confirmed and a new treatment is described using Tetra-IsoPropylorthoTitanate (TIPT) to generate surface coatings containing much less titanium. The difference between the two methods of applying TiO$_2$ has been considered in terms of the oxidation kinetics and the good potential of the TIPT treatment has been shown. Another important finding is the beneficial influence of the treatment on oxide scale adhesion.

* Reprinted from A. Galerie et al.: Influence of a TiO2 surface treatment on the growth and adhesion of alumina scales on FeCrAl alloys. *Materials and Corrosion*. 2008. Volume 59. pp. 423–8. Copyright Wiley-VCH Verlag GmbH & Co. KGaA.

*Table 10.1*  Composition (wt.%) of the studied alloys

|  | Fe | Cr | Al | Y | Hf | Ti | Zr |
|---|---|---|---|---|---|---|---|
| Kanthal AF | Bal. | 20.83 | 5.23 | 0.034 | 0.0003 | 0.094 | 0.058 |
| Aluchrom YHf | Bal. | 19.65 | 5.53 | 0.046 | 0.031 | 0.0098 | 0.054 |
| Aluchrom YHfAl | Bal. | 20.15 | 6.02 | 0.065 | 0.045 | 0.004 | 0.058 |

## 10.2    Experimental

Three FeCrAl alloys were used in this work, Kanthal AF from Kanthal, and Aluchrom YHf and Aluchrom YHfAl both from Thyssen Krupp VDM, the compositions of which are given in Table 10.1.

These alloys were used in the form of thin foils (50 μm in thickness), cut to 25 mm × 15 mm, and were only degreased and rinsed. Two methods were used to enrich the alloy surfaces with TiO₂: slurry deposition and TIPT dipping.

### 10.2.1   Application of a TiO₂ slurry coating

The procedure for applying TiO₂ powder onto the surface of FeCrAl alloys has been described previously [8]. From 1 μm TiO₂ agglomerates suspended in water, attempts were made to obtain various surface coverages by varying the concentration of suspensions. The minimum deposit which led to a visually homogeneous coverage was ~150 μg cm⁻². In this paper, deposits of 170, 220 and 240 μg cm⁻² are described. Figure 10.1 shows the surface appearance of a 170 μg cm⁻² deposit.

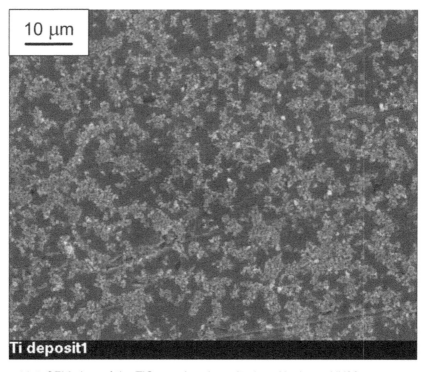

*10.1* SEM view of the TiO₂ coating deposited on Aluchrom YHf from an aqueous slurry (surface coverage: 170 μg cm⁻²)

### 10.2.2    TIPT dipping

TetraIsoPropylorthoTitanate (TIPT $TiC_{12}H_{28}O_4$) from Fluka (Titanium Isopropo-xide) was dissolved in a water/ethanol solution (1:1) adjusted to pH 1.2 with hydro-chloric acid. The TIPT concentration was fixed at 0.4 mol L$^{-1}$. Metallic samples were completely immersed in the solution for 5 min then the solution was slowly (1 mm s$^{-1}$) drained away through a bottom tap and the samples were then allowed to dry at room temperature. This produced a thin Ti-containing film of about 65 µg cm$^{-2}$. When heating samples in oxygen to attain the oxidation temperature, TIPT decomposition occurred according to Eq. 10.1:

$$Ti(C_3H_7O)_4 + 18O_2 \rightarrow TiO_2 + 12CO_2 + 14H_2O \qquad [10.1]$$

This produced a thin $TiO_2$ film of theoretical surface mass 18 µg cm$^{-2}$, 10 times less than that of the slurry deposit.

### 10.2.3    Oxidation procedure and characterisation techniques

For assessment of the influence of $TiO_2$ on oxidation kinetics and alumina formation, treated and non-treated samples were oxidised for 72 h in a dynamic Ar–15% O$_2$ mixture flowing at ~1 mm s$^{-1}$ in a laboratory tube. A microthermobalance or a hori-zontal furnace was used depending on the experiment. To compare the oxidation behaviour, gravimetric curves were recorded and the development of the parabolic rate constant was followed during oxidation, as proposed by Monceau and Pieraggi [10]. As already reported in the literature, parabolic rate constants decrease rapidly before stabilising after a few tens of hours of reaction. In this paper, parabolic rate constants after oxidation for 72 h are given for comparison purposes. X-ray diffrac-tion (Cu Kα radiation with a rear monochromator) and ruby fluorescence (laser wavelength 514.532 nm) were used to characterise the alumina scales that formed. For scale adhesion measurements, all samples were oxidised in laboratory air.

### 10.2.4    Scale adhesion measurements

The mechanical adhesion of oxide scales was quantified in terms of interfacial fracture energy at scale spallation. Oxidised samples were strained in a tensile machine placed in the SEM chamber and the fraction of spalled area was recorded as a function of stress and strain. From the measurements, qualitative comparison was made by observing, for each sample, the fraction of spalled area at a fixed strain of 5%. More quantitatively, the fracture energy $G_i$ (J m$^{-2}$) was determined from the volumetric elastic energy stored in the oxide scale when the first spallation occurred [11–13].

## 10.3    Results and discussion

### 10.3.1    Influence of Ti in the alloy

The influence of titanium on transition alumina inhibition was first observed for industrial alloys containing various Ti contents. As shown in Fig. 10.2, the oxidation kinetics of Kanthal AF (0.1% Ti), Aluchrom YHF (0.01% Ti) and Aluchrom YHfAl (0.004% Ti) at 925 °C were significantly different, the slowest kinetics corresponding to the highest Ti-content. This effect was followed as a function of oxidation

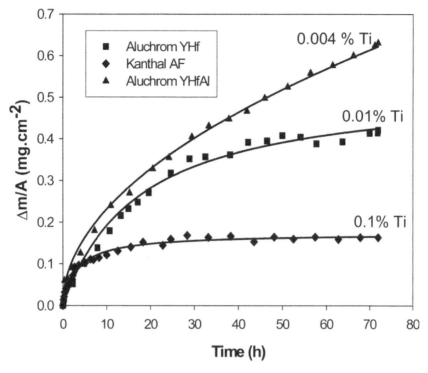

*10.2* Oxidation kinetics of various FeCrAl alloys with different titanium contents. Oxidation temperature: 950 °C, atmosphere: Ar + 15% O$_2$

temperature and the results are presented in Table 10.2. It is observed that the two Aluchrom alloys exhibited increasing scale thickness with increasing temperature after oxidation for 72 h, whereas Kanthal AF showed a quite constant scale thickness resulting from an increasing alpha alumina content in the scale, in good agreement with previously published results [14].

As is known from the literature, the stabilisation of metastable cubic aluminas is favoured by large foreign ions [15], but the effect of small ions is not well documented. Titanium has been cited as promoting alpha alumina [16,17], and Fei *et al.* [9] have clearly demonstrated that a TiO$_2$ coating applied on FeCrAl fibres suppresses the formation of transition aluminas at intermediate temperatures. It should also be

*Table 10.2* Parabolic rate constants ($k_p$) and scale thicknesses ($y$) after oxidation of FeCrAl alloys in Ar + 15% O$_2$ for 72 h

|  |  | 850 °C | 875 °C | 900 °C | 925 °C |
|---|---|---|---|---|---|
| Kanthal AF | $k_p$ (mg$^2$ cm$^{-4}$ s$^{-1}$) | $2.7 \times 10^{-7}$ | $4.4 \times 10^{-7}$ | $3.6 \times 10^{-7}$ | $1.5 \times 10^{-7}$ |
|  | $y$ (μm) | 1.38 | 1.60 | 1.60 | 1.23 |
| Aluchrom YHf | $k_p$ (mg$^2$ cm$^{-4}$ s$^{-1}$) | $1.3 \times 10^{-7}$ | $2.2 \times 10^{-7}$ | $3.9 \times 10^{-7}$ | $7.0 \times 10^{-7}$ |
|  | $y$ (μm) | 0.73 | 1.38 | 1.96 | 2.11 |
| Aluchrom YHfAl | $k_p$ (mg$^2$ cm$^{-4}$ s$^{-1}$) | $3.6 \times 10^{-7}$ | $3.7 \times 10^{-7}$ | $8.1 \times 10^{-7}$ | $15.4 \times 10^{-7}$ |
|  | $y$ (μm) | 1.24 | 2.26 | 2.40 | 2.84 |

noted that γ-TiAl alloys rapidly generate alpha alumina at temperatures where FeCrAl grows transition aluminas [18].

### 10.3.2   Influence of TiO$_2$ slurry coatings

Earlier oxidation experiments with slurry-treated and non-treated samples clearly established that a TiO$_2$ surface coverage as low as 170 μg cm$^{-2}$, close to the smallest slurry deposit achieved, could give a huge reduction in the parabolic rate constant [8]. The X-ray diffraction spectra in Fig. 10.3 showed that characteristic lines of transition alumina totally disappeared when Aluchrom YHfAl was treated with this amount of TiO$_2$ before oxidation. To quantify the influence of the amount of TiO$_2$, the heights of the two X-ray diffraction peaks of quadratic TiO$_2$ rutile ([110] at $d = 322$ pm and [211] at $d = 168.7$ pm) and of rhombohedral α-alumina ([012] at $d = 348.0$ pm and [024] at $d = 174.0$ pm), chosen to be reasonably close, were compared. The results in Fig. 10.4 show that the TiO$_2$/Al$_2$O$_3$ ratio after 72 h oxidation at 925 °C tends to zero with decreasing TiO$_2$ surface coverage. This indicates that all of the deposited Ti$^{4+}$ ions have dissolved into the alumina, promoting the rapid transformation to alpha. Such a coverage, of 170 μg cm$^{-2}$ with no TiO$_2$ excess, was therefore successful, but lower surface coverages were difficult to achieve with water slurries. To investigate the possibility of surface treating with lower quantities of titanium, deposits from TIPT alkoxide Ti(OC$_3$H$_7$)$_4$ were used.

### 10.3.3   Influence of TIPT coatings

TIPT-treated samples were submitted to isothermal oxidation under the same conditions as slurry-treated and non-treated samples (925 °C, 72 h, Ar + 15% O$_2$). Comparative thermogravimetric curves are reproduced in Fig. 10.5, where it is clear

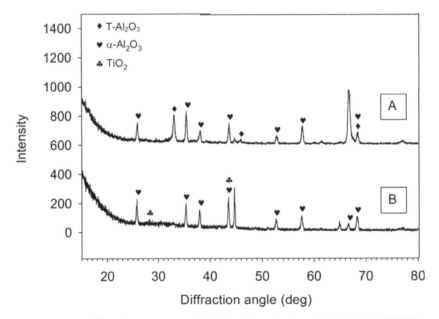

*10.3* X-ray diffraction spectra of YHfAl oxidised in Ar + 15% H$_2$O at 925 °C for 72 h. (A) Non-treated; (B) treated with 170 μg cm$^{-2}$ TiO$_2$ slurry

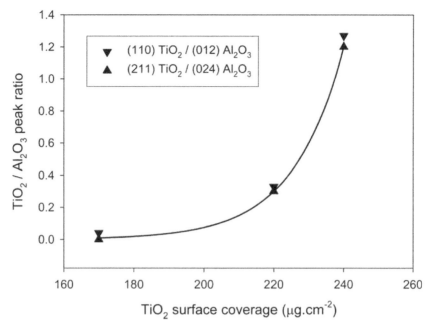

*10.4* Development of the XRD TiO$_2$/Al$_2$O$_3$ peak ratios of scales grown on Aluchrom YHfAl as a function of the amount of pre-deposited TiO$_2$

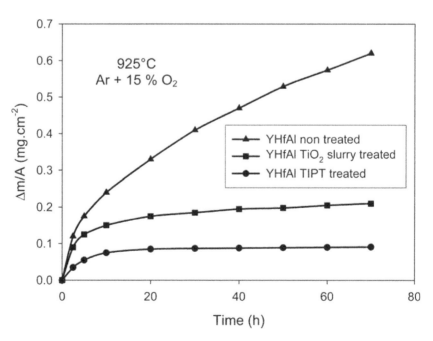

*10.5* Comparative oxidation curves for Aluchrom YHfAl alloy treated with 170 μg cm$^{-2}$ TiO$_2$ slurry or 65 μg cm$^{-2}$ TIPT. Reference: non-treated alloy

*Table 10.3*   Comparative parabolic rate constants for the oxidation at 925 °C of YHfAl samples, treated and non-treated

| Treatment | $k_p$ (mg$^2$ cm$^{-4}$ s$^{-1}$) |
|---|---|
| Non-treated | $15.4 \times 10^{-7}$ |
| Slurry-coated | $1.1 \times 10^{-7}$ |
| TIPT-coated | $0.2 \times 10^{-7}$ |

that the curve for the TIPT sample is far beneath that for the slurry sample, with a parabolic rate constant about five times lower (Table 10.3), although the TiO$_2$ surface coverage is 10 times less. This confirms that a very low amount of Ti is very efficient in promoting the formation of protective alpha alumina.

Surface SEM observation (Fig. 10.6) showed mainly equiaxed surface features characteristic of alpha alumina on treated samples, whereas non-treated samples exhibited the classical plate-like morphology of metastable aluminas. X-ray diffraction gave results identical to those presented in Fig. 10.3 and ruby fluorescence (Fig. 10.7) confirmed that alpha alumina was the major oxide formed on TIPT samples, with no evidence of transition aluminas, like those observed on the slurry samples. As proposed in a previous paper [8], titanium dissolution into growing alumina seems to occur during the early stages of oxidation, when alumina scales are very thin, making low TiO$_2$ amounts very efficient to promote $\alpha$-formation.

### 10.3.4   Influence of TiO$_2$ treatment on scale adhesion

Previous studies using a tensile test [11,19] have shown that the mechanical adhesion of alumina scales on FeCrAl alloys strongly depends on the nature (metastable transition or stable alpha) of the alumina formed. Due to their outward growth, metastable aluminas exhibit poor adhesion because of cavity formation at the metal/scale interface. On the other hand, alpha alumina scales, mainly growing by inward oxygen transport, are strongly adherent to the FeCrAl substrate. Figure 10.8 presents remarkable situations, with extensive local crack formation and spallation at places where transition aluminas predominate. For all FeCrAl alloys, scale adhesion improves during the course of oxidation resulting from the change from transition to alpha, the opposite of what is observed on chromia-forming alloys where a continuous decrease in adhesion is observed with increasing oxidation time [20–22].

In the present experiments, non-treated FeCrAl samples were first oxidised at 850, 950 and 1200 °C for 72 h, before being subjected to tensile testing. Induced spallation recorded at a strain of 5% is shown in Table 10.4.

The results indicate that all alloys form very adherent alpha alumina scales at 1200 °C. At lower temperatures, Kanthal AF exhibits no scale spallation, resulting from the rapid transformation from metastable to alpha alumina reported above. On the other hand, both Aluchrom alloys grow less adherent scales at 850 °C, with noticeable improvement at 950 °C for YHfAl.

Alloy YHfAl, suffering the largest spallation at 850 °C, was chosen for evaluating the effect of TiO$_2$ treatment. The application of TIPT followed the same procedure as described above. Treated and non-treated samples were oxidised in laboratory air at 850 °C for 65 h and subsequently tensile tested. The results are reported in Table 10.5. It was observed that the TIPT treatment greatly increased the strain at which the first

*10.6* Comparative surface morphology of Aluchrom YHfAl samples oxidised at 925 °C for 72 h in Ar + 15% O$_2$. (A) Non-treated, (B) treated with 65 μg cm$^{-2}$ TIPT

spall appeared, a consequence of higher adhesion. That resulted in a very low spalled fraction being measured at 5% strain. Adhesion energy calculations from the strain where the first spall appeared gave more than four times higher adhesion for the scale on the TIPT-treated sample. This could be related to the difference in oxide growth

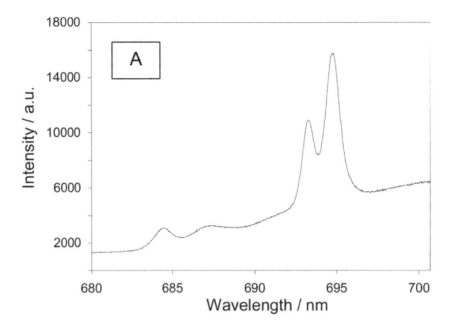

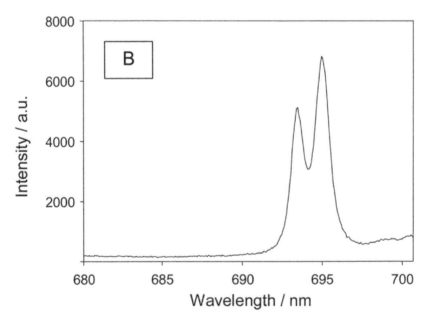

*10.7* Ruby fluorescence spectra of oxide scales in Fig. 10.6. (A) Non-treated. (B) TIPT-treated. Only sample A shows the T doublet near 685 cm$^{-1}$ characteristic of theta alumina

direction as shown by the fracture morphologies in Fig. 10.9, where the scale grown in the presence of TiO$_2$ shows more equiaxed grains and less interface porosity. It should also be noted in Table 10.5 that the scale thickness was lower in the presence of TiO$_2$ as a result of the formation of more protective α-alumina.

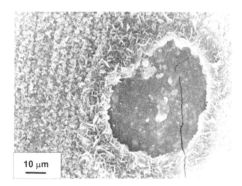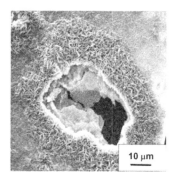

*10.8* Surface observations of scale buckling and spalling at places where needle-shaped oxide indicates the predominance of transition aluminas

*Table 10.4*   Spalled fraction of surface area of FeCrAl alloys oxidised at different temperatures for 72 h in laboratory air, and then tensile strained to 5% at room temperature

| FeCrAl grade | 850 °C | 950 °C | 1200 °C |
|---|---|---|---|
| Aluchrom YHf | 6.23 | 7.86 | 0.00 |
| Aluchrom YHfAl | 7.41 | 0.00 | 0.00 |
| Kanthal AF | 0.00 | 0.00 | 0.00 |

*Table 10.5*   Comparative adhesion testing of non-treated and TIPT-treated Aluchrom YHfAl oxidised at 850 °C in laboratory air for 65 h

| | Oxide thickness (μm) | Spalled fraction at strain = 0.05 (%) | Strain at first spall (%) | Elastic energy at first spall (MJ m$^{-3}$) | First spall adhesion energy (J m$^{-2}$) |
|---|---|---|---|---|---|
| YHfAl non-treated | 1.92 | 7.5 | 3.57 | 248 | 475 |
| YHfAl TIPT-treated | 1.42 | 0.1 | 8.57 | 1500 | 2145 |

## 10.4    Conclusions

Surface treatments of FeCrAl alloys with TiO$_2$ were observed to accelerate the formation of stable alpha alumina compared to metastable transition aluminas. Even low amounts of titanium were shown to have a remarkable effect, allowing the initial slurry deposition process to be modified to a more efficient TIPT process. This modification could reduce the necessary amount of deposited Ti by a factor of 10 and reduce the parabolic rate constant by a factor of five compared to slurry-treated samples and by a factor of 75 compared to non-treated FeCrAl alloys. As a result of the promotion of alpha alumina, the adhesion of scales on treated samples was greatly improved by the inward growth of the scales.

*10.9* FEG-SEM observations of scale fractures. (A) Non-treated. (B) TiO$_2$-treated. Oxidised at 925 °C for 72 h

## References

1. J. Deaking, V. Prunier, G. C. Wood and F. H. Stott, *Mater. Sci. Forum*, 251–254 (1997), 41.
2. W. J. Quadakkers and M. J. Bennett, *Mater. Sci. Technol.*, 10 (1994), 126.
3. E. Andrieu, A. Germidis and R. Molins, *Mater. Sci. Forum*, 251–254 (1997), 357.
4. D. Naumenko, W. J. Quadakkers, A. Galerie, Y. Wouters and S. Jourdain, *Mater. High Temp.*, 20(3/4) (2003), 41.
5. D. Naumenko, L. Singheiser and W. J. Quadakkers, in *Cyclic Oxidation of High Temperature Materials*, 287, ed. M. Schütze and W. J. Quadakkers, EFC Workshop Nr. 27, 1999.
6. J. R. Nicholls, M. J. Bennett and R. Newton, *Mater. High Temp.*, 20(3/4) (2003), 183.
7. E. N'Dah, PhD Thesis, INP-Grenoble, France, (in French), 2005.
8. E. N'Dah, A. Galerie, Y. Wouters, D. Goossens, D. Naumenko, V. Kochubey and W. J. Quadakkers, *Mater. Corros.*, 56(12) (2005), 1.
9. W. Fei, S. C. Kuiry and S. Veal, *Oxid. Met.*, 62(1/2) (2004), 29.
10. D. Monceau and B. Pieraggi, *Oxid. Met.*, 50 (1998), 97.
11. A. Galerie, F. Toscan, E. N'Dah, K. Przybylski, Y. Wouters and M. Dupeux, *Mater. Sci. Forum*, 461–464 (2004), 631.
12. F. Toscan, L. Antoni, Y. Wouters, M. Dupeux and A. Galerie, *Mater. Sci. Forum*, 461–464 (2004), 705.
13. A. Galerie, F. Toscan, M. Dupeux, J. Mougin, G. Lucazeau, C. Valot, A.-M. Huntz and L. Antoni, *Mater. Res.*, 7(1) (2004), 1.
14. R. Cueff, H. Buscail, E. Caudron, C. Issartel and F. Riffard, *Oxid. Met.*, 58 (2002), 439.
15. P. Burtin, J.-P. Brunelle, M. Pijolat and M. Soustelle, *Appl. Catal.*, 34 (1987), 225.
16. B. A. Pint, M. Treska and L. W. Hobbs, *Oxid. Met.*, 47 (1997), 1.
17. K. M. N. Prasanna, A. S. Khanna, R. Candra and W. J. Quadakkers, *Oxid. Met.*, 46 (1996), 465.
18. C. Lang and M. Schütze, in *Microscopy of Oxidation 3*, 265, ed. S. B. Newcomb and J. A. Little. The Institute of Materials, London, 1997.
19. A. Galerie, Y. Wouters, E. N'Dah, A. Crisci and K. Przybylski, *Ceramics*, 92 (2005), 13.
20. J. Mougin, M. Dupeux, L. Antoni and A. Galerie, *Mater. Sci. Eng.*, A359 (2003), 44.
21. G. Bamba, Y. Wouters, A. Galerie, F. Charlot and A. Dellali, *Acta Mater.*, 54 (2006), 3917.
22. S. Chandra-Ambhorn, F. Roussel-Dherbey, Y. Wouters, A. Galerie and M. Dupeux, *Mater. Sci. Technol.*, 23 (2007), 1.

# 11

## A new austenitic alumina forming alloy: aluminium-coated FeNi32Cr20*

### Heike Hattendorf and Angelika Kolb-Telieps

*ThyssenKrupp VDM GmbH, Kleffstraße 23, 58762 Altena, Germany*
*heike.hattendorf@thyssenkrupp.com*

### Ralf Hojda

*DIEHL-Metall, Hönnetalstr. 110, 58675 Hemer, Germany*

### Dmitry Naumenko

*Forschungszentrum Jülich GmbH IEF-2, 52425 Jülich, Germany*

### 11.1  Introduction

Foils of FeCrAl alloys are used as a substrate in metal-supported automotive catalytic converters [1]. Conventionally, they are produced in the form of 50 μm or 30 μm thick foils of Fe–20%Cr–5%Al (by mass) combined with additions of Y and Hf, which previously proved to be superior to other reactive elements [1]. FeCrAl alloys owe their excellent oxidation resistance at high temperature to a slow-growing $\alpha$-$Al_2O_3$ scale. However, in thin foils, this scale growth process gradually consumes the aluminium content of the alloy, so after extended service periods, the protective aluminium oxide layer no longer forms [2]. Increasing exhaust gas temperatures demand improved high-temperature mechanical properties in addition to oxidation resistance. In comparison to austenitic alloys, the ferritic FeCrAl alloys typically have low mechanical strength at higher temperatures. But adding aluminium to an austenitic alloy causes severe workability problems, even at aluminium contents which are below the amount required to form a protective alumina scale.

In an austenitic NiCr alloy, at least 5% Al is necessary in addition to about 10% Cr for a pure alumina scale to form during oxidation [3]. An alloy with such a high aluminium content cannot be produced by ingot melting, hot rolling and cold rolling with intermediate heat treatments because of severe workability problems.

To avoid these problems, foils of an austenitic FeNiCr alloy with an aluminium content higher than 6% can be produced by coating an FeNiCr alloy with pure aluminium, for example by cladding at intermediate thickness followed by cold rolling and diffusion annealing. Such material is easier to work than a homogeneous alloy of high aluminium content. A similar approach has been shown to be successful in increasing the aluminium content to above 7% in FeCrAl alloys [1,4,5].

---

\* Reprinted from H. Hattendorf et al.: A new austenitic alumina forming alloy: aluminium-coated FeNi32Cr20. *Materials and Corrosion.* 2008. Volume 59. pp. 449–54. Copyright Wiley-VCH Verlag GmbH & Co. KGaA.

In this work, the oxidation and mechanical properties of the newly-developed material Nicrofer 3220 PAl (coated FeNiCrAl), which consists of an austenitic FeNi32Cr20 alloy clad with aluminium on both sides, are presented and compared to those of a conventional homogeneous FeCrAl alloy (FeCr20Al6).

## 11.2 Experimental details

### 11.2.1 Production

Since the workability is better for alloys with lower aluminium or no aluminium content, the idea was to start with a strip of an FeNiCr alloy without aluminium, but with additions of yttrium produced by the conventional route. At an intermediate thickness, the strip was clad with aluminium and then further rolled down to final thickness. For the FeNiCr base alloy, a composition of 32% Ni, 20% Cr, minor additions of silicon and manganese and a balance of iron similar to Nicrofer 3220–alloy 800 was chosen. Meanwhile, several industrial heats were produced by cladding with different aluminium contents. Some of their compositions are shown in Tables 11.1 and 11.2. For comparison, heats of FeCrAl alloys are included in Tables 11.1 and 11.2.

The coated samples in Tables 11.1 and 11.2 were produced by melting the FeNiCr alloys with an addition of yttrium or by using FeCrAl alloy of reduced aluminium content and additions of yttrium, zirconium and hafnium, hot rolling followed by cold rolling to an intermediate thickness, coating with aluminium and cold rolling to final thickness. In addition, homogeneous FeCrAl foils were produced via the conventional route, i.e. by melting then hot rolling followed by cold rolling to final thickness with an intermediate anneal.

### 11.2.2 Long-term oxidation tests

The influence of the production route on the oxidation resistance was compared for the heats/trials listed in Table 11.1. After cleaning these foils, model samples simulating catalytic converter bodies were produced and oxidised for 500 to 1000 h in a

*Table 11.1* Chemical composition in mass% of the alloys used for the oxidation tests (balance Fe)

| Material | | Trial | d/µm | Al* | Ni | Cr | Mn | Si | Y | Zr | Hf | |
|---|---|---|---|---|---|---|---|---|---|---|---|---|
| Conventional FeCrAl | H | A1 | 40 | 5.9 | 0.2 | 20.4 | 0.2 | 0.3 | 0.07 | 0.04 | 0.04 | S$_4$ |
| Conventional FeCrAl | H | A2 | 40 | 6.0 | 0.2 | 20.4 | 0.2 | 0.3 | 0.07 | 0.05 | 0.04 | – |
| Conventional FeCrAl | H | A3 | 40 | 6.0 | 0.2 | 20.2 | 0.2 | 0.3 | 0.07 | 0.04 | 0.05 | – |
| Coated FeNiCrAl | C | N1 | 40 | 6.0 | 30.5 | 20.3 | 0.1 | 0.4 | 0.02 | – | – | S$_1$ |
| | | N2 | 40 | 6.8 | | | | | | | | S$_4$ |
| Coated FeNiCrAl | C | N3 | 40 | 6.0 | 30.6 | 20.2 | 0.1 | 0.4 | 0.02 | – | – | |
| | | N5 | 30 | 6.0 | | | | | | | | S$_0$S$_4$ |
| | | N4 | 40 | 10.0 | | | | | | | | – |
| Coated FeCrAl | C | P1 | 50 | 7.8 | 0.2 | 20.2 | 0.2 | 0.3 | 0.05 | 0.05 | 0.03 | S$_4$ |
| Coated FeCrAl | C | P2 | 40 | 8.2 | 0.2 | 20,3 | 0.2 | 0.2 | 0.05 | 0.04 | 0.04 | – |
| | | P3 | 50 | | | | | | | | | – |

H, homogeneous heat; C, coated; S$_0$, cross-section after 1 h; S$_1$, cross-section after 100 h; S$_4$, cross-section after 400 h.
*Total aluminium content.

*Table 11.2*   Chemical composition in mass-% of the alloys used for the hot tensile tests (balance Fe)

| Material | Trial | | d/μm | Al* | Ni | Cr | Mn | Si | Y | Zr | Hf |
|---|---|---|---|---|---|---|---|---|---|---|---|---|
| Conventional FeCrAl | H | A1 | 40 | 5.9 | 0.2 | 20.4 | 0.2 | 0.3 | 0.07 | 0.04 | 0.04 |
| Conventional FeCrAl | H | A4 | 55 | 5.7 | 0.2 | 20.1 | 0.2 | 0.3 | 0.06 | 0.04 | 0.05 |
| Coated FeNiCrAl | C | N2 | 40 | 6.8 | 30.5 | 20.3 | 0.1 | 0.4 | 0.02 | – | – |
| Coated FeNiCrAl | C | N6 | 50 | 8.3 | 30.5 | 20.1 | 0.4 | 0.5 | – | – | – |
| Coated FeNiCrAl | C | N7 | 50 | 5.0 | 30.5 | 20.1 | 0.4 | 0.4 | – | – | – |
| Coated FeCrAl | C | P4 | 50 | 6.9 | 0.2 | 18.2 | 0.2 | 0.6 | 0.05 | 0.01 | 0.05 |
| Coated FeCrAl | C | P5 | 50 | 7.4 | 0.1 | 17.2 | 0.3 | 0.8 | 0.03 | 0.05 | 0.05 |

*Total aluminium content.

furnace in dry laboratory air. The samples were heated to final temperature in 1 h. After every 100 h of exposure the furnace was cooled down to room temperature over a 24-h period. All samples were removed from the furnace and their mass change was determined. No fewer than duplicate samples were annealed for each heat/trial. The mass changes reported are the mean values of these samples. For N1 and N5, additional samples were annealed for 1 h ($S_0$) and 100 h ($S_1$) (see Table 11.1). For these samples and selected samples after 400 h ($S_4$), cross-sections were prepared. These were examined with SEM (Scanning Electron Microscopy), EDX (Energy-Dispersive X-ray Analysis) and FESEM (Field Emission Scanning Electron Microscopy) using a Leo 1530-Gemini equipped with In-Lens detection. The FESEM examinations were carried out on micro-sections, which were polished much longer to etch the grain boundaries of the scale, which could then be detected with an SE In-Lens detector.

### 11.2.3   Mechanical properties

Hot tensile tests were carried out on different heats/trials of conventional FeCrAl, and coated FeCrAl (Aluchrom P) and FeNiCrAl at temperatures up to 1100 °C (compositions are shown in Table 11.2). These tests were conducted at Thyssen-Krupp VDM, at IEHK of RWTH Aachen and at the Fraunhofer IFAM in Bremen [6,7] on foils of 40 to 55 μm thickness after a recrystallisation treatment in a continuous furnace (only for the homogeneous samples), or a diffusion treatment of 1100 °C for 15 min and air cooling, or 1190 °C for 10 min and air cooling.

### 11.3   Results

Figure 11.1 shows the results of the long-term oxidation tests on the coated FeNiCrAl alloy. For comparison, the results for the homogeneous conventional FeCrAl alloy and the coated FeCrAl were included. Figure 11.1 shows that the coated FeNiCrAl has a net mass gain clearly lower than that of the alumina-forming alloys, the conventional FeCrAl and the coated FeCrAl, which indicates that the new material also forms a protective alumina scale. Figure 11.2 shows a picture of a model catalytic converter made of the coated FeNiCrAl after 100 h at 1100 °C in air, which shows the bright grey appearance of α-alumina. Figure 11.3 shows a micro cross-section of FeNiCrAl coated with 6% Al after annealing for 1 h at 1100 °C. A layer of

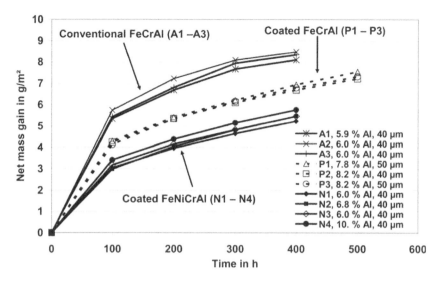

*11.1* Net mass gain of coated FeNiCrAl (N1–N4), conventional FeCrAl (A1–A3) and coated FeCrAl (P1–P3) at 1100 °C in air on foils formed to model catalytic converters

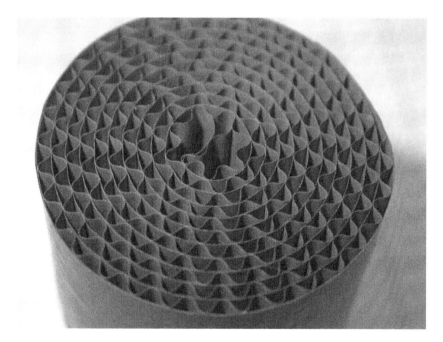

*11.2* Coated FeNiCrAl, N2, 6.8% Al, after 100 h at 1100 °C in air, foil of 40 μm thickness formed to model catalytic converters

an intermetallic AlNi₃ phase can be seen underneath a surface layer of a highly aluminium enriched FeNiCr alloy. After 100 h at 1100 °C, this intermetallic phase had vanished and the alloy was homogeneous and covered with an alumina scale, as shown in Fig. 11.4. Figure 11.5 shows an SEM picture of a micro-cross-section at

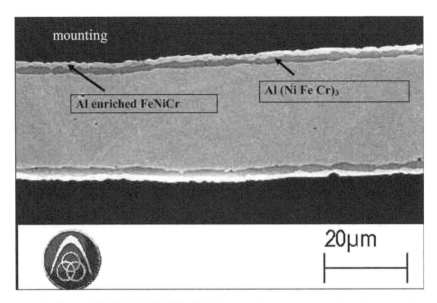

*11.3* Coated FeNiCrAl, N5, 6% Al, 30 µm, micro-cross-section of a model catalytic converter after oxidation at 1100 °C for 1 h in laboratory air

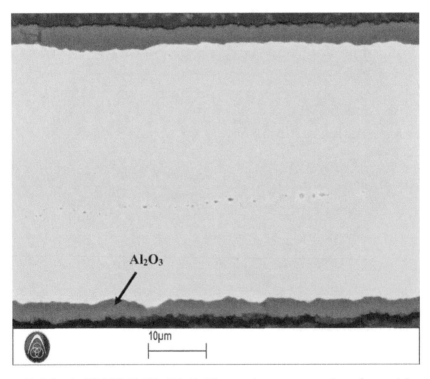

*11.4* Coated FeNiCrAl, N1, 6% Al, 40 µm, micro-cross-section of a model catalytic converter after oxidation at 1100 °C for 100 h in laboratory air

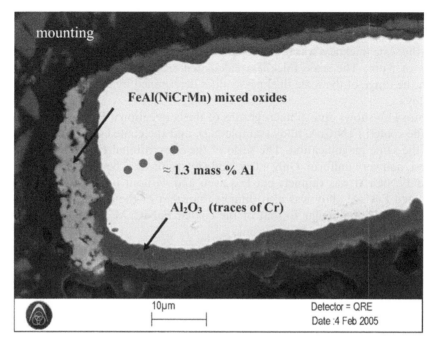

*11.5* Coated FeNiCrAl, N5, 6% Al, 30 μm, cross-section of the edges of a model catalytic converter after oxidation at 1100 °C for 4 × 100 h in laboratory air

the edge of the foil after 4 × 100 h at 1100 °C in air. The edges are not coated with aluminium, so that during the first hours of oxidation, the aluminium content was not sufficiently high to form a pure alumina scale. Instead, FeAl-rich mixed oxides were formed. Later, enough aluminium diffused into the edges so that a closed alumina scale could be formed below the mixed oxides and good protection of the edges was achieved. No spallation of oxide from the foil was observed after oxidation for 400 h.

In Table 11.3, the measurements of scale thickness by optical microscopy after 400 h at 1100 °C in air are summarised. The scale of the conventional FeCrAl had a

*Table 11.3*   Scale thickness after 1100 °C/4 × 100 h in air. The maximum values of scale thickness in SEM/metallography do not take local oxide intrusions into account

|  |  | Sample | d /μm | Al /% | Scale thickness /μm | | |
|---|---|---|---|---|---|---|---|
|  |  |  |  |  | Calc. from mass gain | Optical microscopy | SEM |
| Conventional FeCrAl | H | A1 | 40 | 5.9 | 4.3 | 4–5 | 4.1–5.1 |
| Coated FeNiCrAl | C | N2 | 40 | 6.8 | 2.9 | 2–3, max to 5 | 2.4–3.1 max to 4.2 |
| Coated FeNiCrAl | C | N3 | 40 | 6.0 | 3.0 | 2–3, max to 5 | – |
| Coated FeCrAl | C | P1 | 50 | 7.8 | 3.7 | 2–4, max to 6 | 2.6–3.6 |

H, homogeneous heat; C, coated.

thickness of between 4 and 5 µm whereas that of the coated FeNiCrAl alloy was mostly between 2 and 3 µm, with a maximum value of 5 µm. On the coated FeCrAl alloy, the scale thickness was found to be mostly between 2 and 4 µm, with maximum values of 5 µm. The scale thickness calculated from mass gain measurements is within the range of the scale thickness values determined by optical microscopy for the alloys.

Figure 11.6 shows optical micrographs of the conventional FeCrAl alloy (sample A1), the coated FeNiCrAl alloy (sample N2), and the coated FeCrAl (sample P1), all at the same magnification. The scale of the conventional FeCrAl alloy was the thickest, and very uniform. Only a few oxide intrusions can be seen. The scale of the coated FeNiCrAl was thinner, but less even and without intrusions. The scale of the coated FeCrAl alloy was also thinner than that of the corresponding conventional FeCrAl, but similar to that of the coated FeNiCrAl as well as being more uneven and with some oxide intrusions.

Figure 11.7 shows FESEM pictures of the conventional FeCrAl alloy in comparison with the coated FeNiCrAl alloy and coated FeCrAl alloy. As already mentioned, the scale on a conventional FeCrAl is thicker and more uniform than that on the coated FeNiCrAl. Furthermore, the grain size of the scale on a conventional FeCrAl is smaller than that on the coated FeNiCrAl. Similar to the coated FeNiCrAl, the

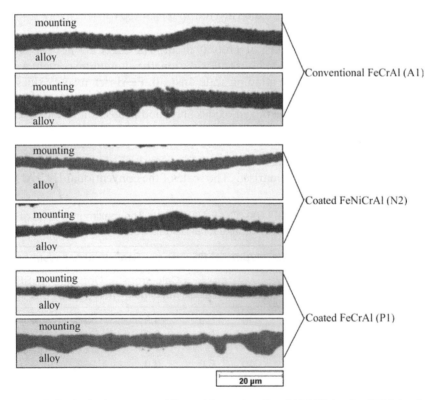

*11.6* Optical microscopy of the oxide scale, after 1100 °C for 4 × 100 h in air, 40 µm foils; coated FeNiCrAl (sample N2), conventional FeCrAl (sample A1) and coated FeCrAl (sample P1)

**Conventional FeCr20Al6 (A1)** — mounting

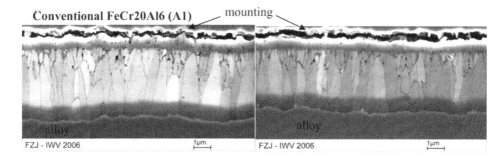

**Coated FeNiCrAl (N2)** — mounting

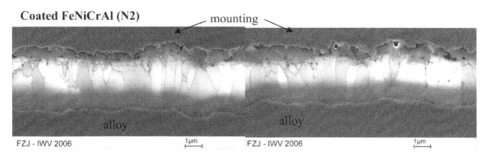

**Coated FeCrAl (P1)** — mounting

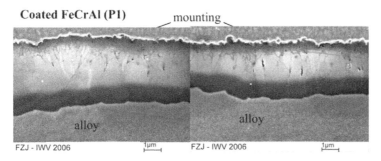

*11.7* FESEM In-lens images of the oxide scale of conventional FeCrAl (sample A1), coated FeNiCrAl (sample N2) and coated FeCrAl (sample P1) after 1100 °C for $4 \times 100$ h in air, 40 µm foils

coated FeCrAl forms a thinner scale, of less uniform thickness compared to the conventional FeCrAl. However, the grain size and thickness variations are far greater for the scale on the coated FeCrAl compared to that on the coated FeNiCrAl. There are conglomerations of very small grains and pores (Fig. 11.8). Figure 11.9 shows the lateral grain size of the oxide scales derived from the SEM images presented in Fig. 11.7 as a function of the distance from the surface of the scale for the coated FeNiCrAl (sample N2), the conventional FeCrAl (sample A1) and the coated FeCrAl (sample P1) [8]. The grain diameter of the scale at the same distance from the surface is largest for the coated FeNiCrAl and smallest for the conventional FeCrAl, with the coated FeCrAl in between.

Figure 11.10 shows the yield strength of the coated FeNiCrAl, the conventional FeCrAl and the coated FeCrAl in relation to temperature. The yield strengths of the coated FeNiCrAl are significantly higher than those of the conventional FeCrAl

## Coated FeCrAl (P1)

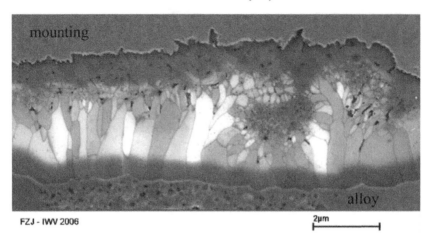

FZJ - IWV 2006                                     2µm

*11.8* FESEM In-lens image of the oxide scale for coated FeCrAl (sample P1) after 1100 °C for 4 × 100 h in air, 40 µm foils, showing agglomerations of small grains

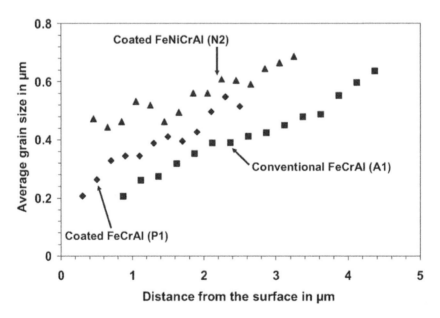

*11.9* Grain diameter of oxide scale after 1100 °C for 4 × 100 h in air in relation to the distance from the surface of the scale for coated FeNiCrAl (sample N2), conventional FeCrAl (sample A1) and coated FeCrAl (sample P1)

and the coated FeCrAl over the whole range of temperatures including the range of special interest above 600 °C. The two FeCrAl alloys have similar yield strengths.

Table 11.4 gives a short summary of the experimental results, pointing out the differences in oxidation resistance and mechanical properties between the studied materials.

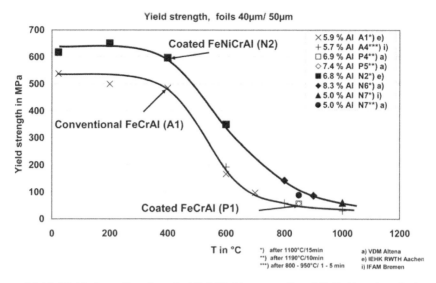

*11.10* Yield strengths of coated FeNiCrAl, conventional FeCrAl and coated FeCrAl for 40 μm and 50 μm foils

*Table 11.4*  Summary of the experimental results

| Conventional FeCrAl | Coated FeCrAl | Coated FeNiCrAl |
|---|---|---|
| FeCrAl, ferritic | FeCrAl, ferritic | FeNiCrAl, austenitic |
| Homogeneous | Coated | Coated |
| Alumina scale | Alumina scale | Alumina scale edges: alumina scale underneath FeAlNiCrMn mixed oxides |
| Higher mass gains, scale thicknesses | Lower mass gains, scale thicknesses | Lower mass gains, scale thicknesses |
| No spallation | No spallation | No spallation |
| More even oxide scale | Uneven oxide scale | Uneven oxide scale |
| Some protuberances | Some protuberances | – |
| – | – | Interim appearance of intermetallic Ni–Al phases |
| Smaller oxide grains | Larger oxide grains | Larger oxide grains |
| Low strength at elevated temperature | Low strength at elevated temperature | High strength at elevated temperature |

## 11.4    Discussion

The coated high-aluminium-containing FeCrAl and FeNiCr alloys show a less uniform thickness of alumina scales (Table 11.3), which most probably is related to the method of manufacture (aluminium coating). The results indicate, however, that the oxidation resistance of the aluminium-coated alloys is higher than that of the conventional wrought FeCrAl. The uncoated edges of the aluminium-coated foils are protected by a closed alumina scale underneath FeAl-rich mixed oxides, so again, there seems to be no enhanced degradation (Fig. 11.5). During service at high temperature, there is an interim appearance of intermetallic Ni–Al phases, which dissolve after a longer exposure time (Figs. 11.3 and 11.4).

As expected for an alloy with an austenitic structure, the coated FeNiCrAl shows higher mechanical properties than the ferritic FeCrAl, especially at temperatures above 600 °C.

Remarkable is the lower net mass gain of the coated samples in comparison to the conventional wrought one (Fig. 11.1), which will be discussed later in more detail.

The presence of chromium oxide is known to enhance the transformation of meta-stable aluminium oxides to α-alumina [9,10]. The concentrations of chromium and iron are always less in the scale of the coated FeCrAl than in the corresponding conventional material for temperatures up to 900 °C and for up to 30 min at 1100 °C [11]. A similar effect is expected for the coated FeNiCrAl alloy. Therefore, in the initial oxidation stages the formation of α-alumina starts earlier in the oxide scale of the conventional FeCrAl with more nuclei and consequently, the scale is more fine-grained compared to the coated FeNiCrAl and the coated FeCrAl. The increase in the oxide grain size has already been reported for the coated FeCrAl [11] and for an FeCrAl alloy with an aluminium coating produced by vapour deposition [12,13]. As is known for yttrium-containing FeCrAl alloys [2,14,15], the α-alumina scale grows mainly by inward diffusion of oxygen along the grain boundaries of the scale. This means that alloys with a larger grain size of the scale – the coated alloys – should have a lower net mass gain and thinner scales in comparison to homogeneous samples, as observed in the present study.

The conventional FeCrAl and the coated FeNiCrAl have a similar grain size at the scale/metal interface after 400 h. However, the scale thickness on conventional FeCrAl is larger because of a finer grain size in the outer part of the scale compared to coated FeNiCrAl. This is consistent with the net mass gain plot in Fig. 11.1. There, the net mass gain of the conventional FeCrAl is clearly larger after 400 h, but the oxidation rate (slope of the mass change curves) for both alloys at this time is quite similar.

For the coated alloys, the scale is more uneven than that of the conventional alloys. As Fig. 11.8 for the coated FeCrAl alloy shows, these thicker parts of the scale can have a smaller grain size. The reason for this effect could be, for example, impurities from the alloy/aluminium layer interface, which upon incorporation into the scale, hinder locally the columnar growth and promote the nucleation of new grains.

The net mass gain of the coated FeNiCrAl is lower than that of the coated FeCrAl. Possible additional causes for this could be an effect of the base material and/or reactive elements (only yttrium in the coated FeNiCrAl and yttrium, zirconium and hafnium in the coated FeCrAl). For example, it has been reported that the alumina scales formed on nickel aluminides can have a very large grain size [16].

## 11.5     Conclusions

- During high-temperature oxidation, the coated FeNiCrAl forms a pure alumina scale on the main surfaces with no oxide spallation.
- The uncoated edges are protected by a closed alumina scale beneath a transient FeAl-rich mixed oxide.
- There is an interim appearance of an intermetallic Ni–Al phase during high-temperature annealing.

- The coated FeNiCrAl and the coated FeCrAl have a lower net mass gain than the conventional FeCrAl alloys due to the larger grain size of the alumina scale in the former materials. Therefore, a lower aluminium consumption prevails in the coated foils.
- The coated FeNiCrAl has higher mechanical strength at elevated temperature than the conventional ferritic FeCrAl and coated FeCrAl.

## Acknowledgements

The authors would like to thank Mr Peters from ThyssenKrupp Duisburg for hot dipping, and Mr Theile from Wickeder Westfalenstahl for cladding. Thanks also go to Mrs R. Jänichen, A. Liebelt, Mr M. Wagner, Mr H. Niecke and Mr D. Siepmann for carrying out the oxidation tests, to Mr M. Schulze for carrying out the hot tensile tests, and to Mrs A. Kalinowski and Dr E. Wessel for metallographic and SEM investigations.

## References

1. J. Klöwer, A. Kolb-Telieps, U. Heubner and M. Brede, in Corrosion 98, NACE Houston, TX, 1998, No. 746.
2. W. J. Quadakkers, H. Holzbrecher, K. G. Briefs and H. Beske, *Oxid. Metals*, 32 (1989), 67.
3. G. R. Wallwork, A. Z. Hed, *Oxid. Met.*, 3 (1971), 171.
4. J. Koepernik, A. Kolb-Telieps, J. Klöwer and U. Heubner, in *Werkstoffwoche 96*, 159–164. Frankfurt, 1997.
5. A. Kolb-Telieps, J. Klöwer, R. Hojda and U. Heubner, in *Metal Supported Automotive Catalytic Converters*, 99–104, ed. H. Bode. Frankfurt, Germany, 1997.
6. G. Leisten and J. Römer, Laboratory Report, IEHK RWTH Aachen, Sept. 2006.
7. M. Brede, Laboratory Reports, Fraunhofer IFAM, Mar. 1997, Jun. 1999, Aug. 2000.
8. S. V. P. Reddy, Master's Thesis, FH Aachen, Jülich, Germany, 2006.
9. M. W. Brumm and H. J. Grabke, *Corros. Sci.*, 33 (1992), 1677.
10. W. C. Hagel, *Corrosion*, 21 (1965), 316.
11. H. Hattendorf, A. Kolb-Telieps, Th. Strunskus, V. Zaporojchenko and F. Faupel, *EFC Publication 34*, 135–147. Published by Maney Publishing on behalf of The Institute of Materials, London, 2001.
12. A. Andoh, S. Taniguchi and T. Shibata, High-temperature oxidation of al-deposited stainless-steel foils, *Oxid. Metals*, 46 (1996), 481.
13. A. Andoh, S. Taniguchi and T. Shibata, Phase transformation and structural changes of alumina scales formed on Al-deposited Fe-Cr-Al foils, in Proc. Int. Symp. on High-Temperature Corrosion and Protection 2000, Hokkaido, Japan, 17–22 Sept. 2000, 297–303, ed. T. Narita, T. Maruyama and S. Taniguchi. Science Reviews, UK, 2000.
14. F. A. Golightly, F. H. Scott and G. C. Wood, *Oxid. Metals*, 10 (1976), 163; 14 (1980), 218.
15. B. A. Pint and L. W. Hobs, *Oxid. Metals*, 41 (1994), 203.
16. V. K. Tolpygo, D. R. Clarke and K. S. Murphy, The effect of grit blasting on the oxidation behaviour of a platinum-modified nickel-aluminide coating, *Metall. Mater. Trans.*, 32A (2001), 1467.

# 12

## Effectiveness of platinum and iridium in improving the resistance of Ni–Al to thermal cycling in air–steam mixtures*

### R. Kartono and D. J. Young

*School of Materials Science & Engineering,*
*University of New South Wales, UNSW, Sydney, NSW 2052 Australia*
*d.young@unsw.edu.au*

### 12.1 Introduction

Water vapour is a product of hydrocarbon fuel combustion, and is invariably present in gas turbine engines. However, its effect on the oxidation of Ni–Al-based systems is not fully understood. For iron-based alloys, Kvernes *et al.* [1] reported that the presence of water vapour adversely affects the selective oxidation of aluminium and chromium. Their results indicated that the increase in oxidation rate with water vapour content was due to gaseous transport across voids and pores within or beneath the scale being facilitated by an $H_2O–H_2–O_2$ carrier gas mixture. Other reports [2,3] have suggested the same effect. The water vapour effect on nickel-based superalloys has been studied by Meier, Pettit and colleagues [4–6]. They proposed that the presence of water vapour caused stress corrosion cracking at the $Al_2O_3$ scale–alloy interface during cyclic oxidation. A study of the oxidation of pure nickel [7] showed that NiO formation in pure water vapour was slower than that in air.

Turbine engine hot stage components are provided with oxidation-resistant coatings, containing high levels of aluminium. New coating compositions containing lower aluminium levels, but with platinum additions, are currently being tested [8–13]. These coatings are based on a $\gamma–Ni+\gamma'–Ni_3Al$-based constitution. The phase diagram for this system is shown in Fig. 12.1, where it is seen that substitution of platinum for some of the nickel leads to partitioning of the platinum between the two phases.

Binary $\gamma/\gamma'Ni–Al$ alloys are known [15] to perform poorly under cyclic oxidation conditions in dry air, and even worse in air-plus-steam mixtures. The aim of the present work was to explore the performance of platinum-modified $\gamma/\gamma'$ nickel aluminide materials. Subsidiary aims were to investigate the effectiveness of hafnium additions and the viability of partial replacement of some of the platinum with iridium.

---

\* Reprinted from R. Kartono and D. J. Young: Effectiveness of platinum and iridium in improving the resistance of Ni–Al to thermal cycling in air-steam mixtures. *Materials and Corrosion.* 2008. Volume 59. pp. 455–62. Copyright Wiley-VCH Verlag GmbH & Co. KGaA.

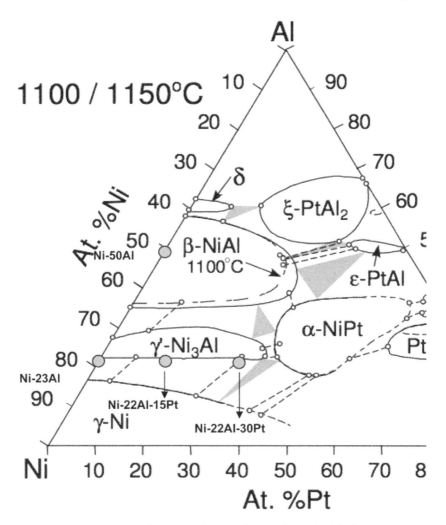

*12.1* Part of the Ni–Al–Pt phase diagram (1150 °C section) [14]

*Table 12.1*  Alloy compositions (at.%) (balance Ni)

| Al | Pt | Ir | Hf, wt.% |
|----|----|----|----------|
| 23 |    |    | 0 and 1 |
| 22 | 15 |    | 0 and 1 |
| 22 | 30 |    | 0 and 1 |
| 20 | 15 | 5  | 0 and 1 |
| 20 | 15 | 10 | 0 and 1 |

## 12.2    Experimental

Alloy compositions are listed in Table 12.1. All compositions are specified in at.% except for hafnium additions, which were at the 1 wt.% level when added. This level corresponds to different molar concentrations in the different alloys, and is quoted in wt.% for simplicity. All alloys containing iridium were provided by Ames National

Laboratory. Iridium-free alloys were prepared by argon arc melting metals [99.99% Ni and Al, 99.999% Pt, 99% Hf (including 0.2% Zr)] using non-consumable electrodes. Alloys were subsequently annealed at 1200 °C for 1 h and 1150 C for 24 h in Ar–5% $H_2$. The iridium-free alloy compositions are marked on the phase diagram of Fig. 12.1. X-ray diffraction (XRD) and metallography were used to verify the phase constitution of these alloys.

Alloy coupons of thickness 1.2 mm and diameter 10 mm were cut from the annealed ingots, ground to a 1200 grit finish and ultrasonically cleaned in acetone before use. These samples underwent cyclic oxidation in flowing gas, either pre-dried air or air–12% $H_2O$. Wet air was produced by flowing air through a saturator–condenser system causing condensation at 51.3 °C, leaving 12% by volume water vapour in the gas. Cyclic oxidation was performed in a vertical tube furnace at 1200 °C, using an automated mechanism for moving samples into and out of the furnace. Each cycle consisted of 1 h at temperature and 10 min cooling above the furnace, still in the reaction gas atmosphere, at a temperature of 80 °C. From time to time, samples were withdrawn and weighed. The precision of the measurement led to an error equivalent to 0.05 mg cm$^{-2}$ in sample weight change. Oxidised samples were examined using conventional metallography. In addition, cross-sections were prepared for field emission scanning electron microscopy using focused ion beam (FIB) milling.

## 12.3    Cyclic reaction in dry air

Weight change kinetic data are shown in Fig. 12.2. The binary $\gamma/\gamma'$ alloy underwent rapid scale spallation and weight loss. The $\beta$-NiAl material also sustained severe damage. Weight change data for the 15% and 30% platinum-containing $\gamma/\gamma'$ alloys were closely similar, and only one is shown on the plot. In each case, no net weight loss was observed over the 1000 cycle test. Nonetheless, spallation did occur, as shown in Fig. 12.3. The difference between the platinum-containing alloys and the binary material was the ability of the former to re-heal after spallation, so as to reform a protective alumina scale.

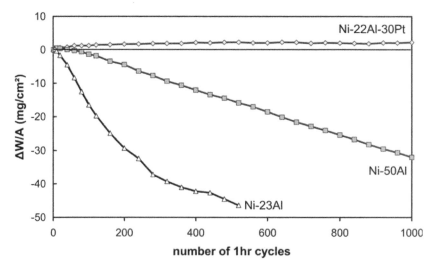

12.2 Effect of Pt on cyclic oxidation kinetics in dry air

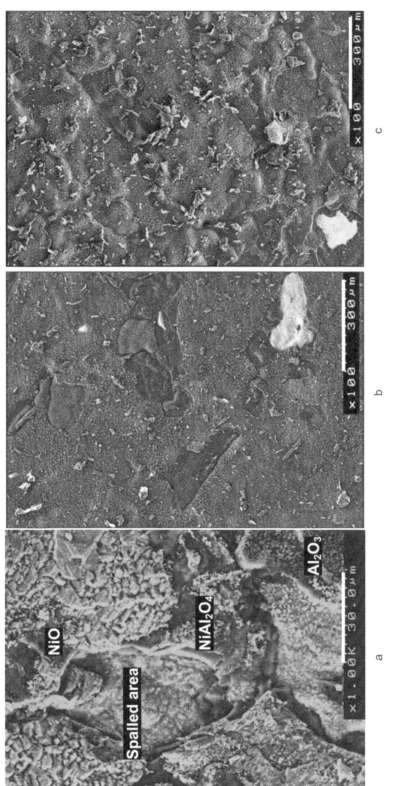

12.3 Scale surfaces after 1000 cycles in dry air. (a) Ni–23Al (60 cycles); (b) Ni–22Al–15Pt; (c) N–22Al–30Pt

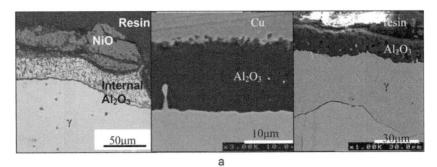

12.4 Cross-sections of alloys after thermal cycling in dry air: (a) scales; left to right: Ni–23Al (520 cycles), Ni–22Al–15Pt, Ni–22Al–30Pt; (b) alloy subsurface zones; left to right: Ni–23Al (60 cycles), Ni–22Al–15Pt, Ni–22Al–30Pt

The benefit of the platinum is clear from Figs. 12.3 and 12.4. Whereas the binary material ceased to be an alumina former after several cycles, growing thick nickel-rich oxide scales which spalled massively, the platinum-containing alloys retained their ability to form alumina scales. Scales on the platinum alloys invariably consisted of α-alumina with a thin outer layer of $NiAl_2O_4$ spinel. The binary alloy had become depleted in aluminium to the extent that $Al_2O_3$ was precipitated internally beneath a thick, nickel-rich oxide scale.

Depletion of aluminium from the alloys led to the formation of a sub-surface, single-phase γ-nickel region in all cases, as seen in Fig. 12.4b. This was very deep in the case of the binary alloy, but only shallow in the case of the platinum-bearing materials. The smaller degree of sub-surface phase transformation in the latter alloys reflected much reduced extents of aluminium depletion. Nonetheless, depletion of the 30% Pt alloy was sufficient to cause the formation of an α+γ sub-surface zone of more than 100 μm depth. The phase transformations occurring within the sub-surface zone correspond simply to aluminium depletion, as is seen from the diffusion paths mapped on the phase diagram of Fig. 12.1.

The somewhat different behaviour of the alloy–scale interface in the case of platinum-containing materials is also evident in Fig. 12.4a. A degree of morphological

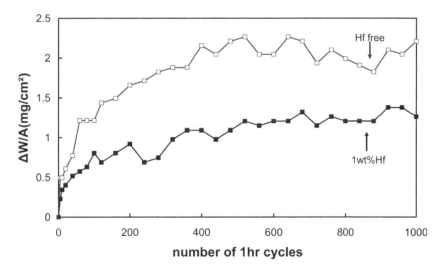

*12.5* Effect of hafnium on cyclic oxidation kinetics of Ni–22Al–30Pt

instability had developed at the interface, resulting in the formation of apparent protrusions of metal into the oxide scale. Some of these had been pinched off, leading to the appearance of Pt–Ni-rich particles within the oxide. In addition, some rumpling of the metal is apparent at the scale–alloy interface in the case of the 30% Pt alloy.

Performance was improved still further with the addition of hafnium to these alloys. Figure 12.5 shows the effect on cyclic oxidation of the 30 Pt $\gamma/\gamma'$ alloy. The slight decrease in total net weight uptake in fact reflected the ability of this alloy to retain its scale and resist spallation, as seen in Fig. 12.6. Whereas the hafnium-free 15% and 30% platinum alloys showed 7% and 4% bare metal, respectively, after 1000 cycles at 1200 °C, and many areas of re-healed spallation, no such damage was seen on the hafnium-bearing versions of these two alloys. Furthermore, the growth of alumina whiskers at the scale exterior observed on all hafnium-free alloys was completely suppressed in the presence of 1 wt.% Hf.

Closer examination of oxide scale cross-sections using FIB milling showed that the oxide formed on the hafnium-containing alloy was pore free and developed a very uniform interface with the metal. In contrast, the oxide formed on the hafnium-free version of the alloy contained voids and developed a convoluted scale–metal interface. The same observations were made on the 15% Pt alloy. The oxide grown on the hafnium-bearing alloys was in all cases $\alpha$-alumina, plus a thin external layer of $NiAl_2O_4$.

## 12.4    Cyclic oxidation in wet air

The introduction of 12% water vapour to the gas had a profound effect on cyclic oxidation weight change kinetics, as seen in Fig. 12.7. Spallation damage to the $\gamma/\gamma'$ Ni–Al alloy was extraordinarily rapid. The addition of platinum to the alloy decreased the rate of spallation markedly, but did not prevent spallation entirely. Inspection of reacted scale surfaces (Fig. 12.8) revealed extensive damage which had been largely re-healed in the case of the platinum-bearing alloys. The oxide observed on the binary alloy consisted of a mixture of $\alpha$- and $\theta$-alumina plus nickel-rich oxides, whereas that formed on the platinum-bearing alloys consisted only of $\alpha$-alumina plus

*12.6* Fine-grained, spall-resistant scale grown during cycling oxidation of Ni–22Al–30Pt + 1 wt.% Hf in dry air

$NiAl_2O_4$. The reaction morphologies shown in Fig. 12.9 reveal internal precipitation of alumina within the binary alloy, underneath an external nickel-rich oxide scale. Platinum-containing alloys formed thin scales, which were mainly $Al_2O_3$ with a thin layer of spinel at the exterior. Relatively deep depletion of aluminium from the alloys (Fig. 12.9b) indicated that the alumina scale had repeatedly formed, spalled and reformed in order to consume the large amount of aluminium. Thus the platinum-containing alloys were still subject to spallation, but were more successful at reforming alumina.

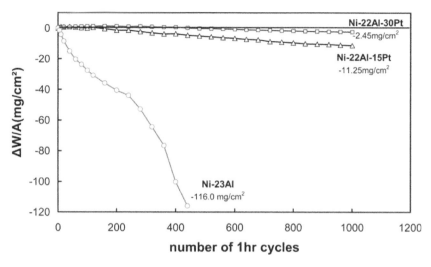

*12.7* Effect of Pt on cyclic oxidation kinetics in air + 12% H₂O

The addition of hafnium was strongly beneficial, even in this aggressive wet air environment. Figure 12.10 shows the weight change kinetics. It is seen that weight loss was largely suppressed in the case of the 30% Pt alloy, and was strongly reduced in the case of the 15% Pt alloy. After 1000 cycles of reaction, the extent of spallation on the 30% Pt + 1 wt.% Hf alloy was negligible, as seen in Fig. 12.11. On the 15% Pt alloy, a similar fine-grained scale was also formed, but some regions had spalled and re-healed. These alloys produced oxide whiskers at the scale exterior in the absence of hafnium, but not when hafnium was present. The improvement in scale behaviour was reflected in much reduced depths of aluminium depletion within the alloys, as shown in Table 12.2.

## 12.5    Effects of iridium on cyclic oxidation in wet air

Weight change kinetics during thermal cycling in wet air are compared in Fig. 12.12 for $\gamma/\gamma'$ Ni–20Al–$x$Pt–$y$Ir, where $x$ is 15% or 30% and $y$ is 5% or 10%. The 15% Pt + 5% or 10% Ir alloys were very much better than iridium-free versions of this alloy and somewhat better than the 30% Pt alloy. The different spallation behaviour is clear in Fig. 12.13. It is also evident in this figure that the addition of 1 wt.% Hf improved scale spallation resistance enormously. Weight change kinetics (not shown) showed no net mass loss for hafnium-containing iridium alloys.

Figure 12.14 shows FIB milled scale cross-sections. In the absence of hafnium, the scale on 15Pt–5Ir $\gamma/\gamma'$ alloy contained substantial numbers of voids and developed to roughly uniform thicknesses. When hafnium was present, the scale developed instead as an extremely thin layer between regions of thicker growth. The thick regions corresponded to the development of ridges in the external scale.

## 12.6    Discussion and conclusions

The effects of water vapour observed at 1200 °C, were very much more severe than have been reported [13] for 1 h cycles at 1100 °C. At the lower temperature, the addition of water vapour to air led to increased rates of weight uptake on $\gamma/\gamma'$

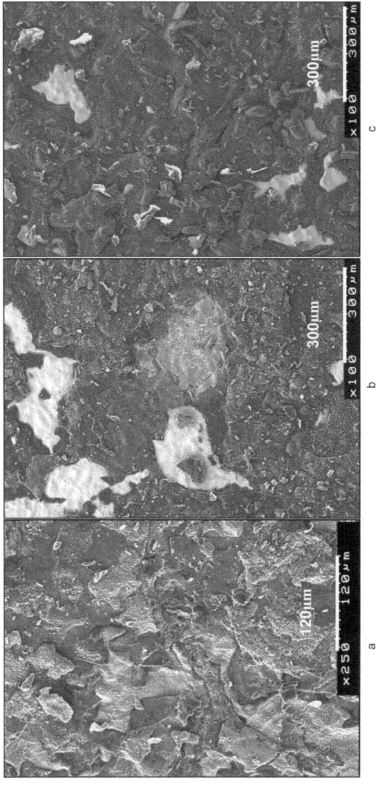

12.8 Scale surfaces after thermal cycling in wet air. (a) Ni–23Al (60 cycles); (b) Ni–22Al–15Pt; (c) Ni–22Al–30Pt

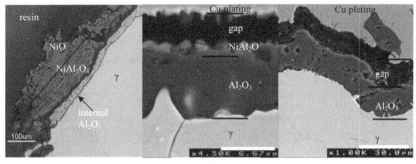

a

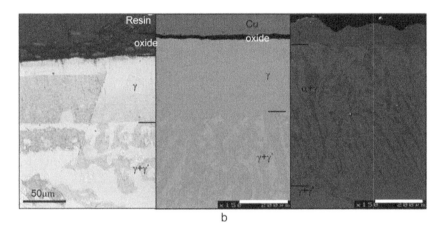

b

*12.9* Cross-section after thermal cycling in wet air: (a) scales; left to right: Ni–23Al (440 cycles), Ni–22Al–15Pt, Ni–22Al–30Pt; (b) alloy subsurface zones; left to right: Ni–23Al (60 cycles), Ni–22Al–15Pt, Ni–22Al–30Pt

platinum-modified alloys. In the present work however, increased scale spallation was the dominant effect observed when water vapour was present.

Because of its relatively low aluminium content, the binary $\gamma/\gamma'$ alloy is a marginal alumina former. Under temperature cycling conditions, it is rapidly depleted in aluminium, to the extent where protective alumina scales cannot be reformed. Loss of protective behaviour is even more rapid in a moisture-containing atmosphere than in dry air. Adding platinum to these alloys greatly improved their resistance to thermal cyclic oxidation.

### 12.6.1  Effects of platinum additions

The Ni–Al–Pt alloys showed no net weight loss when subjected to cyclic oxidation in dry air at 1200 °C (Fig. 12.3). Nonetheless, the platinum-modified alloys were undergoing repeated spallation, as is clear in Fig. 12.3. The superiority of the platinum-modified alloys lay in their ability to reform alumina scales after spallation. Repeated spallation and re-healing events led to quite deep depletion of aluminium from the alloys, as seen in Fig. 12.4 and Table 12.2. It is likely that the increased

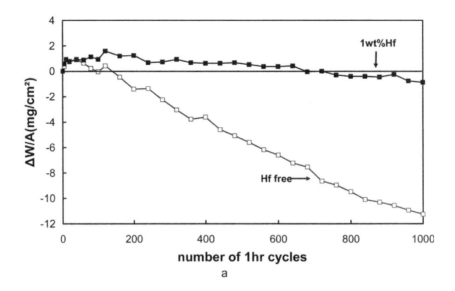

a

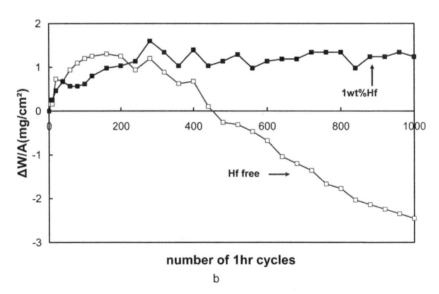

b

*12.10* Effect of hafnium on cyclic oxidation kinetics in wet air. (a) Ni–22Al–15Pt; (b) Ni–22Al–30Pt

Al–Ni ratio in platinum-bearing alloys has the effect of promoting alumina formation and suppressing NiO formation during the transient oxidation period following scale spallation [12]. The beneficial effects of platinum additions to alumina formers have long been known [16–18].

An additional benefit provided by the platinum addition was the promotion of $\alpha$-Al$_2$O$_3$ formation. Whereas the binary alloy produced both $\alpha$- and $\theta$-Al$_2$O$_3$, the platinum-modified alloys produced only the $\alpha$ form. Platinum is well known [19] to be a catalyst for alumina phase transformation, leading to the rapid formation of the stable $\alpha$-phase.

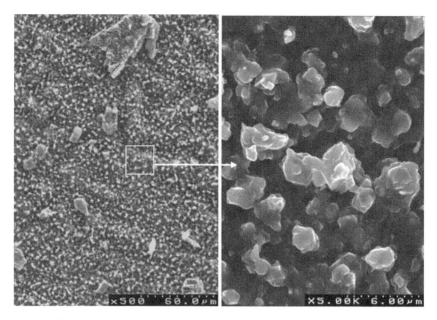

*12.11* Fine-grained, spall-resistant scale grown on Ni–22Al–30Pt + 1Hf during cyclic oxidation in wet air

*Table 12.2*    Reaction morphologies after 1000 cycles in wet air

| Alloy | Scale thickness, μm | | Depleted $\gamma$-zone, μm | |
|---|---|---|---|---|
| | No Hf | 1 wt.% Hf | No Hf | 1 wt.% Hf |
| Ni–23Al | 120* | 60 + 90 | All transformed* | 0 |
| | (+ internal oxide) | (+ protruding oxide) | | |
| Ni–22Al–15Pt | 17 | 6 | 320 | 24 |
| Ni–22Al–30Pt | 20 | 6 | 67 | 0 |
| Ni–20Al–15Pt–5Ir | 12 | 7 | 37 | 6 |
| Ni–20Al–15Pt–10Ir | 12 | 7 | 37 | 0 |

*After 440 cycles.

Platinum additions to the binary alloy are also beneficial to the oxidation perform-ance during cyclic exposure to wet air. However, although the platinum improves performance greatly (Fig. 12.7), the spallation is nonetheless a continuing long-term problem. The extent of damage evident from the weight change kinetics and the observed aluminium depletion from the alloys (Table 12.2) was much greater than reported by Pint *et al.* [13] for reaction at 1100 °C. It seems that although platinum provided essentially the same benefits to the alloy in dry and wet air oxidation, the effect was insufficient to counteract the more severe spallation which occurred during wet air exposure.

### 12.6.2   Effects of hafnium additions

Additions of hafnium to the Ni–Al–Pt alloys brought about several changes to the reaction morphologies. The $\alpha$-Al$_2$O$_3$ scales grown on hafnium-containing alloys were

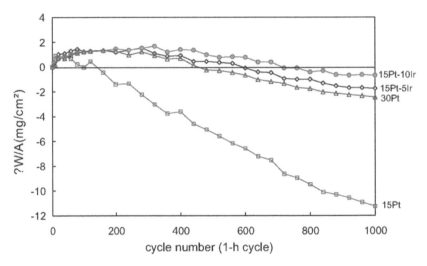

*12.12* Effects of Ir on cyclic oxidation in wet air

thinner and more dense than in the absence of this dopant. Moreover, the growth of alumina whiskers at the scale outer surface was completely suppressed in the presence of hafnium. Evidently, mass transport within the oxide scale was altered by the hafnium, and outward transport of aluminium was greatly decreased. This is to be expected, as the alumina grain boundaries formed on platinum-modified materials have been shown [20] to be enriched with hafnium.

The presence of hafnium is also clearly beneficial to scale adhesion, suppressing or at least reducing greatly the incidence of scale spallation in dry air at platinum levels of both 15 and 30 at.% (Fig. 12.6). In wet air, hafnium prevents scale spallation completely at a platinum level of 30%, but only partially at a platinum level of 15% (Table 12.2). The benefits of hafnium additions have been discussed by others [21,22]. What is evident here is an interaction between platinum and hafnium in the case of wet air cyclic oxidation. Thus, the addition of hafnium prevents spallation when platinum is at the 30% level, but not when platinum is only at the 15% level.

### 12.6.3   Effects of iridium additions

Partial substitution of iridium for platinum in the $\gamma/\gamma'$ alloys improved their resistance to scale spallation, as seen from a comparison of Figs. 12.8 and 12.13. This was reflected in the observed weight change kinetics (Figs. 12.7 and 12.12) and in sub-surface aluminium depletion (Table 12.2). The further addition of hafnium to the platinum–iridium alloys led to the complete prevention of scale spallation under these rather extreme conditions.

The $\alpha$-Al$_2$O$_3$ scale produced on a hafnium-bearing, platinum–iridium modified $\gamma/\gamma'$ alloy (Fig. 12.14) appeared closely similar to the $\alpha$-alumina scale produced during isothermal oxidation of $\beta$-NiAl [23,24]. The thin, flat regions of oxide correspond to the regions where the first formed oxide scale transformed to the $\alpha$-phase. These regions spread laterally from the initial nuclei until neighbouring regions impinged to form high angle grain boundaries. Rapid growth of transient aluminas continued for longer at these grain boundaries, resulting in thicker oxide and the development of

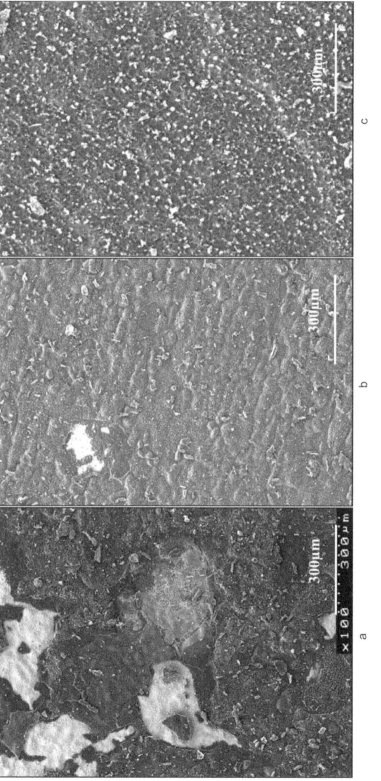

12.13 Scale surfaces after 1000 cycles in wet air, showing effects of Ir and Hf. (a) Ni–22Al–15Pt; (b) Ni–20Al–15Pt–10Ir; (c) Ni–20Al–15Pt–10Ir + 1 wt% Hf

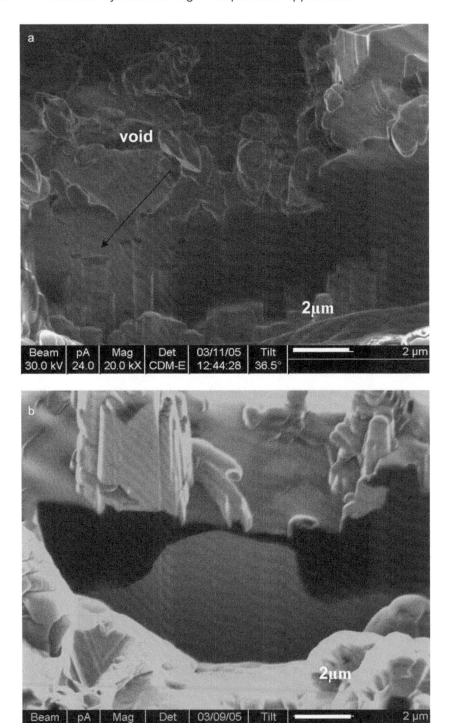

*12.14* FIB milled scale sections on Ni–20Al–15Pt–5Ir alloys after 1000 cycles in wet air. (a) 5Ir hafnium-free; (b) 5Ir + 1 wt. % Hf

externally visible oxide ridges. In short, the oxide morphology achieved on hafnium-bearing platinum–iridium modified $\gamma/\gamma'$ alloys under cyclic reaction conditions at 1200 °C in wet air, was essentially the same as that arrived at for high aluminium content $\beta$-NiAl under the very much less aggressive conditions of isothermal dry air oxidation. Details of the interactions between hafnium, iridium and platinum in the $\gamma/\gamma'$ alloy await clarification.

## Acknowledgements

The authors thank Brian Gleeson of Ames National Laboratory for provision of the iridium-containing alloys and gratefully acknowledge the provision of financial support by NEDO (Japan) and the Australian Research Council.

## References

1. I. Kvernes, M. Oliveira and P. Kofstad, *Corros. Sci.*, 17 (1977), 237.
2. A. Rahmel and J. Tobolski, *Corros. Sci.*, 3 (1965), 333.
3. C. W. Tuck, M. Odgers and K. Sachs, *Corros. Sci.*, 9 (1969), 271.
4. R. Janakiraman, G. H. Meier and F. S. Pettit, in *Cyclic Oxidation of High Temperature Materials/1999*, 38, ed. M. Schütze and W. J. Quadakkers. European Federation of Corrosion Publication No. 27, London, 1999.
5. M. C. Maris-Sida, G. H. Meier and F. S. Pettit, *Met. Trans.*, 34A (2003), 2609.
6. K. Onal, M. C. Maris-Sida, G. H. Meier and F. S. Pettit, *Mater. High Temp.*, 20 (2003), 327.
7. A. Galerie, Y. Wouters and M. Caillet, *Mater. Sci. Forum*, 369–372 (2001), 231.
8. K. Bouhanek, O. A. Adesanya, F. H. Stott, P. Skeldon, D. G. Lees and G. C. Wood, *Mater. Sci. Forum*, 369–372 (2001), 615.
9. B. Gleeson, W. Wang, S. Hayashi and D. Sordelet, *Mater. Sci. Forum*, 461–464 (2004), 213.
10. Y. Zhang, J. A. Haynes, B. A. Pint and I. G. Wright, *Surf. Coat. Technol.*, 200 (2005), 1259.
11. T. Izumi and B. Gleeson, *Mater. Sci. Forum*, 522–523 (2006), 221.
12. S. Hayashi, T. Narita and B. Gleeson, *Mater. Sci. Forum*, 522–523 (2006), 229.
13. B. A. Pint, J. A. Haynes, Y. Zhang, K. L. More and I. G. Wright, *Surf. Coat. Technol.*, 2006, in press.
14. S. Hayashi, S. I. Ford, D. J. Young, D. J. Sordelet, M. F. Besser and B. Gleeson, *Acta Mater.*, 53 (2005), 3319.
15. R. Kartono and D. J. Young, in *High Temp. Corros. Mater. Chem. V*, 43, ed. E. Opila, J. Fergus, T. Maruyama, J. Mizusaki, T. Narita, D. Shifler and E. Wuchina. The Electrochemical Society, Pennington, NJ, 2005.
16. E. J. Felten, *Oxid. Met.*, 10 (1976), 23.
17. G. J. Tatlock and T. J. Hurd, *Oxid. Met.*, 22 (1984), 201.
18. C. W. Corti, D. R. Coupland and G. L. Selman, *Platinum Metals Rev.*, 24 (1980), 2.
19. ICI Ltd., *Catalyst Handbook*, Wolfe Scientific Books, London, 1970.
20. J. A. Haynes, B. A. Pint, K. L. More, Y. Zhang and I. G. Wright, *Oxid. Met.*, 58 (2002), 58.
21. J. L. Smialek, *JOM*, 1 (2000), 22.
22. B. A. Pint, M. Treska and L. W. Hobbs, *Oxid. Met.*, 47 (1997), 1.
23. J. Doychak, J. L. Smialek and T. E. Mitchell, in Int. Cong. Metal Corros., 35. National Research Council of Canada, Ottawa, 1984, vol. 1.
24. J. Doychak, J. L. Smialek and T. E. Mitchell, *Met. Trans. A*, 20A (1989), 499.

# 13

# Effect of manufacturing-related parameters on oxidation properties of MCrAlY bond coats*

M. Subanovic, D. Sebold, R. Vassen, E. Wessel, D. Naumenko,
L. Singheiser and W. J. Quadakkers

*Forschungszentrum Juelich GmbH, IWV, D-52425, Juelich, Germany*

m.subanovic@fz-juelich.de

## 13.1    Introduction

Internally cooled metallic superalloy components protected by a thermal barrier coating (TBC) are commonly used nowadays in high-efficiency gas turbines and aircraft engines. The TBC system applied to the superalloy is usually comprised of an oxidation-resistant MCrAlY (M = Ni, Co) bond coat (BC) beneath a thermal insulator, typically Yttria Stabilised Zirconia (YSZ) deposited by Air Plasma Spraying (APS) or Electron Beam Physical Vapour Deposition (EB-PVD) [1–3]. During TBC deposition and high-temperature exposure, the MCrAlY bond coat forms an alumina scale, which protects the underlying material from oxidation. The main lifetime limiting factor of EB-PVD-TBC systems during service has been shown to be processes related to oxidation of the bond coat. The latter include growth of the oxide scale, phase transformations in the scale and coating, aluminium depletion in the BC, etc.

Thermal barrier coatings deposited by EB-PVD often exhibit very large scatter in lifetime during cyclic oxidation testing. The reason for the variation of lifetime might be related to differences in coating manufacturing parameters, as proposed recently by several authors [4–7]. The manufacturing procedure for EB-PVD-TBC systems commonly used in industrial applications generally consists of four different steps:

1. MCrAlY BC deposition usually by vacuum plasma spraying (VPS)
2. Vacuum heat-treatment
3. A smoothing process to reduce the roughness of the as-sprayed VPS MCrAlY coating surface; and
4. Application of a YSZ top coat by EB-PVD.

The presence of small amounts of reactive elements (yttrium, hafnium, zirconium, etc.) in MCrAl coatings is essential to improve the oxide scale adherence. Although during the past decades, a variety of results were published and a number of hypotheses proposed concerning the reactive element effect, there is still much controversy about the mechanism by which the reactive elements induce their beneficial effect. So far, two mechanisms seem to have gained most experimental support, namely

* Reprinted from M. Subanovic et al.: Effect of manufacturing related parameters on oxidation properties of MCrAlY bondcoats. *Materials and Corrosion.* 2008. Volume 59. pp. 463–70. Copyright Wiley-VCH Verlag GmbH & Co. KGaA.

impurity (sulphur) gettering and a change in the scale growth mechanism [8]. In MCrAlY systems, it has been found that different manufacturing stages can introduce impurities which tie up the reactive elements into stable compounds, thus hindering the positive 'reactive element effect' on scale adherence [9].

In a preliminary study, an EB-PVD-TBC system with a typical MCrAlY bond coat was subjected to cyclic oxidation lifetime testing at 1000 °C in air. The times to failure of the seven cylindrical specimens varied between 324 and 6515 h (Fig. 13.1a). Analytical studies of the oxidised specimens suggested that the TBC lifetime variations were related to differing extents of oxidation of reactive elements during application of the MCrAlY coating by VPS. A large number of tiny reactive element-containing oxide precipitates were found in the bulk MCrAlY bond coat of TBC specimens with short lifetimes, indicating that the reactive elements were tied up into oxides during the coating process. During cyclic oxidation, this resulted in the formation of a thin, uniform, thermally grown oxide (TGO), as shown in Fig. 13.1b, and also in poor oxide scale adherence. In contrast, the TBC specimens with long lifetimes revealed virtually no oxide precipitates in the bulk bond coat. Therefore the incorporation of the reactive elements into the TGO was not hindered, whereby large numbers of reactive element-rich oxide inclusions were found in the scale. The substantial incorporation of reactive elements into the scale indicates that the quantity of reactive elements in the coating was sufficient to impart their beneficial effect on the scale adherence of the TGO (Fig. 13.1c) and consequently much longer TBC lifetimes during cyclic oxidation.

Based on these results in combination with data from the literature, it is evident that, for the same nominal MCrAlY coating composition, the form and distribution of the reactive elements can be significantly different, whereby these variations may substantially affect the scale growth rate, adherence and consequently lifetime of the TBC. Therefore, it is essential to gain knowledge about the effects of different manufacturing steps and/or parameters on the distribution of reactive elements to achieve good reproducibility of the TBC lifetime. As a starting point in the present work, free-standing coatings were produced by VPS, whereby the oxygen partial pressure in the vacuum chamber was deliberately varied to see whether a similar effect on scale growth occurs, as observed in the previously studied TBC specimens. In the industrial manufacturing process of the TBC system, the deposition of the bond coat by VPS is normally followed by a vacuum heat treatment. The heat treatment procedure is commonly optimised for the microstructural and strength properties of superalloys [10]. In some publications [4,5], it was shown that the heat treatment not only affects the microstructure of the superalloy, but also the distribution and incorporation of reactive elements in the TGO during subsequent high-temperature oxidation. In the current work, the vacuum heat treatment was performed on low-oxygen, free-standing coatings under two different conditions, i.e. at 1100 °C for 2 h and at 900 °C for 24 h. The state-of-the-art vacuum heat-treatment step at 1100 °C for 2 h commonly applied in industrial practice results in the formation of reactive element-rich oxide precipitates on the surface and in the sub-surface regions. Typically, the resulting uneven distribution of reactive elements causes local variations in the oxide morphology, which can be detrimental to the spallation resistance of the growing oxide scale. From the oxide dispersion strengthening alumina forming alloys, it is known that the reactive element additions in some cases are more effective as oxide dispersions than as alloy additions [11]. Therefore, an attempt was made to generate a uniform distribution of reactive elements throughout the coating by

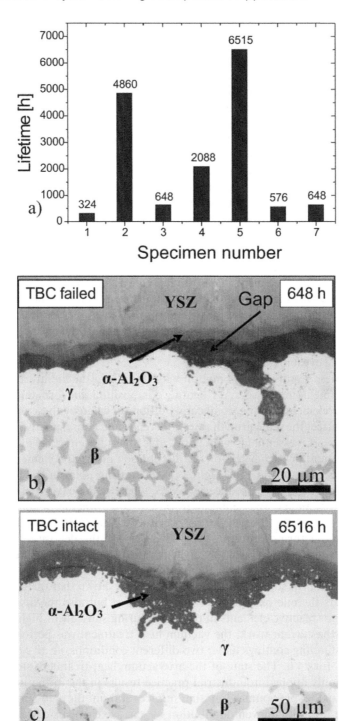

*13.1* (a) Lifetime chart for tested TBC system (Ø10 mm superalloy, 200 μm VPS-MCrAlY-Bond coat, 300 μm YSZ, EB-PVD-TBC); (b) metallographic cross-sections of TBC specimens after cyclic oxidation lifetime testing at 1000 °C in air with short lifetime and (c) with long lifetime

application of a vacuum heat treatment procedure at lower temperatures for a longer time period (24 h at 900 °C). The restricted mobility of the reactive element dopant in the coating was expected to result in the formation of a deep internal oxidation zone consisting of fine, well dispersed reactive element-rich oxide precipitates. Finally, the effect of the distribution of reactive elements induced by the heat treatment on the oxide scale growth during isothermal air oxidation at 1100 °C has been investigated.

## 13.2    Experimental

The material used was a NiCoCrAl coating with nominal composition (wt.%): 20 cobalt, 11.5 aluminium, 17 chromium, 0.28 silicon, 0.6 yttrium and 0.3 hafnium. During the deposition process by vacuum plasma spraying, one set of coatings was manufactured in vacuum with a low partial pressure and another set deliberately with a high partial pressure of oxygen. The low partial pressure of oxygen corresponds to the equilibrium state with the lowest vacuum base pressure achievable by the available vacuum system (~20 mbar). The high partial pressure was obtained by the deliberate admission of oxygen (~20 L min$^{-1}$) controlled by a mass flow control unit.

Free-standing 2 mm thick coatings were sprayed on steel plate substrates. For oxidation testing, specimens with dimensions of 10 mm × 10 mm × 1.5 mm were subsequently prepared by spark erosion and polished to a 1 μm surface finish. Elemental analysis by means of Inductively Coupled Plasma Source Mass Spectrometry (ICP-MS) was performed in order to determine the concentration of major elements in the coating. For the analysis of oxygen and nitrogen, the samples were heated in flowing helium gas in a graphite crucible using resistance heating. The oxygen content was determined by infrared absorption and that of nitrogen by thermal conductivity detection. The results of the chemical analysis are given in Table 13.1.

The free-standing specimens were subjected to two different vacuum heat treatment procedures (vacuum base pressure around 10$^{-6}$ mbar):

1. 1100 °C / 2 h
2. 900 °C / 24 h

The vacuum heat-treated specimens were subjected to isothermal oxidation at 1100 °C for 72 h in air. Extensive characterisation of the specimens in the vacuum heat-treated condition and after oxidation was performed using Sputtered Neutral Mass Spectroscopy (SNMS). Elemental intensity depth profiles obtained from the specimens after various heat treatment and oxidation experiments with a Plasma SNMS facility (INA-5 from Specs GmbH) were converted into atomic concentrations using the procedure described in Ref. 12. Field-emission gun SEM (LEO 1530 Gemini) equipped with a cathodoluminescence detector was used to study the morphology

*Table 13.1*   Chemical analysis of the free-standing coatings

|  | | Al | Cr | Co | Ni | Y | Hf | Si | C | S | N | O |
|---|---|---|---|---|---|---|---|---|---|---|---|---|
| High $pO_2$ | wt.% | 12.80 | 16.50 | 21.50 | 48.00 | 0.49 | 0.29 | 0.36 | 0.013 | 0.001 | 0.027 | 0.428 |
|  | at.% | 23.79 | 15.92 | 17.98 | 41.04 | 0.28 | 0.08 | 0.64 | 0.054 | 0.002 | 0.097 | 1.326 |
| Low $pO_2$ | wt.% | 12.80 | 16.30 | 21.40 | 47.80 | 0.57 | 0.27 | 0.34 | 0.011 | 0.002 | 0.005 | 0.051 |
|  | at.% | 23.94 | 15.83 | 18.01 | 41.12 | 0.32 | 0.08 | 0.61 | 0.046 | 0.003 | 0.018 | 0.161 |

and distribution of the internal oxide precipitates. For analysis of the surface oxide precipitates after different manufacturing stages, confocal Raman and Fluorescence spectroscopy with $Ar^+$ ion laser excitation (wavelength 515.4 nm) were used. For this purpose, the laser with a power density of 18 MW $cm^{-2}$ was focused to a spot of 2–3 μm diameter on the surface of the sample. All Raman spectra were recorded using a grating of 1800 grooves $mm^{-1}$ and a corresponding spectral resolution of 1.5 $cm^{-1}$.

## 13.3    Effect of oxygen partial pressure in the vacuum chamber during BC deposition by VPS on the oxidation behaviour of the BC

In Fig. 13.2, SEM cross-section images illustrate differences between the morphologies of the oxide scales formed on the coatings manufactured under low- and high-oxygen partial pressures after isothermal air oxidation at 1100 °C for 72 h. The oxide scale formed on the high-oxygen specimen appears thin with no visible signs of reactive element-oxide pegs in the scale (Fig. 13.2a). It can be seen that a substantial amount of oxide particles is present in the as-sprayed condition, whereby the tiny dark particles represent the aluminium oxide particles formed at the boundaries between individual sprayed layers. EDX analysis also showed the presence of yttrium- and/or hafnium-rich particles along with aluminium oxide particles in between the sprayed layers. In contrast, the low-oxygen specimen (Fig. 13.2b) is characterised by an inwardly growing oxide scale accompanied by inclusions of the matrix material within the scale. The rapid oxide growth is related to the large amount of reactive elements incorporated into the scale, often referred to as 'overdoping' [13]. The latter results confirm the previous observations regarding the oxide morphology in the TBC system. Furthermore, they indicate that methods should be sought to decrease the sensitivity of the coating oxidation performance to intrinsic variations in the thermal spraying parameters. As will be shown in the next section, one such method is to optimise the heat treatment procedure, which is commonly the one prescribed for the superalloy substrate, but in addition is known [4,5] to affect the distribution of reactive elements in MCrAl coatings.

## 13.4    Effect of heat treatment procedure on the RE-distribution in the MCrAl-coating

Figure 13.3 shows the SNMS depth profile of the studied low-oxygen free-standing coating after vacuum heat treatment for 2 h at 1100 °C. At the very surface, an enrichment of hafnium and to a lesser extent Y is observed. Deeper into the coating, hafnium and yttrium enrichments correspond to an internal oxidation zone with a depth of approximately 15 μm. The observed enrichment of nickel and aluminium in the near surface region relates to beta phase formation, which is apparently caused by chromium evaporation during the vacuum annealing process, as also found by other authors [4].

Raman spectroscopy was applied to identify the oxides formed on the surface of the heat-treated specimen. In Fig. 13.4, the intensity of Raman scattering is plotted against the frequency relative to the incident irradiation (Raman shift). The oxide formed on the surface is mainly yttria, but also hafnia and occasionally some yttrium aluminates could be observed. The broadening and slight shift of the peaks and appearance of a few additional bands for the yttrium- and hafnium-oxides measured on the coating compared to the reference spectra are probably attributable to some

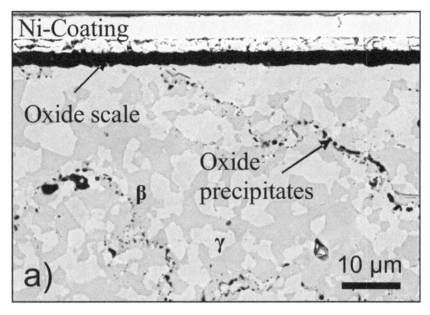

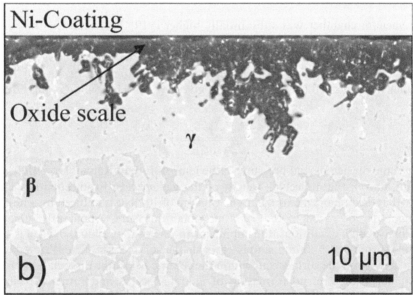

*13.2* SEM cross-section images of the studied free-standing coating after isothermal oxidation for 72 h at 1100 °C: (a) high oxygen coating; (b) low oxygen coating

disorder due to substitution by oxygen interstitials or vacancies and the incorporation of foreign dopant cations into the respective phases, e.g. hafnium into the crystal lattice of yttria, as commonly observed by other authors [14]. No alumina formation could be detected on the coating surface after vacuum heat-treatment for 2 h at 1100 °C using the characteristic [15,16] $Cr^{3+}$-fluorescence signal. The alumina formation was apparently suppressed, in spite of the fact that the oxygen partial pressure

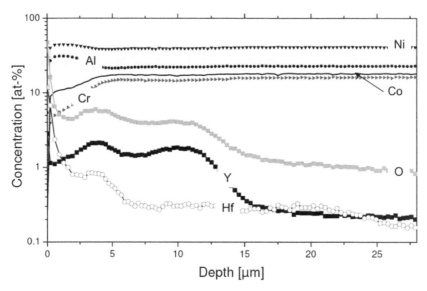

*13.3* SNMS depth profile of low-oxygen free-standing MCrAl, Y + Hf containing coating after vacuum heat treatment at 10⁻⁵ mbar/1100 °C for 2 h

in the vacuum chamber was substantially higher (~$10^{-9}$ bar) than the dissociation pressure of alumina. Therefore, this observation can be explained by kinetic rather than thermodynamic considerations. According to the kinetic gas theory, the ambient partial pressure of oxygen determines the impingement rate of the gas molecules on a surface ($\dot{z}$ in $s^{-1}$ $cm^{-2}$), as given by the Hertz–Langmuir Equation [17]:

$$\dot{z} = \frac{p}{\sqrt{2\pi \cdot m \cdot \mathrm{k} \cdot T}} \qquad [13.1]$$

Here $p$ is the oxygen partial pressure, $T$ the temperature in Kelvin, k the Boltzmann constant, and $m$ the molecular weight of the gas species. From equation 13.1, it follows that the impingement rate in the vacuum of $10^{-9}$ bar is reduced by nine orders of magnitude compared to that at atmospheric pressure. Apparently, under the prevailing vacuum conditions, the impingement rate of the residual oxygen is so low that the yttrium and hafnium supply towards the surface is sufficient to reduce the oxygen activity on the surface and within the coating below the dissociation pressure of alumina. Therefore, no formation of alumina occurs and rather yttrium and hafnium oxide precipitates are formed on the surface and in the sub-surface regions as shown by SNMS analysis in Fig. 13.6.

In agreement with SNMS profiles and Raman spectra, the SEM images revealed enrichment of hafnium and yttrium at the coating surface (light band in Fig. 13.5a) corresponding to oxide formation. Cathodoluminescence imaging was performed to analyse the morphology and distribution of Y-rich internal oxide precipitates in the bond coat. This technique had been used previously to detect yttrium-rich oxide precipitates in alumina scales [18]. Yttria precipitation in a 15 µm deep internal oxidation zone could be clearly observed (Fig. 13.5b) in agreement with the SNMS profiles in Fig. 13.3. The size of the internal oxide precipitates increases with increasing distance from the coating surface. This effect is related to a decrease with time of the penetration velocity of the internal oxidation front, thus promoting precipitate growth,

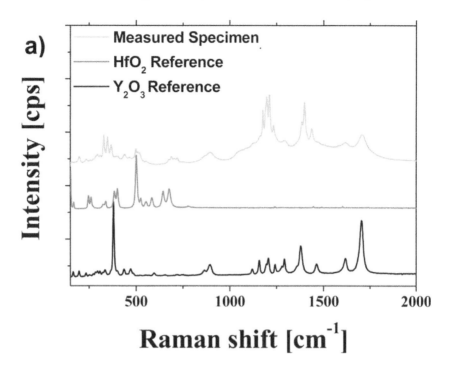

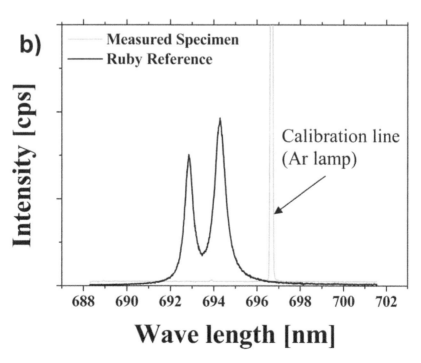

*13.4* (a) Raman and (b) Fluorescence spectra of low-oxygen, free-standing MCrAl(Y/Hf) coating after vacuum heat treatment at 10⁻⁵ mbar/1100 °C for 2 h

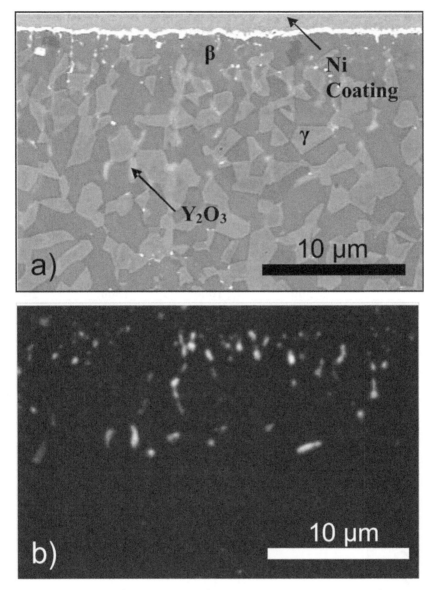

13.5 Low-oxygen free-standing MCrAl, Y + Hf-containing coating after vacuum heat treatment at $10^{-5}$ mbar/1100 °C for 2 h: (a) BSE-SEM image; (b) cathodoluminescence image

as discussed in detail in Ref. 19. It is important to note that the internal oxide precipitates are not laterally homogeneously distributed. This effect is not completely understood. It may be related to the statistical nature of precipitate nucleation or to the distribution of yttrium-rich intermetallics in the sprayed coating. It is believed that this non-homogeneous distribution of the reactive element oxides results in the formation of non-uniform oxide scale, as will be shown in the next section.

The SNMS depth profile of the specimen heat-treated at 900 °C for 24 h in Fig. 13.6 revealed hafnium enrichment at the surface similar to that occurring after heat-treatment at 1100 °C. Contrary to the latter condition, however, at 900 °C, the

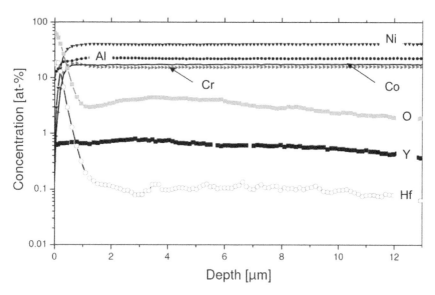

*13.6* SNMS depth profile of low-oxygen free-standing MCrAl, Y + Hf-containing coating after vacuum heat treatment at $10^{-5}$ mbar/900 °C for 24 h

yttrium shows no tendency to enrich at the surface indicating that its mobility at lower temperatures is substantially reduced. Rather, yttrium is tied up by oxygen into internal oxide precipitates in the BC, as indicated by the higher yttrium content (0.6 at.%) measured by SNMS up to a depth of 13 µm, compared to the original yttrium concentration of 0.3 at.%. Furthermore, no β nickel aluminide phase enrichment could be observed at the coating surface in SNMS depth profiles after heat treatment at 900 °C, probably due to less extensive chromium evaporation compared to that at 1100 °C.

This fundamental change of the reactive element surface enrichment at lower temperatures is confirmed by the measured Raman spectra. These suggest hafnia formation on the coating surface, whereas the scattering pattern associated with yttria completely vanished (Fig. 13.7a). Minor amounts of alumina were detected on the surface by means of fluorescence spectroscopy (Fig. 13.7b).

Figure 13.8 shows SEM cross-sections of the specimen heat-treated at 900 °C. The dark precipitates on the coating surface (Fig. 13.8b) were identified as alumina. Along with aluminium oxide, hafnium enrichment on the surface was observed (Fig. 13.8d). An internal oxidation zone contained fine, well distributed yttrium-rich precipitates (Fig. 13.8c).

The observed differences in the reactive element distribution after the different heat treatments can be explained by considerations drawn schematically in Fig. 13.9. Based on the experimental data, the diffusivity of hafnium in the studied NiCoCrAl bond coat considerably exceeds the diffusivity of yttrium. The higher diffusion rates compared to yttrium may be related to the higher solubility of hafnium in the alloy matrix. According to the ASM binary phase diagram database, the solubility of hafnium in γ nickel phase is about 1 at.% at 1200 °C, while the solubility of yttrium is limited to ~0.1 at.% [20]. Therefore, during heat-treatment at the lower temperature of 900 °C, the hafnium diffuses very rapidly to the surface and reacts there with oxygen ($t_1$ in Fig. 13.9a). The ongoing nucleation and growth of hafnia surface

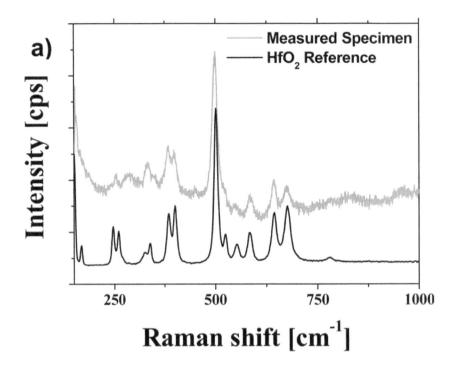

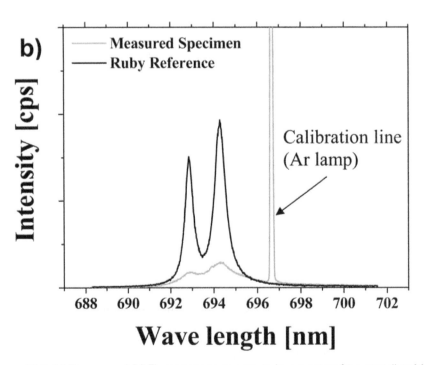

*13.7* (a) Raman and (b) Fluorescence spectrum low-oxygen free-standing MCrAl, Y+Hf-containing coating after vacuum heat treatment at $10^{-5}$ mbar/900 °C for 24 h

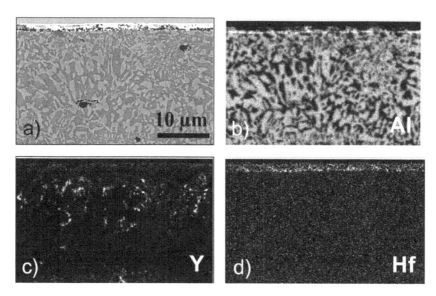

*13.8* (a) BSE image and EDX maps for Al (b), Y (c) and Hf (d) of low-oxygen NiCoCrAl, Y+Hf-containing free-standing coating after heat-treatment at 900 °C for 24 h

precipitates results in hafnium depletion in the bond coat, reducing its supply to the surface. This in turn leads to an increase in the oxygen activity on the coating surface and the formation of alumina precipitates. At this stage ($t_2$ in Fig. 13.9a), the depth of internal oxidation starts to be determined by the competition of diffusion of dissolved oxygen into the coating and counter-diffusion of yttrium. Due to slow diffusion of yttrium at 900 °C, the depth of the internal oxidation is mainly controlled by oxygen diffusion. The considerably higher oxygen permeability ($D_O N_O$) (compared to $D_Y N_Y$) favours higher nucleation rates and small-sized yttrium-rich oxide precipitates [19].

At 1100 °C, the formation of hafnia precipitates on the coating surface is responsible for its rapid depletion in the near-surface coating material. However, unlike the behaviour at lower temperatures, alumina formation does not occur because yttrium diffusion to the surface is sufficiently fast at the higher temperature to result in the formation of yttria precipitates on the surface and in the sub-surface zones ($t_1$ in Fig. 13.9b). The yttria formation reduces the oxygen activity on the surface to a value below that which is necessary for alumina formation. The higher value of $D_Y N_Y$ decreases the penetration rate of the internal oxidation front and reduces the nucleation rate of internal precipitates. The faster diffusion of yttrium at 1100 °C compared to that at 900 °C allows coarsening of the yttria precipitates ($t_2$ in Fig. 13.9b).

## 13.5    Effect of heat treatment on the oxidation behaviour of the coating

The low-oxygen free-standing coatings were isothermally oxidised for 72 h at 1100 °C in air. It was observed that the oxide growth rate was lower for the coating heat-treated at 900 °C compared to that heat-treated at 1100 °C (Fig. 13.10a and b). The depth of internal oxidation was also remarkably different, i.e. ca. 120 μm in the sample heat-treated at 900 °C compared to more than 200 μm for that after heat treatment at 1100 °C.

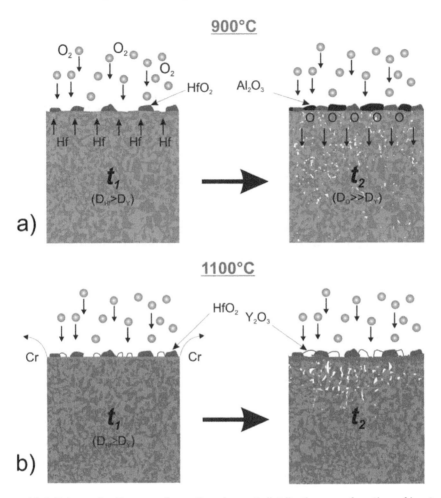

*13.9* Schematic diagram of reactive element distribution as a function of heat treatment procedure: (a) 900 °C and (b) 1100 °C

Figure 13.11a shows the SNMS profile for the freestanding coating heat-treated at 900 °C after oxidation demonstrating that hafnium is mainly enriched in the oxide scale, whereas yttrium is precipitated as internal oxide deeper in the bond coat. The free-standing coating heat-treated at 1100 °C reveals the enrichment of both yttrium and hafnium in the oxide layer (Fig. 13.11b). Therefore, the yttrium and hafnium distributions in the oxidised specimens (Fig. 13.11) qualitatively replicated those found in the specimens after heat treatment (Figs. 13.3 and 13.6). The reason for the faster oxide growth rate of the specimen heat-treated at 1100 °C might be that during the heat treatment, mainly yttria and hafnia precipitates formed at the surface. These precipitates provided short-circuit paths for oxygen penetration into the coating upon incorporation into the alumina scale in the initial stages of oxidation. Mixed hafnium/hafnium oxides also found by other authors [21] in this coating system should exhibit high bulk oxygen diffusivities due to doping with lower valency ions [22]. In contrast, after heat treatment at 900 °C, in addition to hafnia, the near surface regions contain considerable amounts of alumina precipitates, as indicated by

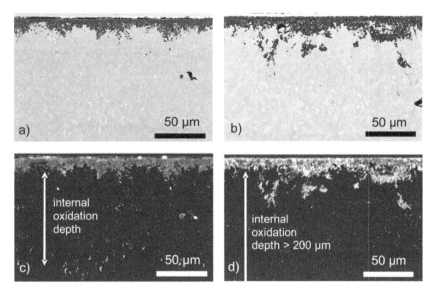

*13.10* SEM cross-sections after isothermal oxidation for 72 h at 1100 °C in air: (a) BSE-SEM image and (c) cathodoluminescence image of the specimen heat-treated in vacuum of 10$^{-5}$ mbar at 900 °C for 24 h; (b) BSE-SEM image and (d) cathodoluminescence image of specimen heat-treated in vacuum of 10$^{-5}$ mbar at 1100 °C for 2 h

Raman spectroscopy (Fig. 13.7). SEM data in Fig. 13.8 indicate that these precipitates cover a large fraction of the surface, hence hindering oxygen penetration and promoting the formation of a more protective alumina-based scale upon oxidation.

## 13.6    Summary and conclusions

It was found that variation of the manufacturing parameters of the MCrAl bond coat for an EBPVD-TBC system has a significant influence on the reactive element distribution in the coating before oxidation in service. After vacuum plasma spraying at high oxygen partial pressure, the reactive elements yttrium and hafnium in the NiCoCrAl coating were tied up in oxide precipitates, thus their beneficial effects on scale adhesion were inhibited. In contrast, spraying at low oxygen partial pressure resulted in an 'over-doped' coating, which showed rapid scale growth and extensive internal oxide formation during oxidation.

During vacuum heat treatment of the low-oxygen coating for 2 h at 1100 °C, yttrium and hafnium oxide surface precipitation was observed. The internal oxidation zone consisted mainly of coarse yttria precipitates with non-uniform distribution. After vacuum heat-treatment for 24 h at 900 °C, however, the surface oxide precipitates mainly consisted of hafnia and alumina. The reduced mobility of yttrium at 900 °C resulted in a deeper internal oxidation zone and a finer precipitate morphology of yttrium oxides compared to that found at 1100 °C.

The distribution of reactive elements established during heat treatment was found to have a substantial effect on the oxide scale growth rate and morphology during subsequent air oxidation at 1100 °C. The coating heat-treated for 2 h at 1100 °C showed stronger inward scale growth and a deeper internal oxidation zone due to a

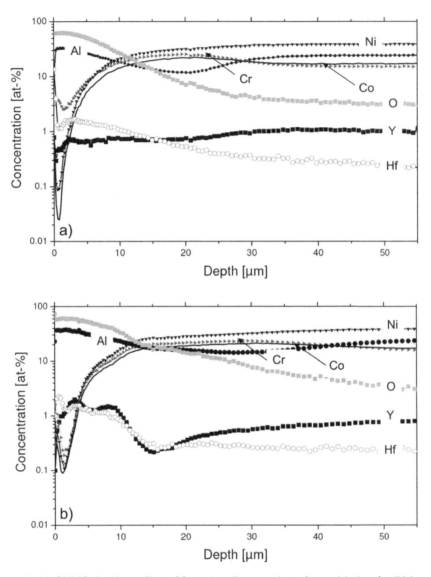

*13.11* SNMS depth profiles of free-standing coating after oxidation for 72 h at 1100 °C in air. Specimens heat-treated in vacuum before oxidation: (a) 900 °C/24 h; (b) 1100 °C/2 h

large number of short-circuit paths for oxygen transport (yttrium/hafnium mixed oxides formed during heat treatment). The coating heat-treated for 24 h at 900 °C exhibits a lower scale growth rate and less internal oxidation compared to that treated at 1100 °C. This effect can be attributed to the formation of alumina precipitates during the heat treatment, thus promoting the formation of a more protective scale during oxidation as well as the absence of the mixed hafnium and yttrium oxide precipitates in the near-surface regions of the BC.

The results demonstrate that the heat treatment procedure has a substantial influence on the oxidation behaviour of the MCrAlY bond coats. It is obvious that the optimisation of the heat treatment procedure offers significant potential for improvement in the MCrAl bond coat oxidation behaviour and, consequently, lifetime extension and better lifetime reproducibility of the TBC coatings.

## Acknowledgement

One of the authors, D. Naumenko, would like to acknowledge the Deutsche Forschungsgemeinschaft (DFG) for funding his work under Grant No. NA 615-1-1. The authors would like to thank Dr L. Niewolak for SNMS measurements and Prof. G. H. Meier from Pittsburgh University for his comments on the manuscript and valuable discussions.

## References

1. U. Schulz, B. Saruhan, K. Fritscher and C. Leyens, *Int. J. Appl. Ceram. Technol.*, 1 (2004), 302.
2. M. Peters, C. Leyens, U. Schulz and W. A. Kaysser, *Adv. Eng. Mater.*, 3 (2001), 193.
3. N. P. Padture, M. Gell and E. H. Jordan, *Science*, 296 (2002), 280.
4. T. J. Nijdam, L. P. H. Jeurgens, J. H. Chen and W. G. Sloof, *Oxid. Met.*, 64 (2005), 355.
5. H. Lau, C. Leyens, U. Schulz and C. Friedrich, *Surf. Coat. Technol.*, 165 (2003), 217.
6. C. G. Levi, E. Sommer, S. G. Terry, A. Catanoiu and M. Rühle, *J. Am. Chem. Soc.*, 86 (2003), 676.
7. N. M. Yanar, F. S. Pettit and G. H. Meier, *Metall. Mater. Trans. A*, 37 (2006), 1563.
8. B. A. Pint, 'Progress in understanding the reactive element effect since the Whittle and Stringer literature review', in Proc. John Stringer Symp., 2001, ASM, Materials Park, OH.
9. B. A. Pint, *J. Am. Chem. Soc.*, 86 (2003), 686.
10. T. Kim, S. S. Chun, X. X. Yao, Y. Fang and J. Choi, *J. Mater. Sci.*, 32 (1997), 4917.
11. B. A. Pint, P. F. Tortorelli and I. G. Wright, *Mater. Corros.*, 47 (1996), 663.
12. M. Göbel, G. Borchardt, S. Weber and S. Scherrer, *J. Anal. Chem.*, 358 (1997), 131.
13. J. Klöwer, *Mater. Corros.*, 51 (2000), 373.
14. M. Yashima, H. Takahashi, K. Ohtake, T. Hirose, M. Kakihana, H. Arashi, Y. Ikuma, Y. Suzuki and M. Yoshimura, *J. Phys. Chem. Solids*, 57 (1996), 289.
15. D. M. Lipkin, D. R. Clarke, M. Hollatz, M. Bobeth and W. Pompe, *Corros. Sci.*, 39 (1997), 231.
16. Q. Ma and D. R. Clarke, *J. Am. Chem. Soc.*, 76 (1993), 1433.
17. H. Lüth, *Surfaces and Interfaces of Solid Metals*, 3rd edn. Springer Verlag, Berlin, Heidelberg, New York, 1995.
18. A. Gil, V. Shemet, R. Vassen, M. Subanovic, J. Toscano, D. Naumenko, L. Singheiser and W. J. Quadakkers, *Surf. Coat. Technol.*, 201 (2006), 3824.
19. N. Birks, G. H. Meier and F. S. Pettit, *Introduction to High Temperature Oxidation of Metals*, Cambridge University Press, 2006.
20. *ASM Binary Alloy Phase Diagrams*, ASM International, 1996.
21. C. Mercer, S. Faulhaber, N. Yao, K. McIlwrath and O. Fabrichnaya, *Surf. Coat. Technol.*, 201 (2006), 1495.
22. M. F. Trubelja and V. S. Stubican, *J. Am. Chem. Soc.*, 74 (1991), 2489.

# 14

# Effect of creep-deformation on oxidation behaviour of nickel-based alloy with a rhenium-based diffusion barrier coating*

## Toshio Narita

*Research Group of Advanced Coating, Hokkaido University,*
*Kita-13 Nishi-8 Kita-Ku, Sapporo 060-8628, Japan*
*narita@eng.hokudai.ac.jp*

## Mikihiro Sakata

*Graduate student, Graduate School of Engineering, Hokkaido University, Sapporo, Japan*

## Takumi Nishimoto and Takayuki Yoshioka

*Research associate in Hokkaido University, Sapporo, Japan*

## Shigenari Hayashi

*Graduate School of Engineering, Hokkaido University, Sapporo, Japan*

## 14.1    Introduction

Heat-resistant alloys for high-temperature applications have been used along with coatings such as overlay MCrAlY (M = Ni and/or Co) coatings and/or Al diffusion coatings of β-NiAl, β-Ni(Pt)Al, and γ'-(Ni,Pt)$_3$Al [1,2]. Loss of Al from the coatings occurs by the formation of Al$_2$O$_3$ scales and also by mutual diffusion between the coating and substrate. The degradation modes of coatings can be divided into two groups [2,3]. One is spallation of Al$_2$O$_3$ scales due to thermal and/or mechanical stresses, and the other is changes in the microstructure of the alloy substrate due to mutual diffusion with the Al-reservoir layer, resulting in the precipitation of topologically close-packed (TCP) phases. Furthermore, it has been accepted that Ni-aluminide coated alloys can suffer significant degradation of mechanical properties under creep conditions [4,5].

In addition to suppression of the loss of Al to the substrate, suppression of the diffusion of Ni and other alloying elements such as S, Ti, Mo, and Ta from the substrate into the coating would be desirable [3]. To achieve this, barrier compounds such as TiN, AlN, SiC, AlON, ZrO$_2$, Al$_2$O$_3$, Ir, Re, and W have been investigated. However, overall, almost all of the diffusion barrier layers proposed lose their protective function after relatively short exposure.

---

\* Reprinted from T. Narita et al.: Effect of creep-deformation on oxidation behaviour of nickel-based alloy with a rhenium-based diffusion barrier coating. *Materials and Corrosion.* 2008. Volume 59. pp. 471–5. Copyright Wiley-VCH Verlag GmbH & Co. KGaA.

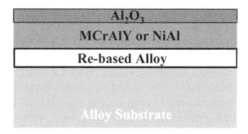

*14.1* Schematic representation of the diffusion barrier coating, with an external $Al_2O_3$ scale; an outer Al-reservoir (MCrAlY or β-NiAl), and inner diffusion barrier (Re-based alloy) on the alloy substrate

In previous investigations [6–11], a Re-based alloy was selected as a diffusion barrier because, at 1423 K, the Ni–Cr–Re ternary system [12] has a σ-phase which has a melting temperature around 2635 K, higher than that of $Al_2O_3$ (2323 K). As shown in Fig. 14.1, a diffusion barrier-type coating with a duplex layer structure, an inner Re-based alloy (σ-phase in the Re–Cr–Ni system) as a diffusion barrier and an outer Al reservoir of MCrAlY or Ni-aluminides (β-NiAl) have been formed on nickel-based single crystal superalloys, nickel-based alloys, and Nb-based alloys.

The oxidation behaviour of these alloys with diffusion barrier coatings was investigated in air under thermal cycling between room temperature and temperatures between 1373 and 1573 K. It was found that the structure and composition of the Re-based alloy layer was retained with little change. Further, there was little Al in the Re-based alloy. It could be concluded that Re-based alloys such as σ (Re (W), Cr, Ni) are very promising candidates as a diffusion barrier between alloy substrates and the Al reservoir layer of the MCrAlY, β-NiAl or γ'-$Ni_3Al$.

In the present work, a diffusion barrier coating with a duplex layer structure was formed on a Ni–Mo-based alloy and the effect of creep deformation on the oxidation behaviour of the coated alloy was investigated at 1243 K in air. For comparison purposes, creep tests were also carried out for a sole Ni-aluminide coated alloy and uncoated alloys with different alloy grain sizes. Particular attention was paid to the structure and composition of the coatings before and after the high-temperature oxidation under creep deformation.

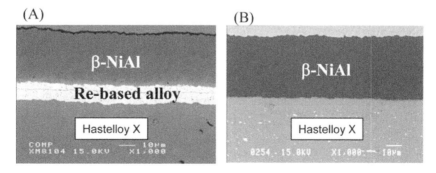

*14.2* Cross-section microstructures of coated Ni-based alloy; (A) diffusion barrier coating and (B) β-NiAl coating

## 14.2    Experimental

The Ni–Mo-based alloy (Hastelloy X) used as the alloy substrate for the coating has a nominal composition (at.%) of (53–56)Ni, (20–23)Cr, (17–20)Fe, (8–10)Mo, (0.5–2.5)Co, 0.1 > (W, Si, Mg), and 0.1 > (C, B, S). A 1.2 mm thick alloy sheet was cut into creep specimens. Creep tests were carried out under a constant stress of 22.5 MPa at 1243 K in air using conventional creep-testing equipment, and creep strains were monitored by CCD (charge coupled device) equipment with a precision better than 1 μm. In the creep test, the furnace temperature was raised to 1243 K at a heating rate of 0.108 K $s^{-1}$ without external loading and after a dwell time of 3.6 ks at 1243 K, the loading was applied to the specimen and maintained for specified time intervals, followed by furnace cooling without the external loading.

The creep specimen was polished with 150-grit waterproof sand paper, followed by ultrasonic degreasing in a methanol–benzene solution. The formation procedure of the diffusion barrier coating with a duplex structure, an inner Re-based alloy and outer Ni-aluminide layers, has been described in detail elsewhere [9–11] and will be briefly introduced here.

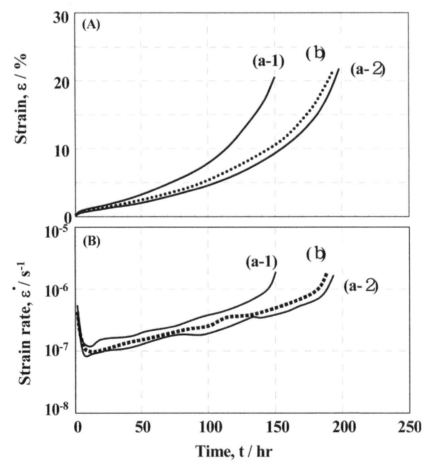

*14.3* Changes in strain (A) and strain rate (B) of the bare alloy (b) and the alloy with β-NiAl coating (a-1) and (a-2) with time during creep tests under 22.5 MPa at 1243 K in air. The average alloy grain size is several tens of micrometres

1. A 10 µm-thick Re–Ni film containing approximately 70 at.% Re was electro-plated from a pH 3 aqueous solution using a Ni anode at a temperature of 323 K. The solution contained $ReO^{4-}$ (0.1 mol $L^{-1}$), nickel sulphate (0.1 mol $L^{-1}$), and citric acid (0.1 mol $L^{-1}$).
2. Cr-pack cementation was then carried out in a vacuum of $10^{-3}$ Pa at 1523 K for 36 ks using an $Al_2O_3$ crucible containing a specimen buried in a powder mixture of Ni–30 at.% Cr alloy and $Al_2O_3$ powders, with trace additions of $Ni_2Al_3$ powder.
3. A 30 µm-thick Ni film was coated on this specimen using a Ni Watts solution
4. Then Al-pack cementation was carried out at 1073 K for 1.2 ks in argon, with the specimens buried in a mixture of Al, $NH_4Cl$, and $Al_2O_3$ powders.

For comparison purposes, the present investigation used the as-received and heat-treated alloys with a Ni-aluminide coating prepared by using processes (3) and (4). Figure 14.2(A) and (B) show cross-section microstructures of the alloy with a diffusion barrier coating and with a conventional β-NiAl coating, respectively.

During the Cr pack cementation process (2), the grain size of the alloy substrate increased significantly, from several tens of micrometres for the as-received alloy to several hundreds of micrometres for the coated alloy, and the creep behaviour of the alloy was found to change significantly depending on the alloy grain size [13]. In the present investigation, therefore, the alloy specimen was heat-treated under the same conditions as in the process (2) above to control the alloy grain size.

The oxidised specimens with or without external loading were examined using an electron probe micro analyser (EPMA) to determine the concentrations of each element and their concentration profiles (compositions expressed in at.%, unless

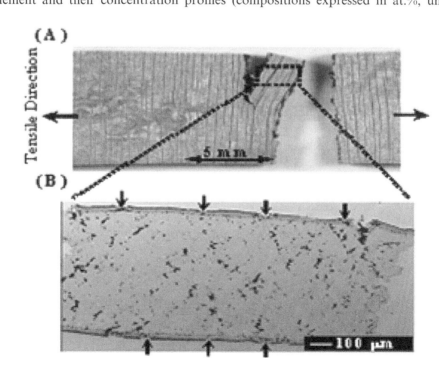

*14.4* Surface morphology (A) and cross-section (B) of the β-NiAl coated alloy after creep rupture at 1243 K in air

otherwise indicated); X-ray diffraction analysis (XRD) was used to identify the oxide products. An X-ray element analyser was used to measure the concentrations of each element on the oxidised surface as a function of oxidation time. In this measurement, the X-ray penetration depth was estimated to be 5–10 μm through the external oxide scale to the alloy surface layer.

## 14.3    Results and discussion

Figure 14.3(A) shows creep-curves for the alloy with and without β-NiAl coating at a stress level of 22.5 MPa at 1423 K in air, and changes in creep-strain rates with time are given in Fig. 14.3(B). The average grain size was several tens of micrometres for the alloy substrate with and without β-NiAl coating. As shown in Fig. 14.3(A), the rupture time for the bare alloy (b) was about 180 h, while rupture times for the β-NiAl coated alloys (a-1) and (a-2) varied between 150 and 190 h. Strain rates for both alloys with and without a β-NiAl coating decreased rapidly at the beginning of the creep process and then strain rates tended to increase gradually, followed by a

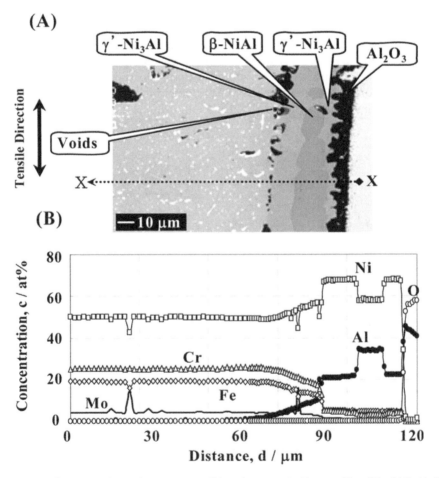

14.5 Cross-section microstructure (A) and concentration profiles (B) of Ni, Al, Cr, Fe, and Mo for the β-NiAl coated alloy after creep rupture at 1243 K in air

rapid increase near the fracture time. From these results, it may be concluded that the creep life of the β-NiAl coated alloy is shorter than that of the bare alloy.

Figure 14.4(A) and (B) show the surface morphology and cross-section microstructure, respectively, of the β-NiAl coated alloy (b) in Fig. 14.3 at creep rupture after 190 h at 22.5 MPa stress; the fracture strain was about 22%. As seen in Fig. 14.4(A), cracks formed periodically perpendicular to the applied tensile stress axis. There are many cracks, voids and cavities within the alloy substrate and these flaws are arrayed along alloy grain boundaries and their nodal points, because sliding of alloy grain boundaries seems to be a major mechanism of creep deformation for the alloy at 1243 K. Arrows in Fig. 14.4(B) indicate cracks perpendicular to the applied stress axis, running through the external oxide scale.

Figure 14.5(A) and (B) show cross-section microstructures and concentration profiles of each element, respectively, for the β-NiAl coated alloy (b in Fig. 14.3) at creep rupture after 190 h at 22.5 MPa. It was found that γ'-$Ni_3Al$ formed in the mutual diffusion zone between the alloy substrate and the β-NiAl and also that γ'-$Ni_3Al$ appeared under the external oxide scale due to the selective oxidation of Al to form an Al-rich oxide scale. Many voids formed near the original interface between the alloy substrate and the β-NiAl coating and a bright, Mo-rich phase precipitated in the alloy substrate below the voids. The β-NiAl coating layer was deformed by creep to a strain of 22% without the development of flaws such as cracks and voids because β-NiAl is ductile at the creep temperature of 1243 K.

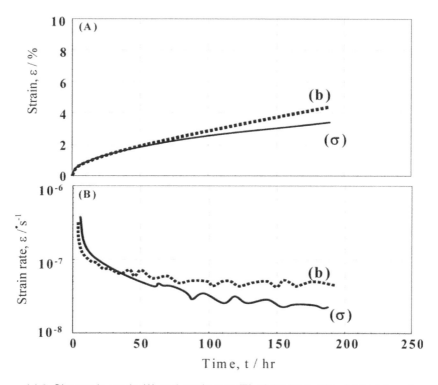

*14.6* Change in strain (A) and strain rate (B) of the bare alloy (b) and the alloy with the diffusion barrier coating (σ) with time during creep tests under 22.5 MPa at 1243 K in air. The average alloy grain size is several hundreds of micrometres

Figure 14.6(A) and (B) show changes in strain and strain rate with creep time, respectively, for the diffusion barrier coated alloy and bare alloy under a stress of 22.5 MPa at 1243 K in air. The average grain size was several hundreds of μm for the alloy substrate with and without the diffusion barrier coating. Creep tests were discontinued at 190 h with strains of 3.5 and 4.5% for the alloys with and without a diffusion barrier layer, respectively. The strain rates for the bare alloy (b) decreased rapidly and then became constant for up to 190 h, while strain rates for the diffusion barrier coated alloy ($\sigma$) continued to decrease slowly. It was concluded that the strain rates for the barrier coated alloy were smaller than those of the bare alloy.

Figure 14.7 shows surface morphologies of the diffusion barrier coated alloy, in which the creep test was discontinued at a strain of 3.5% and strain rate of $8 \times 10^{-8}$ $5^{-1}$ under a stress of 22.5 MPa after 190 h at 1243 K in air. There is a crack indicated by an arrow, and the external oxide scale was partially exfoliated during creep. The external scale consisted mainly of $\theta\text{-Al}_2\text{O}_3$ during the initial stage of creep, and with further oxidation, the external scale became a duplex layer: outer plate-like $\theta\text{-Al}_2\text{O}_3$ and inner equiaxed $\alpha\text{-Al}_2\text{O}_3$. The outer $\theta\text{-Al}_2\text{O}_3$ scale exfoliated significantly. From

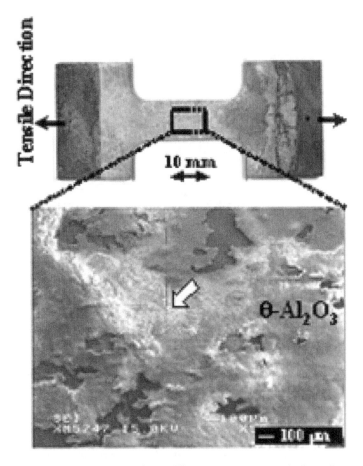

14.7 Surface morphologies of the diffusion barrier coated alloy, discontinued after 190 h. An arrow shows the crack

these results, it was concluded that cracking and spallation of the external scale due to creep deformation at a strain rate of $8 \times 10^{-8}$ $5^{-1}$ resulted in rapid consumption of Al from the Al-reservoir layer.

Figure 14.8 shows a cross-section at low magnification of the coated alloy following a test that was discontinued after 190 h, indicating few flaws such as cracks, voids, and cavities in the diffusion barrier layer over the whole creep specimen. From this result, it was concluded that the Re-based alloy layer used as a diffusion barrier could be creep-deformed along with the creep-deformation of the alloy substrate at least up to a strain of 3.5%.

Figure 14.9 shows the cross-section microstructure and concentration profiles of each element for the diffusion barrier coated alloy, for a test that was discontinued at a strain of 3.5% under a stress of 22.5 MPa after 190 h at 1243 K in air. As shown in Fig. 14.9, the Al-reservoir layer consisted of a single β-NiAl layer with a uniform aluminium content of about 36 at.% Al after creep under a stress of 22.5 MPa for 190 h. There is little penetration of Al through the barrier layer towards the Hastelloy-X substrate. Furthermore, there are few flaws such as cracks, voids, and cavities within the barrier layer.

## 14.4    Conclusions

A diffusion barrier coating with a duplex layer structure consisting of an inner Re-based alloy and outer β-NiAl layer, was formed on a Ni-based alloy, and the coated alloy was oxidised at 1243 K in air for up to about 190 h with an external tensile stress of 22.5 MPa. For comparison, the alloys, as-received, heat-treated, and with the Ni-aluminide coating, were oxidised under creep deformation. The results obtained may be summarised as follows.

- The creep rupture time for the diffusion barrier coated alloy was longer than those for the bare alloy and the β-NiAl coated alloy. Under creep deformation, the Re-based alloy layer acts effectively as a barrier against inward diffusion of aluminium and outward diffusion of elements from the substrate.
- The structure and composition of the diffusion barrier and β-NiAl layers were similar before and after creep deformation for 190 h and there were few cracks and flaws within the inner, Re-based alloy layer following creep deformation to a strain of 3.5%.

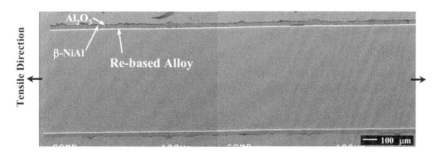

*14.8* Cross-section structure of the coated alloy after the creep deformation shown in Fig. 14.7

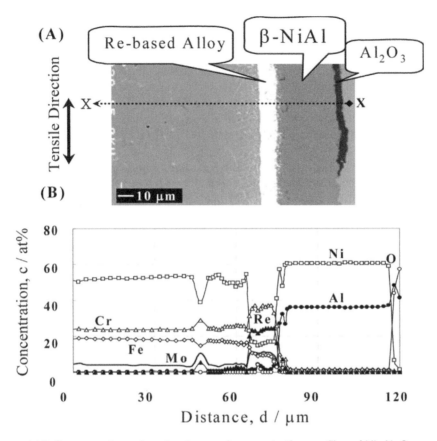

*14.9* Cross-section microstructure and concentration profiles of Ni, Al, Cr, Re, Fe, and Mo for the diffusion barrier coated alloy after creep deformation in Fig. 14.8

- The as-received and β-NiAl coated alloys creep deformed rapidly and fractured within 180 h under a stress of 22.5 MPa with strains of about 20–25%.
- The external scale consisted mainly of θ-$Al_2O_3$ during the initial stage of creep and, with further oxidation became a duplex layer with plate-like θ-$Al_2O_3$ on the outside and equiaxed α-$Al_2O_3$ on the inside. The outer θ-$Al_2O_3$ scale exfoliated significantly.

### Acknowledgements

The investigation was carried out with financial support from Specially Promoted Research under contract number 16001004, from a Grant-in-Aid for Scientific Research, Ministry of Education, Culture, Sports, Science and Technology, Japan.

### References

1. A. G. Evans, D. R. Mumm, J. W. Hutchinson, G. H. Meier and F. S. Pettit, Mechanism controlling the durability of thermal barrier coatings, *Prog. Mater. Sci.*, 46 (2001), 505.

2. B. Gleeson, Thermal barrier coating for aero-engine applications, *J. Propul. Power*, 22(2) (2006), 375–383.
3. T. Narita, S. Hayashi, L. Fengqun and K. Zaini Toshin, The role of bond coat in advanced thermal barrier coating, *Mater. Sci. Forum*, 502 (2005), 99–104.
4. T. Narita, M. Shoji, Y. Hisamatsu, D. Yoshida, M. Fukumoto and S. Hayashi, Rhenium coating as a diffusion barrier on a nickel-based superalloy at high temperature oxidation, *Mater. High Temp.*, 18(S) (2001), 245–251.
5. A. Sato, Y. Aoki, M. Arai and H. Harada, Effect of aluminide coating on creep properties of Ni-based single crystal superalloys, *J. Japan Inst. Metals*, 71 (2007), 320–325.
6. T. Narita, S. Hayashi, M. Shoji, Y. Hisamatsu, D. Yoshida and M. Fukumoto, 'Application of rhenium coating as a diffusion barrier to improve the high temperature oxidation resistance of nickel-based superalloy', in Corrosion 2001, paper 01157. NACE International, Houston TX, 2001.
7. M. Fukumoto, Y. Matsumura, S. Hayashi, T. Narita, K. Sakamoto, A. Kasama and R. Tanaka, 'Coating formation on the Nb-based alloys using electrolytic process and its oxidation behaviour', in Report of the 123rd Committee on Heat Resisting Materials and Alloys, *Japan Society for Promotion of Science*, 43(3) (2002), 383–390.
8. T. Narita, M. Fukumoto, Y. Matsumura, S. Hayashi, A. Kasama, I. Iwanaga and R. Tanaka, 'Development of Re-based diffusion barrier coatings on Nb-based alloys for high temperature applications', in Proceedings of Niobium for High Temperature Applications, 99–112, ed. Y.-Won Kim and T. Carneiro, TMS, 2004.
9. Y. Matsumura, M. Fukumoto, S. Hayashi, A. Kasama, I. Iwanaga, R. Tanaka and T. Narita, Oxidation behaviour of a Re-base diffusion-barrier/β-NiAl coating on Nb-5Mo-15W at high temperatures, *Oxid. Met.*, 61(1/2) (2004), 105–124.
10. Y. Katsumata, T. Yoshioka, K. Zaini Thosin, S. Hayashi and T. Narita, 'Effect of diffusion barrier coating on oxidation behaviour of Hastelloy-X at high temperature', Report of the 123rd Committee on Heat Resisting Materials and Alloys, *Japan Society for Promotion of Science*, 46(2) (2005), 183–189.
11. W. Huang and Y. A. Chang, A thermodynamic description of the Ni-Al-Cr-Re system, *Mater. Sci. Eng.*, A259 (1999), 110–119.
12. Y. Aoki, M. Arai, M. Hosoya, S. Masaki, Y. Koizumi, T. Kobayashi and H. Harada, 'Turbine blade material development for air craft engine rotor application, status and perspective', Report of the 123rd Committee on Heat Resisting Materials and Alloys, *Japan Society for Promotion of Science*, 43(3) (2002), 257–264.
13. Y. Yoshioka, D. Saito, K. Fujiyama and N. Okabe, Effect of microstructure on creep resistance of Hastelloy X, *J. Iron Steel*, 80(10) (1994), 55–60.

# 15

## Hot corrosion of coated and uncoated single crystal gas turbine materials*

N. J. Simms, A. Encinas-Oropesa and J. R. Nicholls

*Cranfield University, Cranfield, Bedfordshire, MK43 0AL, UK*

*n.j.simms@cranfield.ac.uk; a.encinas-oropesa@cranfield.ac.uk;*
*j.r.nicholls@cranfield.ac.uk*

### 15.1    Introduction

Industrial gas turbines (IGTs) are at the heart of many current power stations [1–4], usually as part of combined cycle systems fired on natural gas. It is anticipated that IGTs will be used even more widely in the future, with increasing numbers of natural gas fired combined cycle systems being installed and as new solid fuel (coal, biomass, waste) combined cycle power generation systems move from development to commercialisation [1–4]. The development of economically viable power systems relies on their IGTs giving high efficiencies together with acceptable performance and availability. In particular, the vane and blade materials for the hot combustion gas paths must give acceptable and predictable in-service lifetimes.

IGTs have been developed to use increasingly higher firing temperatures and pressures to increase their efficiency of power generation [1–4], with current models firing at temperatures up to ~1400 °C and at pressures of up to ~30 bar. The requirements for materials in the hot-gas paths of gas turbines are very demanding [5–8], with materials needing to be capable of operating at bulk temperatures up to ~950 °C under both high and fluctuating stresses, whilst also withstanding the surrounding environments. The environments produced can be both physically and chemically aggressive, with particles producing erosion or deposition whilst the mixture of gaseous and vapour phase species can produce different forms of deposition, as well as oxidation and hot corrosion. During the last 40 years, these topics have been the subject of many investigations and the potential problems that may be encountered in gas- and oil-fired gas turbines have been well characterised [5,6,8].

Gas turbine materials will oxidise in the combustion gases produced in all gas turbine systems, but the rate of oxidation at metal temperatures below ~900 °C is sufficiently low so as not to be life limiting. However, hot corrosion of turbine materials can occur much more rapidly and is potentially life limiting. For hot corrosion to occur, a liquid (usually sulphate) deposit is required on the surface of components. The formation of this deposit depends on trace metal species (e.g. sodium compounds) in the gas streams and other reactive gas species (e.g. $SO_2$, $SO_3$, HCl). The rates of corrosion will depend (among other factors) on the rate of deposit formation, deposit

---

* Reprinted from N. J. Simms et al.: Hot corrosion of coated and uncoated single crystal gas turbine materials. *Materials and Corrosion.* 2008. Volume 59. pp. 476–83. Copyright Wiley-VCH Verlag GmbH & Co. KGaA.

composition, metal temperature and the composition of the surrounding environment. Two general types of hot corrosion in gas turbine environments have been identified to date [e.g. 5,6]: Type I hot corrosion, typified by internal oxidation/sulphidation, at ~750–900 °C; and Type II hot corrosion, typified by pitting, at ~600–800 °C. Similar types of materials degradation can be expected in gas turbines using solid-derived fuels, as some of the contaminant species are the same as for oil- and/or gas-fired systems [e.g. 9–12]. The contaminant species of particular interest for corrosive degradation are those containing sulphur, chlorine, alkali metals (i.e. Na and K) and other trace metals (e.g. Pb and Zn). The levels of these contaminants are significantly different in the various potential solid fuels [9,11,12]. These variations persist to differing degrees for each element/fuel utilisation process by the time the gas streams reach the gas turbine.

This paper reports results from a series of eight tests carried out to characterise quantitatively the corrosive effects on single crystal alloys/coatings of gases/deposits anticipated in a gas turbine operated on solid fuel derived fuel gases. Single-crystal superalloys have been developed for use with clean fuels but are now being deployed in industrial gas turbines. The performance of these materials, with coatings, has to be determined before they can be used with confidence in combustion environments that contain high levels of potentially damaging elements. The tests were targeted at the effects of alkali sulphate deposits containing both sodium and potassium; the mixture chosen had a melting point of ~823 °C – significantly lower than the melting points of either sodium sulphate (~883 °C) or potassium sulphate (1071 °C) [13]. A critical part of the test programme was the generation of materials performance data in terms of metal loss (as well as the more conventional mass change information). The measured metal losses for each sample provide the basis for corrosion modelling activities, the first stages of which are reported herewith, using the data obtained for a CMSX-4 substrate.

## 15.2    Experimental procedures

### 15.2.1    Materials tested

The materials and coatings exposed in the tests included:

* Single crystal base alloys (compositions in Table 15.1):
  * CMSX-4 [7]
  * SC$^2$-B [14]
* Coating: Pt–Al (a Pt electroplate–Al diffusion coating).

Samples for these tests were manufactured as cylindrical bars (~10–16 samples per bar, each sample 10 mm long and 10 mm diameter) for ease of machining and later coating, where required. Before testing, individual samples were cut from the bars.

*Table 15.1*   Nominal alloy compositions (weight %)

| Alloy | Ni | Cr | Co | Al | Ti | W | Mo | Ta | Re | C | Others |
|---|---|---|---|---|---|---|---|---|---|---|---|
| CMSX-4 | Bal. | 6.5 | 10 | 5.6 | 4.9 | 6 | 0.6 | 6 | 2.9 | 0.1 | 0.1 Hf |
| SC$^2$-B | Bal. | 11.8 | 5.0 | 4.0 | 4.6 | 3.9 | 1.0 | 2 | | 0.019 | <5 ppm S |

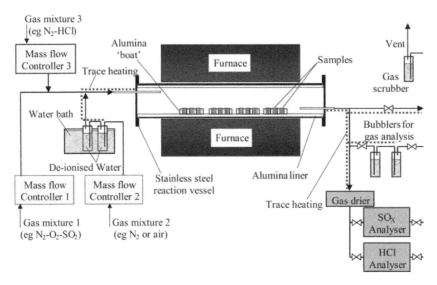

*15.1* Schematic diagram of controlled atmosphere test rig used at Cranfield University for hot corrosion testing [15]

### 15.2.2  Test method

The test method selected for the laboratory hot corrosion experimental programme was the 'deposit recoat' method [10,15,16]. This method has proved to be an effective way of simulating 'in-service' deposition conditions within a laboratory test and now forms the basis for EU-recommended guidelines for hot salt corrosion testing [16].

In this type of corrosion test, samples are coated with a controlled amount of deposit and then exposed in a controlled atmosphere furnace (Fig. 15.1). Periodically, the samples are cooled and recoated with a controlled amount of mixed-salt deposit. Details of this test method are available elsewhere [15]. A matrix of tests was devised to investigate the effects of deposition flux, temperature, gas composition and deposit composition on the hot corrosion of the selected materials. The test programme carried out is summarised in Table 15.2.

The severity of the test depends on the quantity of deposit used (in $\mu g\ cm^{-2}$) and the cycle periodicity (in hours), which combine to give the deposition flux parameter (in $\mu g\ cm^{-2}\ h^{-1}$) that has been identified as a key variable for hot corrosion in gas

*Table 15.2*  Test conditions for hot corrosion tests

| | |
|---|---|
| Deposit composition | 4/1 mol% $(Na/K)_2SO_4$ |
| Deposition flux | 1.5, 5.0 and 15 $\mu g\ cm^{-2}\ h^{-1}$ |
| Deposit recoat interval | 50 h |
| Gas compositions: | Air + 50 vpm $SO_2$ (tests 1 and 2) |
| | Air + 500 vpm $SO_2$ (tests 3 and 4) |
| | Air + 50 vpm $SO_2$ + 500 vpm HCl (tests 5 and 6) |
| | Air + 500 vpm $SO_2$ + 500 vpm HCl + 5% $H_2O$ (tests 7 and 8) |
| Test temperatures | 700 °C (tests 1, 3, 5, 7) and 900 °C (tests 2, 4, 6, 8) |
| Test duration | 500 h |

turbines (rather than hot corrosion in laboratory tests, which have traditionally used excessive levels of deposits) [6]. For example, the most severe conditions used in this test programme, 15 μg cm$^{-2}$ h$^{-1}$, required a 2.4 mg deposit recoat on each sample every 50 h. The deposit recoat test procedure was characterised for the eutectic alkali sulphate (4/1 mol% (Na/K)$_2$SO$_4$) deposit mix. The results demonstrated that the deposit recoat procedure produced a normal distribution; the 95% confidence limits for the three nominal deposition fluxes of 1.5, 5.0 and 15 μg cm$^{-2}$ h$^{-1}$ encompassed the ranges: 0.87–1.96; 2.89–6.55 and 8.68–19.64 μg cm$^{-2}$ h$^{-1}$, respectively. This gives three distinct levels of deposition with a known degree of variability in each level to permit the stochastic nature of the corrosion process to be investigated.

### 15.2.3  Materials performance assessment

During the corrosion tests, the progress of the hot corrosion reactions was monitored by mass-change data generated from the measurements required for the 'deposit recoating' procedure. However, the metal-loss data required for the modelling process must be generated by dimensional metrology [10,15]. Before exposure, the dimensions (especially diameters) of the samples were measured using accurate contact metrology. After exposure, cross-sections of each sample were dry prepared (using paraffin lubricant for cutting, an oil lubricant for grinding/polishing and a solvent for cleaning) and measurements were carried out on these sections using a semi-automated measurement system. At least 24 sets of measurements were made on each sample section to determine the position of the metal surface, depth of internal corrosion, coating thickness, thickness of interdiffusion zone, etc. [15]. Metal losses were calculated by comparison of the measurements made before and after the sample exposure. These data allowed determination of the sensitivity of hot corrosion damage to changes in exposure conditions and formed the basis of the hot corrosion modelling process investigated for both coated and uncoated single crystal materials.

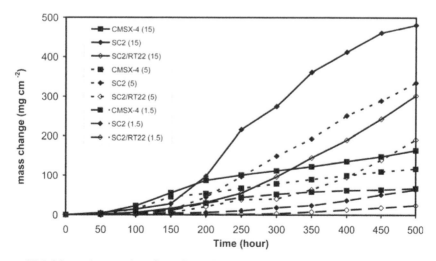

*15.2* Mass change data for selected materials exposed at 900 °C in air–500 vpm SO$_x$ atmosphere with 1.5, 5 or 15 μg cm$^{-2}$ h$^{-1}$ flux of 4/1 (Na/K)$_2$SO$_4$

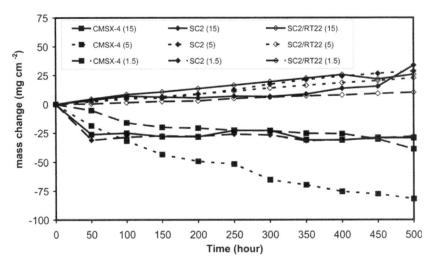

*15.3* Mass change data for selected materials exposed at 700 °C in air–500 vpm SO$_x$ atmosphere with 1.5, 5 or 15 µg cm$^{-2}$ h$^{-1}$ flux of 4/1 (Na/K)$_2$SO$_4$

## 15.3    Results and discussion

### 15.3.1    Mass change data

The mass gains due to the hot corrosion reactions were generally greater for the uncoated single crystal materials than for the coated materials (e.g. Figs. 15.2 and 15.3). The mass change curves showed that incubation times using the 'deposit recoat' test method were short (generally less than 50 h) at 700 °C, whereas at 900 °C, incubation times were much longer, especially for coated samples at lower deposition fluxes. As expected from results on conventional nickel-based superalloys [15], the mass changes were sensitive to the deposition flux for both coated and uncoated materials. It was found that the mass change was much more sensitive to changes in SO$_2$ levels than to HCl levels in the exposure environment, with the high SO$_2$ tests showing a lot more damage than the low SO$_2$ tests with and without HCl. The presence of water vapour also had the effect of increasing the mass changes observed. The uncoated single-crystal materials showed very high degradation rates under many of the test conditions, emphasising the need for appropriate coatings to be used with these materials.

### 15.3.2    Optical microscopy

Following preparation of the cross-sections through the exposed samples, all of the materials were examined by optical microscopy. Figures 15.4–15.7 illustrate the forms of damage observed on cross-sections of some of the samples exposed in the tests. These illustrations have been selected from the large numbers of samples tested to show the sensitivity of the different forms of corrosion to changes in the exposure conditions. (Detailed information on the chemistry of the corrosion layers observed was obtained by SEM/EDX analyses of the sample cross-sections, but is beyond the scope of this paper. However, the compositions of these layers are consistent with the well-known characteristics of Types I and II hot corrosion [e.g. 4,5].)

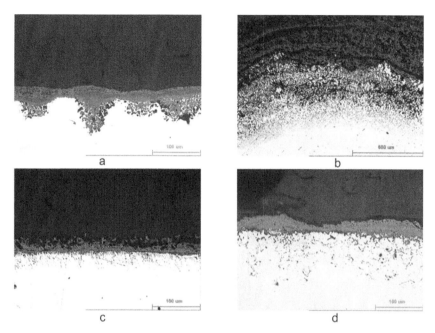

*15.4* Effect of gas composition and temperature on uncoated SC²-B exposed with a deposition flux of 5 µg cm⁻² h⁻¹ for 500 h; (a) air–500 vpm SO₂ at 700 °C; (b) air–500 vpm SO₂ at 900 °C; (c) air–50 vpm SO₂ + 500 vpm HCl at 700 °C; (d) air–50 vpm SO₂ +500 vpm HCl at 900 °C

For uncoated SC²-B and CMSX-4, the corrosion damage increases significantly with increasing deposition flux levels from 1.5 to 5 and from 5 to 15 µg cm⁻² h⁻¹. Figure 15.4 illustrates some of these features for uncoated SC²-B exposed with a deposition flux of 5 µg cm⁻² h⁻¹ for 500 h at 900 °C and 700 °C. At 900 °C, the damage increases significantly on increasing the contaminants from 50 to 500 vpm (volumes per million) $SO_X$ (Fig. 15.4(b)); however, the effect of HCl addition to 50 vpm $SO_X$ is more subtle, increasing the extent of internal oxidation/sulphidation damage that is typical of Type I hot corrosion (Fig. 15.4(d)). At 700 °C, the characteristic Type II pitting damage increases on changing the contaminants from 50 to 500 vpm $SO_X$ (Fig. 15.4(a)). However, the effect of HCl addition to 50 vpm $SO_X$ appears to suppress this form of pitting damage and leads to a broader front attack with internal damage as well (Fig. 15.4(c)). This effect is much more pronounced in the single crystal materials than in conventional nickel-based superalloys with higher chromium contents [15,19]. For the lower chromium content single-crystal superalloys, it is believed that the HCl addition to the gas atmosphere is sufficient to change the acid/base balance of the deposit (by dissolution of the HCl and release of $SO_X$) such that the hot corrosion attack mode changes from a gas induced acidic fluxing mode to a basic fluxing mode [19].

For Pt–Al coated SC²-B exposed with a deposition flux of 5 µg cm⁻² h⁻¹ for 500 h at 700 °C (Fig. 15.5), the extent of pitting increases on changing the contaminants from 50 to 500 vpm $SO_X$. However, the addition of HCl to 50 vpm $SO_X$ appears to reduce this form of damage significantly. Figure 15.6 shows the effect of two deposition fluxes on Pt–Al coated SC²-B exposed in an air–500 vpm $SO_X$ environment for 500 h

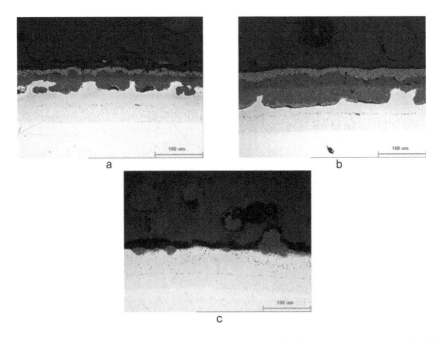

15.5 Effect of gas composition on Pt–Al coated SC²-B exposed at 700 °C with a deposition flux of 5 μg cm⁻² h⁻¹ for 500 h [18]; (a) air–50 vpm SO₂; (b) air–500 vpm SO₂; (c) air–50 vpm SO₂ + 500 vpm HCl

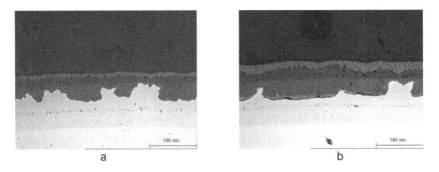

15.6 Effect of deposition flux on Pt–Al coated SC²-B exposed at 700 °C with a gas composition of air–500 vpm SOₓ for 500 h [18]; (a) 1.5 μg cm⁻² h⁻¹; (b) 5 μg cm⁻² h⁻¹

at 700 °C; the characteristic type II pitting damage increases significantly with increasing deposition flux. At the highest deposition flux used (15 μg cm⁻² h⁻¹), the damage locally penetrated through the whole coating thickness after 500 h.

Figure 15.7 shows the influence of the substrate alloy on the performance of Pt–Al coatings exposed at 700 °C for 500 h with a deposition flux of 15 μg cm⁻² h⁻¹ and gas composition of air–50 vpm SOₓ. With a CMSX-4 substrate, the Type II pitting attack is more localised and has penetrated through the Pt–Al coating and into the underlying alloy. However, with a SC²-B substrate, the attack of the Pt–Al is more broad-fronted and has generally penetrated through less than half the coating thickness.

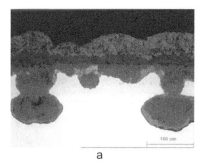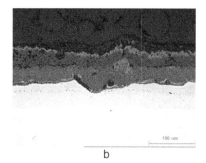

a                                                    b

*15.7* Effect of substrate composition on hot corrosion of Pt–Al coating at 700 °C with deposition flux of 15 µg cm$^{-2}$ h$^{-1}$ and gas composition of air–50 vpm SO$_X$ for 500 h [18]. (a) CMSX-4/Pt–Al. (b) SC$^2$-B/Pt–Al

The corrosion performance of the 'Pt–Al' coating on these two different composition substrates is different because the 'Pt–Al' coating composition will not be the same on both substrates. As 'Pt–Al' is applied as a platinum electro-deposition process followed by a diffusion aluminising process and recovery heat treatments, the composition of the substrate material will affect (by interdiffusion) the coating composition that develops during both the aluminising process and the recovery heat treatment. Details of the complex microstructures that develop are beyond the scope of this paper, but are available elsewhere [20]. The interdiffusion processes will continue during the corrosion exposures, though at lower rates than during processing of the coating due to the lower exposure temperatures.

### 15.3.3  Dimensional metrology

The cross-sections were all measured using the procedure outlined above. The data generated were combined with pre-exposure measurements to give change in sound metal results (negative values represent metal losses). The data sets from each sample have been ordered and are best viewed on probability type plots [15,17]. In this form of presentation, change in sound metal is plotted as a function of the probability of that damage level being exceeded. Figure 15.8 illustrates this for three samples of Pt–Al coated SC$^2$-B exposed at 700 °C for 500 h in an air–500 vpm SO$_X$ environment with three deposition fluxes of 4/1 (Na/K)$_2$SO$_4$.

### 15.3.4  Sensitivity of hot corrosion damage to exposure environments

From the ordered sound metal data, it is possible to determine the sensitivity of hot corrosion damage at particular probability levels to changes in the various environmental parameters, i.e. deposition flux, SO$_X$, HCl and H$_2$O concentrations, deposit composition and temperature. In most figures related to this section, a sound metal loss with a 4% probability of exceedance has been selected (and for simplicity, this has been labelled as 'maximum' corrosion damage).

Figure 15.9 illustrates the effects of deposition flux (of 4/1 (Na/K)$_2$SO$_4$) on sound metal loss for the four materials systems at 700 °C with a gas containing 50 vpm SO$_X$; the sound metal loss has an approximately logarithmic dependence on the deposition flux, as has been observed previously for conventional nickel-based superalloys in the flux limited hot corrosion regime [10,15].

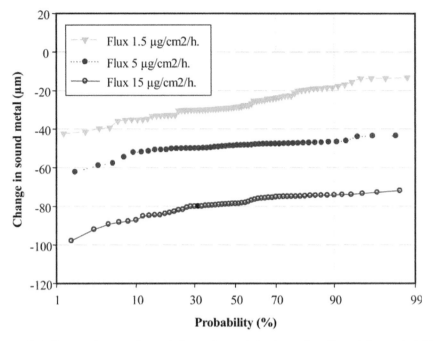

*15.8* Change in sound metal dimensions plotted as a function of the probability that it is exceeded for three samples of Pt–Al coated SC$^2$-B exposed at 700 °C for 500 h in an air–500 vpm SO$_x$ environment at three deposition fluxes of 4/1 (Na/K)$_2$SO$_4$ [18]

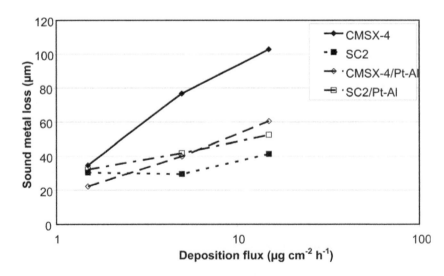

*15.9* Effect of deposition flux of 4/1 (Na/K)$_2$SO$_4$ on the hot corrosion performance of selected materials in terms of sound metal loss (with a 4% probability of being exceeded) after exposure at 700 °C in air–50 vpm SO$_x$ for 500 h

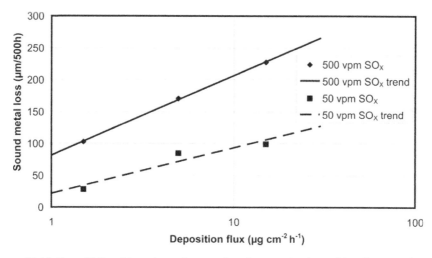

*15.10* Sensitivity of 'maximum' corrosion damage to deposition flux as a function of gas $SO_x$ content for CMSX-4 at 700 °C

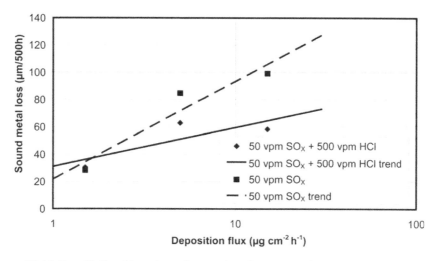

*15.11* Sensitivity of 'maximum' corrosion damage to deposition flux as a function of gas HCl content (with 50 vpm of $SO_x$) for CMSX-4 at 700 °C

Figures 15.10–15.13 illustrate the sensitivity of CMSX-4 'maximum' corrosion damage to changes in gas composition and deposition flux at 700 °C. This measure of corrosion damage increases with gas $SO_x$ and moisture content, as well as deposition flux; however, the measured damage decreases with higher gas HCl content. This is consistent with the damage observed during microscopic examinations, and is believed to be related to the addition of HCl to the gas atmosphere being sufficient to alter the acidic/base balance of the deposits (by dissolution of HCl and release of $SO_x$) enough to change the predominant mode for hot attack on the lower chromium-content single-crystal materials [19]. From the data presented in Figs. 15.10–15.13, a quantitative relationship between corrosion damage data and the exposure conditions can be derived. After several trials of alternative equations,

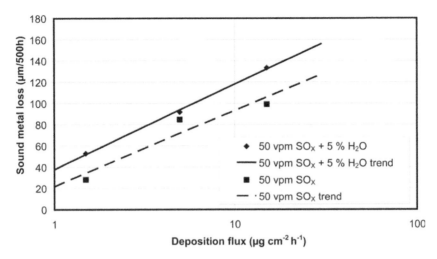

*15.12* Sensitivity of 'maximum' corrosion damage to deposition flux as a function of gas moisture content (with 50 vpm of SOx) for CMSX-4 at 700 °C

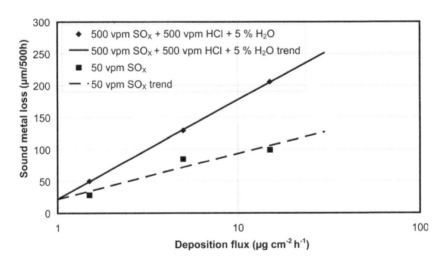

*15.13* Sensitivity of 'maximum' corrosion damage to deposition flux as a function of gas $SO_x$, HCl and moisture content for CMSX-4 at 700 °C

the following was found to be a good framework model for predicting this hot corrosion damage:

$$\text{Corrosion damage} = y_o + A.\ln\left(1 + (\text{alkali sulphate flux/flux}_o)\right) \qquad [15.1]$$

Where:

- $A = A_o.[SO_X+1]^k.[HCl+1]^m.[H_2O+1]^n$      [15.2]
- $\text{flux}_o = F_o.[SO_X+1]^r.[HCl+1]^s.[H_2O+1]^t$      [15.3]
- $A_o$, k, m, n, $F_o$, r, s, t and $y_o$ are constants for each alloy/coating system, temperature and damage probability level.

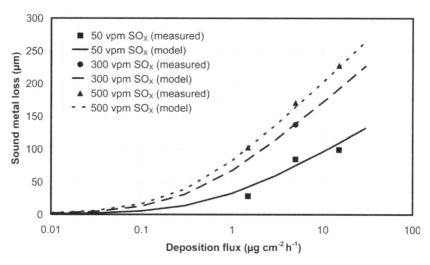

*15.14* Lines are model predictions for 'maximum' corrosion damage as a function of deposition flux for three different gas $SO_x$ contents; points are measured data for CMSX-4 at 700 °C

Previous studies on a conventional nickel-based superalloy (IN738LC) moved towards the use of a sigmoidal relationship between corrosion damage and deposition flux [e.g. 15]. However, these studies were based on a much more extensive set of materials performance data (in terms of the scope of the conditions tested, the numbers of conditions and the numbers of samples) and it was not possible to follow the same methodology in this study, which was based on a much more focused set of exposure conditions.

The need for the '+1' factor in the gas composition parameters is to keep the expressions valid even if a particular gas species is not present. Figures 15.14 and 15.15 illustrate predictions made using this form of framework model with constants derived for CMSX-4 at 700 °C, and compare them to measured data points; the measured data points and model predictions follow the same trends in all cases and the predictions are generally close to the measured values (within 20%). Figure 15.16 compares measured and predicted values for this model over a wider range of conditions (gas compositions and deposition fluxes); again there is a generally good agreement between predicted and measured values.

The effects of evaluating the corrosion damage at different probability levels are illustrated in Fig. 15.17, when 'maximum' corrosion damage model predictions and measured data are compared to the predictions of a median corrosion damage model and measured data for CMSX-4 at 700 °C. As the damage to CMSX-4 in these tests is relatively uniform around the samples (after 500 h, it has progressed from pitting to broad-front corrosion damage), there is relatively little difference between the median and 'maximum' measurements of corrosion damage. However, this difference does increase with increasing deposition flux.

Further work is in progress to evaluate potential improvements to the modelling expressions linking corrosion damage to the various exposure parameters for all of the materials and both exposure temperatures. In addition, more hot corrosion tests are being carried out to address some of the outstanding issues raised by the analysis

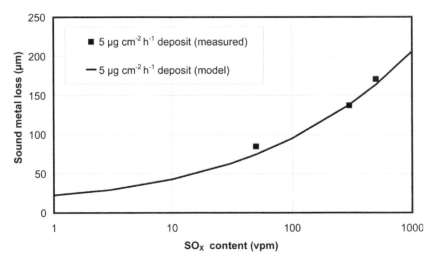

15.15 Line is model predictions for 'maximum' corrosion damage as a function of gas $SO_x$ content for one deposition flux (5 μg cm$^{-2}$ h$^{-1}$); points are measured data (for CMSX-4 at 700 °C)

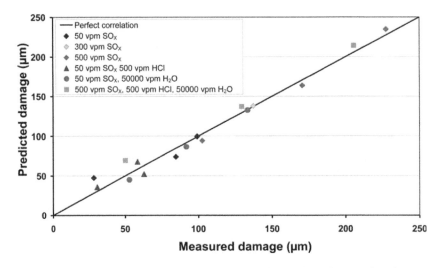

15.16 Comparison of measured and predicted 'maximum' corrosion damage data for CMSX-4 at 700 °C

of the data, e.g. the effects of HCl and whether there is a critical value of HCl needed to produce the reduction in corrosion damage/change in mechanism observed; and the effect of water vapour on the hot corrosion reactions.

## 15.4    Conclusions

This paper reports results from a series of laboratory tests carried out using the 'deposit replenishment' technique to investigate the sensitivity of candidate materials to exposure conditions anticipated in future gas turbines. The materials investigated

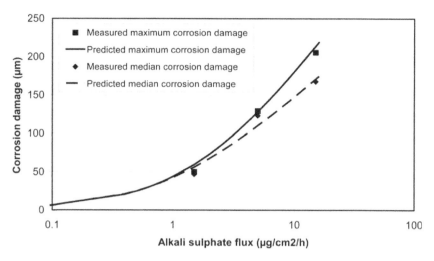

*15.17* Comparison of models and measured data for maximum and median corrosion damage data for CMSX-4 at 700 °C

have included CMSX4 and SC²-B, both bare and with Pt–Al coatings. The exposure conditions within the laboratory tests have covered ranges of $SO_X$ (50 and 500 vpm), HCl (0 and 500 vpm) and moisture (0 and 5% $H_2O$ as steam) in air, as well as 4/1 $(Na/K)_2SO_4$ deposits, with deposition fluxes of 1.5, 5 and 15 $\mu g\,cm^{-2}\,h^{-1}$, for periods of up to 500 h at 700 and 900 °C. Data on the performance of materials have been obtained using dimensional metrology. These measurement methods have allowed distributions of damage data to be determined (as required for future materials performance model developments). In addition, the types of damage observed have been characterised using standard microscopy techniques.

The damage rates of the single crystal materials without coatings are too high for them to be used with confidence in gas turbines fired with fuel gases containing high levels of potentially damaging elements. Under the more severe combinations of gas composition, deposition flux and metal temperature, the corrosion rates of these materials with Pt–Al coatings are also excessive. The data produced from these tests have allowed the sensitivity of hot corrosion damage to changes in the exposure environment to be determined for the single-crystal alloys and coating systems examined. These data have in turn formed the basis for initial hot corrosion models for these materials.

## Acknowledgements

This work was funded as part an EU Project on 'Advanced Long Life Blade Turbine Coating Systems' (ALLBATROS), contract ENK5-CT2000-00081.

## References

1. D. H. Allen, J. E. Oakey and B. Scarlin, in *Materials for Advanced Power Engineering 1998*, 1825, ed. J. Lecomte-Beckers et al., 1998.
2. S. T. Scheirer and R. Viswanathan, presented at ASM International's Materials Solutions Conference, St. Louis, Missouri, 9–10 October 2000, ASM International Materials Park, Ohio, USA.

3. T. Schulenberg, in *Materials for Advanced Power Engineering 1998*, 849, ed., J. Lecomte-Beckers et al., 1998.

4. 'Life cycle issues in advanced energy systems', in *Materials at High Temperature*, 23, ed., J. F. Norton et al., 2003.

5. C. T. Sims, N. S. Stoloff and W. C. Hagel, *Superalloys II*, Wiley, 1987.

6. 'Hot corrosion standards, test procedures and performance', *High Temp. Technol.*, 4 (1989), 7.

7. N. J. Simms, D. W. Bale, D. Baxter and J. E. Oakey, in *Materials for Advanced Power Engineering 2002*, 73, ed., J. Lecomte-Beckers et al. Forschungszentrum Jülich, 2002.

8. J. Stringer and I. G. Wright, *Oxid. Met.*, 44 (1995), 265.

9. M. Decorso, D. Anson, R. Newby, R. Wenglarz and I. Wright, presented at ASME Int. Gas Turbine and Aeroengine Congress 1996, ASME, 1996, ASME Paper 96-GT-76.

10. N. J. Simms, J. R. Nicholls and J. E. Oakey, *Mater. Sci. Forum*, 369–372 (2001), 833.

11. P. Kilgallon, N. J. Simms and J. E. Oakey, in *Materials for Advanced Power Engineering 2002*, 903, ed., J. Lecomte-Beckers et al., Forschungszentrum Jülich, 2002.

12. C. E. Neilson, *Biomass Bioenerg.*, 15 (1998), 269.

13. R. S. Roth, T. Negas and L. P. Cook, *Phase Diagrams for Ceramists–Volume IV*, The American Ceramic Society, 1981.

14. S. Mercier, F. Iozzelli, M.-P. Bacos and P. Josso, *Mater. Sci. Forum*, 461–464 (2004), 949.

15. N. J. Simms, P. J. Smith, A. Encinas-Oropesa, S. Ryder, J. R. Nicholls and J. E. Oakey, in *Lifetime Modelling of High Temperature Corrosion Processes*, 246, ed., M. Schütze et al., EFC No. 34, Maney, 2001.

16. *Draft Code of Practice for Discontinuous Corrosion Testing in High Temperature Gaseous Atmospheres*, European Commission Project SMT4-CT95-2001 'TESTCORR', ERA Technology, UK, 2000.

17. J. R. Nicholls and P. Hancock, in *High Temperature Corrosion*, 198, ed., R. Rapp, NACE, 1983.

18. N. J. Simms, A. Encinas-Oropesa and J. R. Nicholls, *Mater. Sci. Forum*, 461–464 (2004), 941.

19. A. Encinas-Oropesa, PhD Thesis, Cranfield University, UK, 2006.

20. N. Vialas, PhD Thesis, l'Institut National Polytechnique de Toulouse, France, 2004.

# 16

## Oxidation and alkali sulphate-induced corrosion of aluminide diffusion coatings with and without platinum*

Pavleta A. Petrova and Krystyna Stiller

*Department of Applied Physics, Chalmers University of Technology, Gothenburg,
SE – 41296, Sweden*

*pavleta.petrova@fy.chalmers.se*

### 16.1    Introduction

Ni-based superalloys are extensively used as components in stationary gas turbines. These materials are exposed at temperatures in the range of 680–950 °C to high pressure and hot gas atmospheres, i.e. mixtures of inlet gas, combusted fuel and injected water/steam. As a result, the durability of the turbine materials is reduced. One of the processes that is considered decisive for their failure is salt-induced hot corrosion. The most frequently encountered form of such corrosion is alkali sulphate hot corrosion. The salts are formed from the reaction between sulphur, present as an impurity in liquid fuel and coal, and alkali metal ions found in the inlet gas. Once the temperature in the turbines is above the melting point of the eutectic sulphate mixtures, they can cause accelerated corrosion attack on the exposed components.

The corrosion process consists of two main stages; an initiation stage, very similar to oxidation without any deposit and a propagation stage, when the effect of the deposit is more evident [1,2]. The key to continuous protection against hot corrosion is the formation of a protective oxide scale that maintains the hot-corrosion process in its initial, incubation stage. If this incubation stage is not maintained, and the propagation stage begins, a rapid loss of the material occurs.

Aluminide coatings are widely used to improve the oxidation resistance of Ni-base superalloy turbine components during high-temperature oxidation and/or corrosion. The coatings consist basically of a β-NiAl layer deposited on the surface of a γ-Ni/ γ'-Ni₃Al-based substrate. During service, slowly growing aluminium oxide is formed and prevents further oxidation. Although protective against high-temperature oxidation and corrosion, alumina scale is susceptible to cracking and spallation during service.

For high-temperature oxidation (above 900 °C), it has been shown that modifying aluminide coatings by the addition of platinum can substantially improve their performance. Even though the beneficial effects of platinum on the oxidation of the coating have been extensively studied [3–13], very little is known about its effect on the hot corrosion process [14–18]. There is thus a need for a comprehensive study of how the coating performance during corrosion is influenced by the presence of platinum.

---

* Reprinted from P. A. Knutsson and K. Stiller: Oxidation and alkali sulphate-induced corrosion of aluminide diffusion coatings with and without platinum. *Materials and Corrosion.* 2008. Volume 59. pp. 484–8. Copyright Wiley-VCH Verlag GmbH & Co. KGaA.

In the present work, the environmental conditions of turbines have been simulated by salt deposition and subsequent tube-furnace laboratory experiments. The observations from these studies have then been compared with those obtained in high-temperature oxidation. Apart from the gravimetric studies, the microstructures of the materials have been evaluated by means of SEM, EDX and XRD. This paper reports the initial results from the investigation.

## 16.2    Experimental materials and conditions

Two commercial CVD coatings, namely a platinum-free coating (MDC210) and a platinum-rich coating (MDC150L) (Table 16.1), were chosen for this investigation. Both coatings were produced using the same process parameters as those used for commercial products. They were deposited by the vendor on a single crystal CMSX4 substrate (Table 16.2). The samples produced were in the form of cylinders with a diameter of 12 mm and height of 4 mm. The <001> direction of the substrate was parallel to the cylinder axis.

Horizontal tube furnaces were used both for oxidation and corrosion experiments. Samples were placed individually in ceramic crucibles and exposed at 900 °C for two periods of 50 h. Before exposure, the samples were cleaned in acetone and ethanol. Two samples of each material were used for each experimental condition.

In the case of corrosion, the salt deposit was applied by the 'deposit recoat technique' [19]. This technique was chosen since it assures the composition of the deposited salt mixture. The salt mixture was deposited by spraying an aqueous 0.01 M solution of 4:1 $Na_2SO_4 + K_2SO_4$ at a rate of 5 $\mu g$ cm$^{-2}$ h$^{-1}$ onto one side of the samples. This creates more aggressive conditions than those occurring in reality and causes accelerated corrosion so that the effect of the salt can be studied after only 100 h. The salt mixture was deposited in two intervals of 50 h to ensure its presence on the sample surface throughout the entire exposure time. A mixture of 500 ppm $SO_2$ plus air was used as the inlet gas. The flow rate was 20 L h$^{-1}$. In the oxidation experiment, still laboratory air was used. The samples were taken out after 50 h, left to cool down

Table 16.1   Typical structure of the platinum-free and platinum-rich coatings in the as-coated condition

| Coating | Thickness, μm | Outer zone, μm[a] | Interdiffusion zone, μm | Grain size, μm |
|---|---|---|---|---|
| Platinum-free | $42 \pm 2$ | $22 \pm 2$ | $20 \pm 2$ | $105 \pm 8$ |
| Platinum-rich | $47 \pm 2$ | $24 \pm 2$ | $23 \pm 2$ | $113 \pm 8$ |

[a]Outer zone – zone from the coating surface to the interdiffusion zone.

Table 16.2   Nominal composition of CMSX-4 superalloy in weight percent and atomic percent

| Composition | Ni | O | Cr | Al | Ti | Ta | Mo | W | Re | Hf |
|---|---|---|---|---|---|---|---|---|---|---|
| Wt.% | 61.7 | 9.0 | 6.5 | 5.6 | 1.0 | 6.5 | 0.6 | 6.0 | 3.0 | 0.10 |
| At.% | 63.7 | 9.3 | 7.6 | 12.6 | 1.3 | 2.2 | 0.4 | 2.0 | 1.0 | 0.03 |

in a desiccator to room temperature and, after weighing, were put back into the furnace for a further 50 h period.

Before and after the exposures, the samples were weighed on a Sartorius R160P balance with a resolution of 10 μg. In order to account for the spalled products, the samples were weighed in the crucibles. Surface (top-view) and cross-section images were obtained using a Zeiss Ultra 55 FESEM and an FEI Quanta 200 environmental scanning electron microscope. Chemical analysis in the SEM was performed with an Oxford Inca Energy Dispersive X-ray system (EDX). Two types of cross-sections were produced using an FEI Strata 235 Dual Beam (focused ion beam plus SEM) microscope, one beam for SEM imaging and the other for EDX analysis in the SEM. The specimens for SEM imaging were produced with the ion beam perpendicular to the sample surface, while the samples for EDX analysis were prepared with the ion beam inclined at 45° to the surface. To produce both cross-sections, platinum strip was deposited to protect the surface of the sample from the ion beam. Then the chosen area was ion-milled by a focused Ga ion beam operated at 30 kV using a current that was adjusted successively from 20 nA to 500 pA. Since the ion and electron columns in the Dual Beam microscope are 52° apart from each other, viewed by the SEM detector, the imaged cross-section represents a projection of that produced by ions. Therefore, a correction for the measured distances was applied.

Glancing angle X-ray diffraction (XRD) was used to investigate the phases present on the oxidised and corroded samples.

## 16.3    Results and discussion

SEM top-view surface micrographs of the samples before and after the exposures are shown in Figs. 16.1 and 16.2, respectively. Even though the deposit recoating technique is a very reproducible method for salt deposition in terms of mass of salt, it gives uneven salt distribution on the sample surface before the exposure (Fig. 16.1b). However, since the samples were exposed to temperatures higher than the melting temperature of the salts, it was assumed that, in liquid form, the salt deposit will redistribute evenly on the specimen surface.

Two different forms and sizes of crystals were found on the salt-sprayed samples before the exposure: elongated and round structures (Fig. 16.3a). After the exposure, smaller and facetted crystals appeared (Fig. 16.3b). This indicates that recrystallisation of the salt crystals from the melt had occurred during cooling. Moreover, the

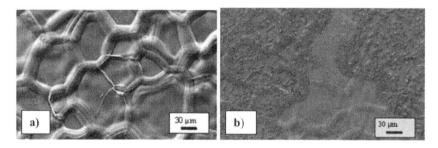

*16.1* SEM surface micrographs of the coatings before exposure: (a) oxidation experiment; an as-received sample – a typical surface structure with visible ridges at the grain boundaries of the coating; (b) corrosion experiment; uneven distribution of salt deposit on the surface of a salt-sprayed sample

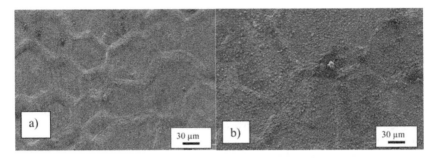

*16.2* (a) SEM surface image of platinum-free coating after exposure in air; thin oxide, grain boundaries of the coating are still visible; (b) sample surface after exposure in sulphur dioxide plus air; less visible grain boundaries due to a thick oxide layer

*16.3* SEM micrographs showing salt crystals on the specimen surface; (a) the salt-sprayed sample before exposure; two different forms of crystals – elongated and round – are apparent; (b) after the exposure, crystal form is different indicating recrystallisation of salts

presence of salt crystals on the surface after the exposure confirms the salt presence during the whole 50 h period of exposure.

The weight change of the samples was used to study the kinetics of the corrosion and the oxidation process. Since weight measurements of samples were performed at 50 h intervals, only two points in the weight change diagram were obtained for each material after 100 h of exposure. Thus, based on this information, it is impossible to determine when exactly the incubation stage ends and the corrosion propagation stage starts. However, after 100 h, both corrosion weight change curves lay above the oxidation ones, which indicates that the propagation stage had already initiated. For the platinum-rich coating, the corrosion and oxidation processes follow the same kinetics during the first 50 h of exposure, i.e. the corrosion process is repressed in the presence of platinum.

The SEM surface images (Figs. 16.2 and 16.4) reveal a difference in the surface appearance of the oxidised and corroded samples. The less visible coating on grain boundaries on the SEM surface micrographs of the corrosion-exposed samples on both platinum-rich and platinum-free coatings (Fig. 16.2b) compared to the oxidised

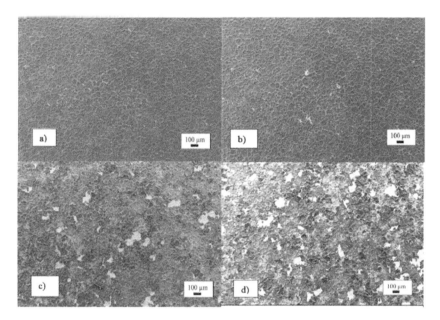

*16.4* SEM surface images showing spalled areas on (a) platinum-rich coating after the oxidation experiment; almost no spallation observable; (b) platinum-free coating after the oxidation experiment; separate spallation areas; (c) platinum-rich coating after the corrosion experiment; spalled spots observed on the surface of the specimen; (d) platinum-free coating after the corrosion experiment; many spalled areas

*Table 16.3*   Thickness of the scale measured using FIB-SEM after 100 h of oxidation (air) and corrosion (salt, sulphur dioxide plus air) at 900 °C

|  | Oxidation, coating, µm | | Corrosion, coating, µm | |
| --- | --- | --- | --- | --- |
|  | Platinum-rich | Platinum-free | Platinum-rich | Platinum-free |
| Thickness | 1.7 ± 0.3 | 1.3 ± 0.4 | 1.9 ± 0.4 | 3.8 ± 0.5 |

ones (Fig. 16.2a), imply the presence of a thicker oxide scale on the corroded materials. This statement is also confirmed by the gravimetrical results and the measurement of the scale thickness using cross-section images (Table 16.3). The values in Table 16.3 represent the average oxide thickness measured at several locations on the cross-section SEM images. Here it should be mentioned that the corrosion product was much more uneven in thickness than that of oxidation. The heaviest spallation was observed to take place in the case of corrosion exposure of the platinum-free coating (Fig. 16.4). In this material, the cross-section micrographs reveal the occurrence of a few places where pitting corrosion was observed, confirming that the propagation stage had already initiated (see arrow in Fig. 16.5). The calculations of the scale thickness from the gravimetric analysis were also performed for comparison, but were considered to be unreliable in the corrosion case because of the salt residues that influence the calculations. The above results show that the presence of platinum slows down both corrosion and oxidation processes.

*16.5* SEM view of cross-sections of platinum-rich coating; (a) after oxidation exposure; a thin singe-layer oxide is apparent; (b) after corrosion exposure; thicker oxide is observed. The arrow indicates a pitting area

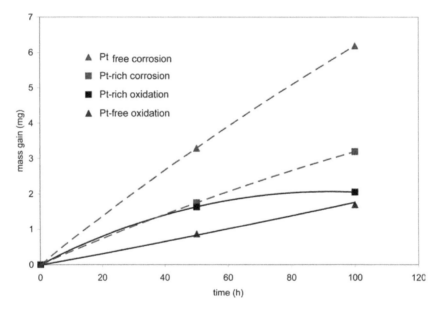

*16.6* Specific weight gain for platinum-free and platinum-rich coatings for corrosion (dashed lines) and oxidation (continuous line) experiments

Due to the complex morphology of the surface of the corroded samples, it was difficult to find places where pitting attack occurred and to produce cross-sections suitable for EDX analysis at these locations. Within the cross-sections produced for EDX, no pitting corrosion was found. The maps and the line-scan profiles in those samples indicate the presence of alumina alone (Fig. 16.7) on both platinum-free and platinum-rich coatings. According to the XRD analyses, the oxide scale on both types of coatings consists of stable $\alpha$-$Al_2O_3$. However, on platinum-rich samples, traces of transient $Al_2O_3$ were also found. Sulphur was not detected by SEM-EDX in the produced cross-sections, neither within the oxide scale nor at the metal–oxide interface (Figs. 16.7 and 16.8). Since it is clear that the propagation stage has already started, the manufacture of samples with pitting areas is necessary for further analysis.

EDX line-scan profiles (Fig. 16.8) show that both after oxidation and corrosion the aluminium content in the coatings below the oxide scale was about 40 at.%. This amount of aluminium is in agreement with that in the as-coated condition.

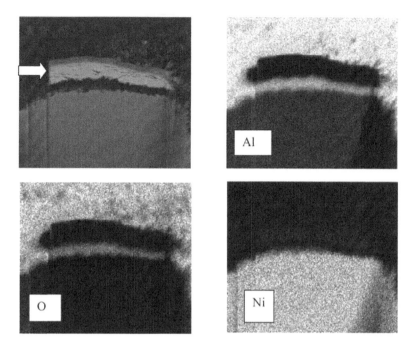

*16.7* EDX elemental maps showing the typical elemental distribution in the scale on the surface of the coatings after the corrosion experiment. The arrow indicates the Pt-strip deposited as a protective layer when producing the cross-section. Note that only maps with sufficient contrast for the corrosion product have been included. Alumina is the predominant scale product. No sulphides were detected by means of EDX

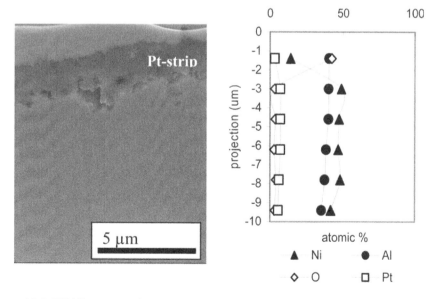

*16.8* EDX line-scan, showing depth profiles in atomic percent for nickel, aluminium, oxygen and platinum of platinum-rich coating after corrosion exposure

In the platinum-rich coating, the depletion of the aluminium reservoir to 32% was first observed after 500 h of oxidation at 1050 °C [9]. In this investigation, the thickness of the oxide scale does not change much between 100 h and 500 h of exposure [9]. This shows that the main reason for the decrease in aluminium in the coating after 100 h exposure is the process of interdiffusion between the coating and the substrate and not consumption of Al by the formation of the oxide scale.

## 16.4    Summary and conclusion

Corrosion and oxidation exposures were performed on platinum-rich and platinum-free coatings. It was found that after exposures totalling 100 h at 900 °C, aluminium oxide was the main product on all of the samples. Traces of transient alumina were detected on platinum-rich samples during corrosion but were difficult to quantify. The aluminium oxide scale is thicker in the case of corrosion, which causes more spallation. A large deviation from the parabolic law was observed in the case of corrosion for both types of coating. However, for platinum-rich material, corrosion and oxidation followed almost the same trend during the first 50 h. Platinum also proved to be beneficial against the corrosion process by prolonging the incubation stage and thereby decreasing the scale thickness and spallation.

## Acknowledgements

Thanks are due to the High Temperature Corrosion Centre, Sweden, for providing financial support, and to Siemens – Sweden for cooperation and for financial support. Professor Gunnar Johansson from the Department of Inorganic Chemistry at CTH and Dr Xin-Hai Li at Siemens are specially acknowledged for their fruitful discussions.

## References

1. T. S. Sidhu, R. D. Agrawal and S. Prakash, *Surf. Coat. Technol.*, 198 (2005), 441.
2. Ch. T. Sims, N. S. Stolloff and W. C. Hagel, *Superalloys II*, Wiley, New York, 1987.
3. J. A. Haynes, K. L. More, B. A. Pint, I. G. Wright, K. Cooley and Y. Zhang, presented at 5th International Symposium on High Temperature Corrosion, 679–686, Les Embiez, France, 22–26 May 2000.
4. J. A. Haynes, B. A. Pint, K. L. More, Y. Zhang and I. G. Wright, *Oxid. Met.*, 58 (2002), 513.
5. Y. Zhang, J. A. Haynes, B. A. Pint, I. G. Wright and W. Y. Lee, *Surf. Coat. Technol.*, 163–164 (2003), 19.
6. J. Angenete and K. Stiller, *Oxid. Met.*, 60 (2003), 83.
7. J. Angenete, K. Stiller and V. Langer, *Oxid. Met.*, 60 (2003), 47.
8. J. Angenete, K. Stiller and E. Bakchinova, *Surf. Coat. Technol.*, 176 (2004), 272.
9. J. Angenete and K. Stiller, *Surf. Coat. Technol.*, 150 (2002), 107.
10. Y. Zhang, J. Allen Haynes, W. Y. Lee, I. G. Wright, B. A. Pint, K. M. Cooley and P. K. Liaw, *Metall. Mater. Trans. A*, 32A (2001), 1727.
11. J. Allen Haynes, K. L. More, B. A. Pint, I. G. Wright, K. M. Cooley and Y. Zhang, *Mater. Sci. Forum*, 369–372 (2001), 679.
12. Y. Zhang, W. Lee, J. Allen Haynes, I. Wright, B. Pint, K. Cooley and P. Liaw, *Metall. Mater. Trans. A*, 30A (1999), 2679.
13. J. Allen Haynes, Y. Zhang, W. Lee, B. Pint, I. Wright and K. Cooley, *Sci. Technol. III*, (1999), 185.

14. B. Bordenet and H. P. Bobmann, *Schriften des Forschungszentrums Juelich, Reihe Energietechnik*, 871, 2002.
15. G. P. De Gaudenzi, A. Colombo, G. Rocchini and F. Uberti, presented at Symposium, Anaheim, CA, 4–8 February, 1996, 301.
16. R. E. Malush, P. Deb and D. H. Boone, *Surf. Coat. Technol.*, 36 (1988), 13.
17. B. Waschbuesch and H. P. Bossmann, *Eur. Fed. Corros. Publications*, 34 (2001), 261.
18. E. Rocca, L. Aranda, M. Vilasi and P. Steinmetz, *Mater. Sci. Forum*, 461–464 (2004), 917.
19. N. J. Simms, J. R. Nicholls and J. E. Oakey, *Mater. Sci. Forum*, 369–372 (2001), 833.

# 17

# Improving the oxidation resistance of $\gamma$-titanium aluminides by halogen treatment*

## Alexander Donchev and Michael Schütze

*DECHEMA e.V. Karl-Winnacker-Institut*
*D-60486 Frankfurt, Germany*
*donchev@dechema.de*

## 17.1 Introduction

Alloys based on titanium and aluminium with an aluminium content of about 40–50 at.%, the so-called gamma-TiAl alloys, are of great interest for several structural high-temperature applications in, for example, aero or automotive engines due to their good high-temperature strength and their low specific weight [1]. The use of TiAl-alloys would lead to an increase in the efficiency of these engines if the heavier superalloys which are presently used could be replaced. The specific weight of $\gamma$-TiAl is about 4 g cm$^{-3}$ while the density of nickel-based superalloys is up to 9 g cm$^{-3}$. The mechanical properties of TiAl-alloys would allow their use at temperatures above 800 °C [2]. However, the poor oxidation resistance of these alloys limits their application to approximately 800 °C [3]. The oxidation resistance has to be improved to reach envisaged usage temperatures of 900 °C and above. There are several ways to improve the oxidation behaviour, e.g. alloying with niobium [4]. Alloying of the bulk of the component can cause problems depending on the required mechanical properties. Surface methods are therefore a better way to improve the oxidation resistance. The addition of small amounts of halogens into the surface zone of TiAl alloys reduces the oxidation kinetics drastically and improves the oxidation resistance [5]. The positive effect of halogens is due to the formation of a protective, well-adherent alumina scale [6]. Untreated TiAl alloys form a fast growing mixed oxide scale during high-temperature oxidation in air which finally leads to the degradation of the metallic compound [7]. This work presents the results of isothermal and thermocyclic oxidation experiments with different cycle durations. These results were obtained with small coupons, larger complex samples and real components. Additionally, the results of creep tests on untreated and fluorine-treated samples are included to elucidate the influence of halogen treatment on mechanical properties. The results are discussed with regard to the service conditions of TiAl components.

## 17.2 Experimental

Small coupons were cut from bars into 1 mm thick slices or from a 1 mm thick sheet into 10 mm × 10 mm pieces. The alloys investigated are listed in Table 17.1. These

* Reprinted from A. Donchev and M. Schütze: Improving the oxidation resistance of $\gamma$-titanium aluminides by halogen treatment. *Materials and Corrosion*. 2008. Volume 59. pp. 489–93. Copyright Wiley-VCH Verlag GmbH & Co. KGaA.

*Table 17.1*   Compositions of the alloys

| Alloy | Titanium, at.% | Aluminium, at.% | Additional elements, at.% |
|---|---|---|---|
| γ-MET | 49.5 | 46.5 | 4(Cr, Nb, Ta, B) |
| TNB | 48.5 | 41.5 | 10Nb |
| MoCuSi 1 | 53.8 | 45 | 0.65Cu–0.35Mo–0.2Si |
| TNB V | 47 | 45 | 8Nb |

coupons were polished to 1200 grit using SiC-paper, ultrasonically cleaned in ethanol for 10 min and dried before further treatment. Some samples were exposed without any treatment to study the oxidation behaviour of untreated materials. Halogen treatment was done either by ion implantation at the Rossendorf Research Centre or by liquid phase treatment, i.e. spraying of a fluorine-containing compound (Treatment I) or by gas phase treatment, i.e. an 'out of pack' procedure (Treatment II). Larger samples were exposed in the as-received condition and after fluorine treatment with fluorine applied by a liquid phase process. TiAl turbocharger rotors were used as real components and were treated in the same way as the large samples. Isothermal oxidation tests were performed in a vertical tube furnace under dry and wet synthetic air mostly at 900 °C. The samples were connected to a thermobalance so that the mass change could be detected continuously (thermogravimetric measurements). Thermocyclic oxidation tests were done in an open chamber furnace under laboratory air and synthetic air with 10% water vapour at 900 °C with a 24 h cycle. The samples were kept at temperature for 24 h, cooled to room temperature within 30 min, weighed and returned into the furnace after 1 h. Shorter cycles were applied to larger round specimens in a movable tube furnace at 1050 °C. The hot dwell time was 1 h and the cooling time was 30 min in accordance with the COTEST procedure [8]. Samples were weighed after each 100 cycles (= time at temperature). The TiAl turbocharger rotors were oxidised at 1050 °C in an open chamber furnace. The samples were removed from the furnace after each 100 h and investigated optically. The total quasi-isothermal oxidation time was 1200 h. The creep testing was performed according to ISO 204:1997(E) with cylindrical specimens (Fig. 17.1). The temperatures were 700 and 850 °C and the load was 170 MPa in both cases.

## 17.3    Results and discussion

Untreated TiAl samples displayed a brownish/grey surface after high-temperature oxidation in air while fluorine-implanted samples were grey, and fluorine gas and

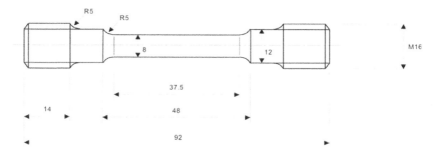

*17.1* Geometry of the creep samples (ISO 204)

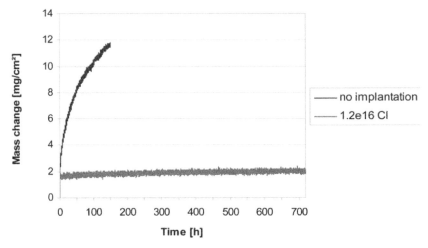

*17.2* Thermogravimetric results for the alloy γ-MET without chlorine treatment and with beamline chlorine implantation (35 keV, 1.2 × 10$^{16}$ Cl cm$^{-2}$) at 900 °C in synthetic air plus 10% H$_2$O (144 and 720 h isothermal oxidation time)

liquid phase-treated samples were white/yellow. The success of the halogen treatment was visible to the naked eye. Thermocyclic exposure led to the spallation of parts of the oxide scale on untreated samples. The spallation of oxide flakes was due to the mismatch of the thermal expansion coefficients of the oxide and the metallic substrate during cooling. Stresses caused by this mismatch were relieved by cracking and finally spallation if the oxide scale grew too thick, which was the case for untreated TiAl-alloys. They formed a mixed oxide/nitride scale (TiO$_2$/TiN/Al$_2$O$_3$) during high-temperature oxidation in air, which was non-protective and fast growing [9]. The oxide scale after halogen treatment was much thinner and consisted of a protective Al$_2$O$_3$ scale [10]. This scale was formed by the selective transport of Al via gaseous halides which were finally oxidised to Al$_2$O$_3$ [11]. The isothermal thermogravimetric results for the industrial alloy γ-MET (Table 17.1), with and without chlorine beam-line implantation at 900 °C in synthetic air with 10% water vapour, are presented in Fig. 17.2. Water vapour in the atmosphere led to enhanced oxidation of TiAl [12] but the halogen effect worked also in wet environments which could be seen in the slow kinetics of the chlorine implanted sample. Thermocycling leads also to failure of untreated alloys [13]. Figure 17.3 shows the mass change behaviour of the alloy γ-MET during a thermocyclic 24 h cycle test at 900 °C, again in synthetic air with 10% H$_2$O. A positive effect was only achieved by fluorine. Chlorine was effective only for the first day; increased mass gain was observed after extended exposure times. However, the treatment with fluorine led to slow mass change behaviour (Al$_2$O$_3$ kinetics) over the whole testing period (50 days = 1200 h). In laboratory air, the 24 h cycle test was extended to 1 year (365 days = 8760 h). The mass change behaviour of a beamline (24 keV, 2 × 10$^{17}$ F cm$^{-2}$) implanted sample, a liquid phase treated sample, and a gas phase treated sample can be seen in Fig. 17.4. The gas phase treated sample is still under investigation. The beamline implanted sample showed a slow mass gain for the first 50 days roughly but after that, mass losses occurred so that the final mass was negative. These mass losses were due to the spallation of oxide flakes at the unprotected edges and corners of the sample, indicating that the beamline technique

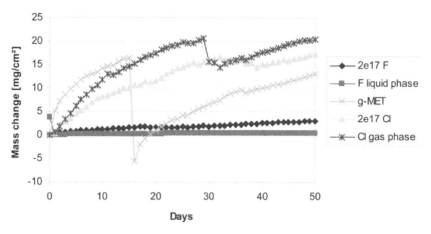

*17.3* Thermocyclic (24 h cycle test) mass change behaviour of the alloy γ-MET without any treatment and with beamline chlorine (45 keV, $2 \times 10^{17}$ Cl cm$^{-2}$) implantation, beamline fluorine implantation (25 keV, $2 \times 10^{17}$ F cm$^{-2}$), chlorine treatment via the gas phase (Treatment II) and liquid phase (Treatment I) fluorine treatment at 900 °C in synthetic air plus 10% H$_2$O for 50 days (= 1200 h)

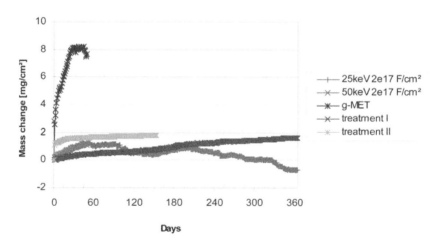

*17.4* Thermocyclic (24 h cycle test) mass change behaviour of the alloy γ-MET without any treatment, with beamline fluorine implantation (25 and 50 keV, $2 \times 10^{17}$ F cm$^{-2}$), liquid phase (Treatment I) and gas phase (Treatment II) F-treatment at 900 °C in laboratory air for up to 365 days (= 8760 h)

is limited to flat sample geometries. The faces were covered with grey alumina but the edges were covered with yellow mixed oxide which partly spalled. The mass gain of the liquid phase treated sample (Treatment I) was steady. This sample showed no spallation and no TiO$_2$ formation. The mass loss after the first day was caused by the decomposition of the fluorine-containing compound. The gas phase treated sample (Treatment II) revealed a higher mass gain after the first day which could be explained by the uptake of fluorine via the gas phase and reaction with the Al of the substrate.

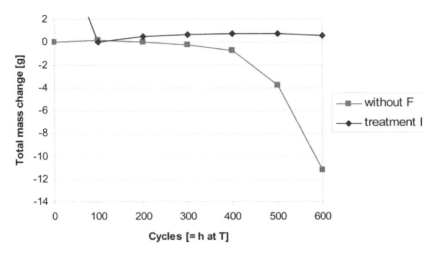

*17.5* Thermocyclic (1 h cycle test) mass change behaviour of near-component samples of the alloy TNB V without and with F-treatment via the liquid phase (Treatment I), at 1050°C in laboratory air (600 cycles = 600 h at temperature)

Samples with a more complex geometry were cut from turbocharger rotors. These round specimens with a varying diameter were exposed thermocyclically at 1050 °C in air. One sample was treated with fluorine via the liquid phase process (Treatment I). The cycle was 1 h hot and 30 min cold. In Fig. 17.5, the mass changes of these samples are shown. The mass loss of the untreated sample was dramatic, obviously the whole oxide scale spalled during each cooling cycle. Figure 17.6a and b show photographs of the samples after 600 cycles (= time at temperature in hours).

To determine the influence of the halogen treatment on the mechanical high-temperature properties, high-temperature creep tests were performed on untreated and treated samples. Samples of the alloys TNB and MoCuSi 1 (Table 17.1) were tested without any treatment at 700 and 850 °C under a constant load of 170 MPa in synthetic air and air plus 10% $H_2O$. The results at 700 °C have been presented

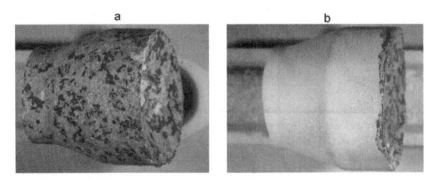

*17.6* a, b: Photographs of the near-component samples after 600 h thermocyclic oxidation (1 h cycle test) at 1050 °C (a) without F, (b) with F on the circumference, Treatment I

**Creep (850°C/laboratory air/170 MPa)**

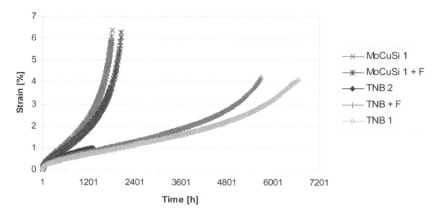

*17.7* Creep curves of the alloys MoCuSi 1 and TNB with and without fluorine at 850 °C in air under a constant load of 170 MPa (TNB 2 failure of the test equipment)

elsewhere [14]. At 850 °C, the samples of the alloy TNB all failed at nearly the same strain of about 4.2% while the alloy MoCuSi 1 failed at a strain of about 6.2% but after shorter times (Figure 17.7). During one test, the equipment failed so the test could not be run to fracture of the sample. The time of failure of MoCuSi 1 was shorter because its oxidation rate was also much higher at this temperature. The metal was consumed by fast oxidation and internal nitridation so that the metallic cross-section decreased. This was shown by plain oxidation tests. Metallographic cross-sections of alloys MoCuSi 1 and TNB after oxidation at 900 °C in air are shown in Fig. 17.8a and b. There was almost no metal left after 90 h of oxidation of a 1 mm thick coupon of MoCuSi 1 while the mixed oxide on the surface of the TNB sample consumed only a little of the metallic substrate. The creep curves of untreated and F-treated samples were almost the same. The F-treatment only affected the surface zone so that the mechanical properties of the bulk material such as the creep behaviour remained unchanged.

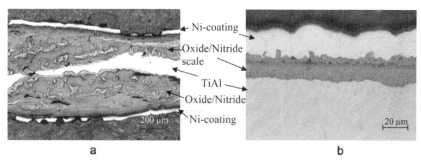

*17.8* a, b: Metallographic cross-sections of the alloy MoCuSi 1 (a) and TNB (b) after oxidation in air at 900 °C for 90 h (MoCuSi 1) and 100 h (TNB)

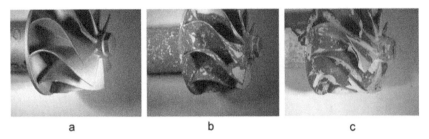

17.9 a–c: Photographs of the untreated TiAl turbocharger rotor before oxidation (a) and after 100 (b) and 1200 (c) hours at 1050 °C in air

Turbocharger rotors made of alloy TNB V (Table 17.1) were tested as examples of real components. One rotor was treated with fluorine via the liquid phase (Treatment I). The rotors were exposed to air in an open furnace at 1050 °C. The samples were removed from the furnace after each 100 h and cooled to room temperature outside the furnace. After optical inspection and documentation, the samples were returned to the furnace. The test was stopped after 1200 h which corresponds roughly to 120 000 km of operation of an automotive engine. The untreated sample can be seen in Fig. 17.9a–c. This sample showed massive spallation of the oxide layer during each cooling period. The thickness of the blades had decreased because the metal was consumed by accelerated oxidation of the spalled surface during each heating step. The fluorine treated sample showed no spallation. The colour of the sample had changed after the first heating but remained the same during the whole test (Fig. 17.10a–c). The thickness of the blades remained the same as before the test.

### 17.4    Conclusions

Compounds made of TiAl alloys can be used at temperatures above approximately 800 °C after surface treatment with fluorine. The oxidation resistance of TiAl alloys can be improved by the fluorine effect in such a way that these alloys withstand thermocyclic exposure in wet environments at 900 °C and thermocyclic exposure in laboratory air up to temperatures of at least 1050 °C up to envisaged lifetimes of several thousand hours. In creep tests, the mechanical high-temperature properties were not affected because the halogen effect is restricted to the subsurface zone.

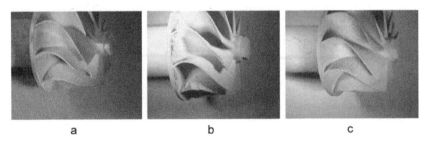

17.10 a–c: Photographs of the F-treated (treatment I) TiAl turbocharger rotor before oxidation (a) and after 100 (b) and 1200 (c) hours at 1050 °C in air

Industrial components can, therefore, be designed for specific applications. The fluorine treatment would be the last step in the processing route.

## Acknowledgements

The authors would like to thank the German Ministry of Economy for funding the work via the German Federation of Industrial Research Associations (AiF) under contract nos. 104 and 176 ZBG. Additionally, the authors would like to thank DaimlerChrysler AG Stuttgart/Germany, GfE GmbH Nuremberg/Germany, PLANSEE AG Reutte/Austria, Rolls Royce Deutschland AG Dahlewitz/Germany and Tital GmbH Bestwig/Germany for providing the alloys and components in the frame of the projects. The implantation work of the Rossendorf Research Centre is also acknowledged.

## References

1. H. Kestler and H. Clemens, in *Titanium and Titanium Alloys*, 351, ed. C. Leyens and M. Peters, Wiley-VCH, Weinheim, 2003.
2. F. Appel and M Oehring, in *Titanium and Titanium Alloys*, 89, ed. C. Leyens and M. Peters, Wiley-VCH, Weinheim, 2003.
3. C. Leyens, in *Titanium and Titanium Alloys*, 187, ed. C. Leyens and M. Peters, Wiley-VCH, Weinheim, 2003.
4. Y. Shida and H. Anada, *Oxid. Met.*, 45/1–2 (1996), 197.
5. M. Hald and M. Schütze, *Mater. Sci. Eng. A*, 240 (1997), 847.
6. M. Schütze, G. Schumacher, F. Dettenwanger, U. Hornauer, E. Richter, E. Wieser and W. Möller, *Corros. Sci.*, 44 (2002). 303.
7. A. Rahmel, W. J. Quadakkers and M. Schütze, *Mater. Corros.*, 46 (1995), 271.
8. M. Malessa and M. Schütze, *COTEST – Code of Practice, Test Method for Thermal Cycling Oxidation Testing*, Final Version (V5), 2005.
9. F. Dettenwanger, E. Schumann, J. Rakowski, G. H. Meier and M. Rühle, *Mater. Sci. Forum*, 251–254 (1997), 211.
10. G. Schumacher, F. Dettenwanger, M. Schütze, U. Hornauer, E. Richter, E. Wieser and W. Möller, *Intermetallics*, 7 (1999), 1113.
11. A. Donchev, B. Gleeson and M. Schütze, *Intermetallics*, 11 (2003), 387.
12. A. Zeller, F. Dettenwanger and M. Schütze, *Intermetallics*, 10 (2001), 59.
13. T. Shimizu, T. Ikubo and S. Isobe, *Mater. Sci. Eng. A*, 153 (1992), 111.
14. A. Donchev, E. Richter, M. Schütze and R. Yankov, *Intermetallics*, 14 (2006), 1168.

# 18

# Effect of surface preparation on the durability of NiCoCrAlY coatings for oxidation protection and bond coats for thermal barrier coatings*

E. M. Meier Jackson, N. M. Yanar, M. C. Maris-Jakubowski[1], K. Onal-Hance, G. H. Meier and F. S. Pettit

*Department of Materials Science & Engineering, University of Pittsburgh, Pittsburgh, PA 15261, USA*

*[1]Currently with Siemens Power Corp., Orlando, FL, USA*

## 18.1    Introduction

Current state-of-the-art thermal barrier coating (TBC) systems consist of a nickel-base superalloy onto which is deposited an aluminium rich bond coat (BC). The BC is either a diffusion aluminide coating or an overlay coating, usually consisting of the elements Ni, Co, Cr, Al, and Y. The NiCoCrAlY coatings are deposited either by electron beam physical vapour deposition (EBPVD) or plasma spraying (vacuum, argon-shrouded, etc.). A thermally grown oxide (TGO), which ideally consists of α-alumina, forms on the bond coat during deposition of an yttria-stabilised zirconia (YSZ) TBC, which is fabricated either by EBPVD or air plasma spraying (APS). The TGO serves both as a bonding layer for the TBC and as the oxidation barrier to protect the underlying metallic layers.

The effect of various BC surface treatments on the lifetimes of NiCoCrAlY bond coats, containing 0.5 wt.% Y, under an EBPVD TBC during cyclic oxidation has recently been studied [1]. A range of behaviours was observed based on the type of surface treatment. Those treatments which resulted in significant improvements in life (particularly deposition of Pt overlayers or fine polishing) were proposed to decrease or eliminate the effect of 'defects' which exist in state-of-the-art systems, particularly near the TGO. Defects have previously been proposed to decrease the life of NiCoCrAlY bond coats, containing 0.3 wt.% Y, under EBPVD TBCs [2]. Some-what contrary to the results in Ref. 1, an earlier study [3] found that fine polishing produced only limited increases in coating life for a NiCoCrAlY coating containing approximately 0.1 wt.% Y.

The current study was directed at ascertaining how Y-content and surface preparation affect the oxidation of NiCoCrAlY coatings.

* Reprinted from E. M. Meier Jackson et al.: Effect of surface preparation on the durability of NiCoCrAlY coatings for oxidation protection and bond coats for thermal barrier coatings. *Materials and Corrosion*. 2008. Volume 59. pp. 494–500. Copyright Wiley-VCH Verlag GmbH & Co. KGaA.

*Table 18.1*  Alloy and coating composition (wt%)

| Alloy | Ni | Cr | Al | Co | Ta | W | Mo | Ti | Hf | Zr | Y | Re |
|---|---|---|---|---|---|---|---|---|---|---|---|---|
| N5 | Bal. | 7 | 6.2 | 7.5 | 6.5 | 6 | 0.6 | 1 | 0.1 | | | 3 |
| NiCoCrAlY | Bal. | 16.5 | 12.6 | 21.8 | | | | | | | 0.1 | |
| NiCoCrAlY | Bal. | 16.5 | 12.6 | 21.8 | | | | | | | 0.5 | |

## 18.2    Experimental

### 18.2.1    Alloys and coatings

The alloy substrate used in this investigation was the Ni-based superalloy René N5, the composition of which is listed in Table 18.1. The superalloy was a cast single crystal with a gamma matrix and gamma prime precipitates.

Two bond coats were studied. Both were NiCoCrAlY overlay coatings deposited by argon-shrouded plasma spraying. The nominal compositions were identical (Table 18.1) except that one coating contained 0.1 wt.% Y, the other 0.5 wt.% Y. These coatings were exposed with a variety of surface preparations including media finishing (tumbling the specimens in a bowl containing alumina pellets and water), light grit blasting (30 min with 700 µm alumina at a pressure of 0.17 MPa), light grit blasting followed by a NaOH etch to remove embedded grit, and hand polishing to a 3 µm diamond finish. The surfaces of coatings following these treatments are shown in Fig. 18.1. All specimens were ultrasonically cleaned in soapy water, and then ultrasonically cleaned in acetone.

TBC specimens had EBPVD thermal barrier coatings deposited onto coupons consisting of the N5 substrate overlaid with the 0.5 wt.% Y NiCoCrAlY coating. The thickness of the yttria-stabilised zirconia TBC was approximately 100 µm. Before

**Media Finish    R$_a$≈2-5µm**          **Hand Polished    R$_a$≈0.3µm**

**Lightly Grit Blasted    R$_a$≈2µm**    **Lightly Grit Blasted & Etched  R$_a$≈2µm**

*18.1* Surfaces of NiCoCrAlY coatings with various surface conditions deposition of the TBC, bond coats were subjected to a variety of surface preparations including heavy grit blasting (with 2 mm alumina at a pressure of 0.5 MPa), media finishing, and hand polishing to a 3 μm diamond finish. Selected specimens had the bond coat aluminised or platinum coated before TBC deposition.

### 18.2.2   Cyclic oxidation exposures

Cyclic oxidation tests of specimens with and without TBCs were performed in a bottom-loading furnace in laboratory air. The thermal cycle used was 10 min heating to 1100 °C, holding at 1100 °C for 45 min, and cooling to approximately 100 °C in 10 min.

### 18.2.3   Microscopy

After high-temperature exposure of the coatings, electron and optical microscopy were used to describe qualitatively the performance of the materials. The scanning electron microscope (SEM) used in this investigation was a Philips XL30 FEG (field emission gun) instrument. This SEM is equipped with secondary electron and backscattered electron detectors for imaging along with an X-ray detector for chemical analysis by energy dispersive spectroscopy (EDS).

### 18.3   Results and discussion

### 18.3.1   TBC degradation

The effect of bond coat surface finish on TBC life for systems with NiCoCrAlY bond coats has been discussed in detail in a previous paper [1]. Several of the key results will be reviewed here to provide a context for the present work. Figure 18.2 presents average spallation lifetimes for various surface preparations ('state-of-the-art' refers to bond coats that were given a heavy grit blasting before TBC deposition). All of the treatments resulted in some increase in coating life but the use of Pt-overlays and hand polishing produced the largest effects.

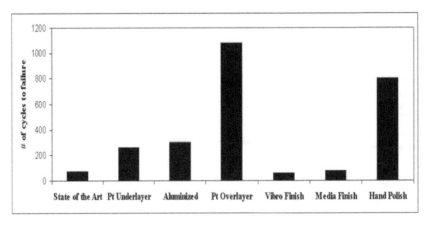

*18.2* Average failure times for TBC systems with modified NiCoCrAlY bond coats exposed to cyclic oxidation at 1100 °C

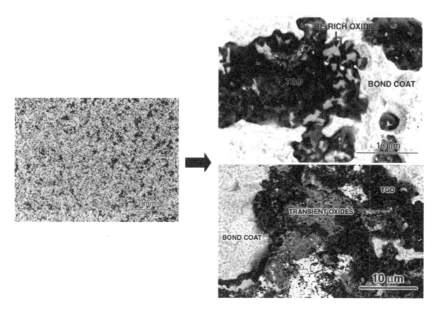

*18.3* Fracture surfaces of TBC systems with NiCoCrAlY bond coats (heavily grit blasted) after failure in cyclic oxidation at 1100 °C

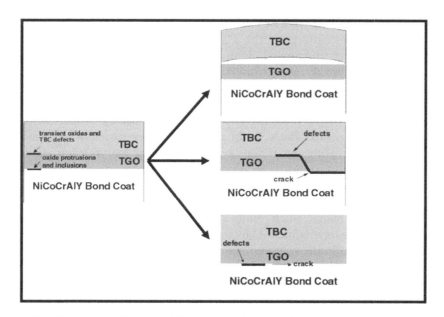

*18.4* Schematic diagram of the cyclic oxidation failure process for TBC systems with NiCoCrAlY bond coats (heavily grit blasted)

The effect of the Pt-overlays was essentially one of 'burying' coating defects that had been found to initiate coating failure in state-of-the-art systems. Examples of several of these defects are presented in Fig. 18.3. These included transient Ni-rich oxides and concentrations of oxides of reactive elements, e.g. Y, which were present in depressions in the bond coat. Figure 18.4 shows a schematic diagram of how the

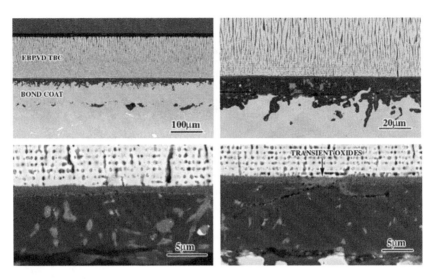

*18.5* Cross-sections of a TBC system with NiCoCrAlY bond coat (hand polished) that had not failed after 1520 cycles at 1100 °C

systems with NiCoCrAlY bond coats failed. In some instances, there were gaps between the TBC and TGO resulting from the deposition process. In these cases, fracture typically followed the TBC/TGO interface. In the majority of cases, defects at the upper or lower surfaces of the TGO initiated cracks which, in the former, travelled down to and propagated along the TGO/BC interface. In the latter, the cracks propagated directly along the TGO/BC interface. In the presence of the Pt-overlayer, the TGO formed on the top of a Pt–Al layer formed by interdiffusion and was thus separated from the defects.

The mechanisms for the dramatic effects of hand polishing in increasing TBC lifetimes were less obvious. The smoothness of the bond coat might be expected to result in fewer pre-existing separations between the TBC and TGO as shadowing during TBC deposition would be reduced. Figure 18.5 presents cross-sections of a specimen with a hand polished bond coat which survived 1520 cycles without failure. As can be seen, there are very few defects along the TBC–TGO interface and this interface is very planar. Such results indicate that the development of TGOs on NiCoCrAlY bond coats is influenced by the condition of the bond coat surface. The TGO contains a significant volume fraction of Y-rich oxides, some of which have resulted in oxide fingers growing to a considerable depth in the alloy.

The present study was undertaken, in part, to understand how the oxidation of NiCoCrAlY coatings is influenced by surface preparation. Since Y oxidation is clearly evident in the oxidation of these coatings, Y concentration was included as a variable.

### 18.3.2    Bond coats without a TBC

Figure 18.6 presents the cross-section of an as-processed NiCoCrAlY coating. The micrograph to the left shows alumina inclusions (dark particles) formed during coating deposition. The higher magnification micrograph to the right shows the presence of Ni-yttrides at the $\gamma/\beta$ phase boundaries. Figure 18.7 presents cross-sections

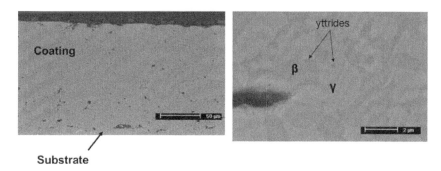

*18.6* Cross-sections of an as-processed NiCoCrAlY coating on René N5 showing oxide inclusions (dark particles) and the presence of Ni-yttrides at the $\gamma/\beta$ phase boundaries

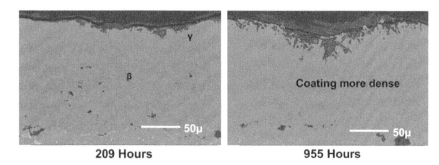

*18.7* Cross-section of a heavily grit blasted NiCoCrAlY coating after 209 and 955 cycles at 1100 °C. The formation of Y-rich oxides in the TGO increases with exposure time and the density of inclusions in the coating decreases as the Al is depleted

of a heavily grit blasted NiCoCrAlY coating after 209 and 955 cycles at 1100 °C. The formation of Y-rich oxides in the TGO increases with exposure time. Comparison of Figs. 18.6 and 18.7 indicates that the density of alumina inclusions in the coating decreases as the Al is depleted and there are fewer voids visible in the coating. Figure 18.8 presents cross-sections of a heavily grit blasted NiCoCrAlY coating

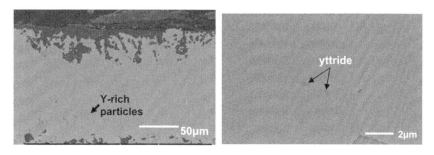

*18.8* Cross-section of a heavily grit blasted NiCoCrAlY coating after 3031 cycles at 1100 °C. The high magnification micrograph to the right indicates the continued presence of yttrides in the bulk of the coating

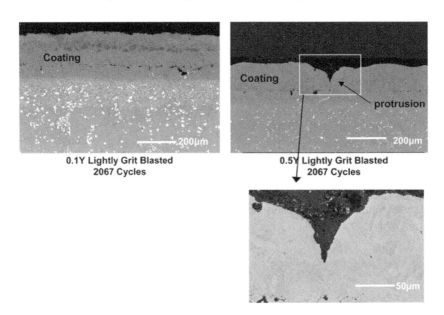

*18.9* Cross-section of lightly grit blasted NiCoCrAlY coatings after 2067 cycles at 1100 °C. The coating to the left contained 0.1 wt.% Y and that to the right contained 0.5 wt.% Y

after 3031 cycles at 1100 °C. The extent of yttrium-rich protrusions growing into the coating has increased. Nevertheless, the high magnification micrograph to the right indicates the continued presence of yttrides in the bulk of the coating. There are also considerable numbers of yttrides in the coating just above the coating/substrate interface. The yttrides are more apparent after exposure than in the as-processed coating. This observation is not understood but is believed to be the result of coarsening of yttrides that are too small to be observed in the as-processed coating.

Figure 18.9 presents a comparison of lightly grit blasted coatings containing 0.1 and 0.5 wt.% Y after more than 2000 cycles at 1100 °C. The coating/oxide interface for the coating with 0.1 wt.% Y has remained relatively planar and a significant amount of the Al-rich β-phase (dark grey phase) is still evident in the coating. The coating with 0.5 wt.% Y has formed deep oxide protrusions into the coating. Some β-phase was still present in this coating. Specimens exposed with media finished surfaces oxidised in the same manner as the lightly grit blasted specimens. It should be pointed out that the 0.5 wt.% Y coatings with either lightly grit blasted or media finished surfaces were observed to develop a network of blue coloured oxide on their surfaces after only 200 cycles, which corresponded to the protrusions observed, whereas the 0.1 wt.% Y coating maintained a uniform grey surface even after nearly 3700 cycles.

Figure 18.10 presents a cross-section of the lightly grit blasted coating with 0.5 wt.% Y after almost 3700 cycles. The oxide protrusions have penetrated completely through the coating and into the interdiffusion zone. Figure 18.11 presents higher magnification micrographs of one of these protrusions. The protrusion is filled with layers of alumina which are separated by cracks. In addition, the protrusion has formed 'sidearms' associated with alumina forming around Y-rich oxide particles. The source

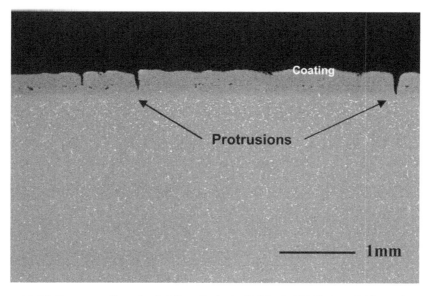

*18.10* Cross-section of a lightly grit blasted NiCoCrAlY (0.5 wt.% Y) coating after 3674 cycles at 1100 °C

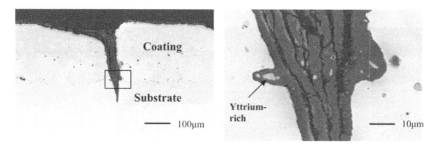

*18.11* Higher magnification micrographs of the coating in Fig. 18.10. The micrograph on the right is the area in the box on the left and shows more detail of the growth of the sidearms

of the yttrium for the 'sidearms' appears to be yttrides that are concentrated near the coating–substrate interface. The β-phase had been completely consumed from this coating.

Figure 18.12 presents a cross-section of the hand polished coating with 0.5 wt.% Y also after almost 3700 cycles. This coating contains no indication of the formation of deep protrusions. This coating did not develop the blue surface oxide observed on the more roughly polished surfaces. Figure 18.13 presents a higher magnification micrograph of the coating/oxide interface for this coating showing one of the few deviations from planarity. The alumina in this depression contains some Y-rich oxide but the penetration into the coating is only of the order of 20 μm.

Figure 18.14 presents an etched cross-section of a hand polished NiCoCrAlY (0.1 wt.% Y) coating after the same number of cycles at 1100 °C. The coating/oxide interface is still planar and β-phase is still present in the coating. The NiCoCrAlY (0.1 wt.% Y) coatings had not undergone significant degradation even after 5000

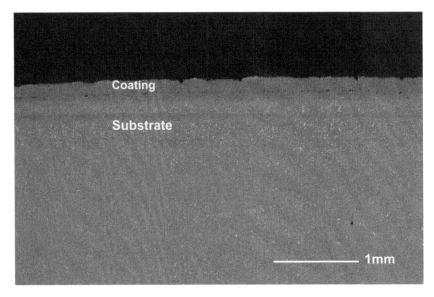

*18.12* Cross-section of a hand polished NiCoCrAlY (0.5 wt.% Y) coating after 3674 cycles at 1100 °C

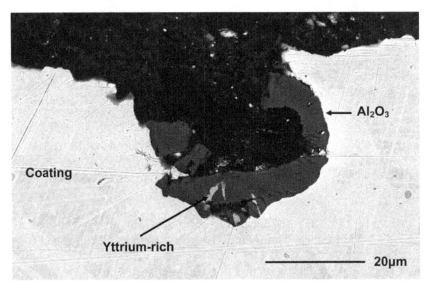

*18.13* Higher magnification view of the coating/oxide interface for the coating in Fig. 18.12

cycles. Figure 18.15 shows the surface of a coating after 5000 cycles. The surface is covered with alumina and some spinel. No oxide protrusions are evident in the cross-section. The β-phase had been completely depleted from this coating but islands of γ′ still remained in the centre of the coating, which otherwise consisted of the γ-phase.

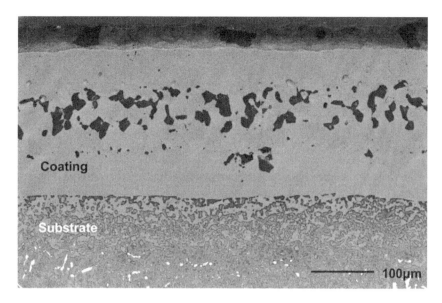

*18.14* Cross-section of a hand polished NiCoCrAlY (0.1 wt.% Y) coating after 3674 cycles at 1100 °C (the coating has been etched)

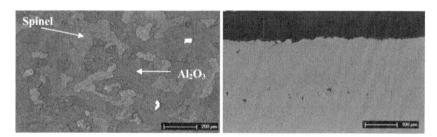

*18.15* Surface (left) and etched cross-section (right) of a NiCoCrAlY coating with 0.1 wt.% yttrium after 5000 cycles at 1100 °C

Figure 18.16 presents a cross-section of a lightly grit-blasted NiCoCrAlY (0.5 wt.% Y) coating after 3674 cycles at 1100 °C. This micrograph shows Ni- and Co-rich oxides forming above the oxide protrusions which cause the network of blue oxide to be apparent on the coating surfaces.

### 18.3.3   Mechanism of formation of oxide protrusions

The above data show two clear results:

1.  The formation of deep oxide protrusions occurs in the high Y-content coatings with rough surfaces but does not occur in the low Y-content coatings for any surface preparation
2.  Fine polishing of the high Y-content coating suppresses the formation of deep oxide protrusions.

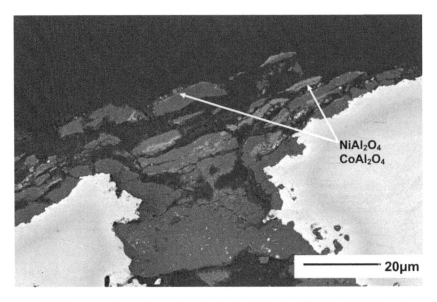

NiAl$_2$O$_4$
CoAl$_2$O$_4$

20µm

*18.16* Cross-section of a lightly grit-blasted NiCoCrAlY (0.5 wt.% Y) coating after 3674 cycles at 1100 °C showing Ni- and Co-rich oxides forming above the oxide protrusions

The first result indicates that the amount of Y in the coating that is available to oxidise near the surface scale is an important factor. Figures 18.17 and 18.18 show observations made of the oxidation of a cast alloy with a very high Y-content (3 wt.% Y). Figure 18.17(a) shows the as-cast microstructure of the alloy with Ni-yttrides outlining the grain boundaries. Figure 18.17(b) shows the surface of a specimen after oxidation for 30 min in air at 1200 °C. A Ni-rich oxide has formed over the grain boundary network of yttrides. Figures 18.17(c) and (d) are cross-sections after 1 and 16 h of oxidation, respectively, showing preferential oxidation of the yttrides. After 16 h, the oxidation has penetrated more than 100 µm into the alloy. Figure 18.18 shows more detail of one of the oxidised yttrides. There is a Ni-rich cap that has formed on the surface, consistent with Fig. 18.16(b). This type of oxidation of Ni (as well as of Co) is believed to contribute to the Ni- and Co-rich oxides observed in Fig. 18.16 and the blue colour observed on the surfaces of specimens that developed protrusions. The X-ray maps in Fig. 18.18 indicate that the oxide protrusion is a Y-rich oxide enveloped by alumina. The Cr map indicates that there is a gap between the receding yttride and advancing oxide (area marked with an arrow in Fig. 18.18a). Apparently, the yttride is dissolving and the yttrium is diffusing to the tip of the oxide protrusion which keeps growing even though encapsulated by alumina. This is also seen in Fig. 18.17d. These results indicate that the NiCoCrAlY coating with 0.1 wt.% Y does not have a large enough volume fraction of yttrides to supply yttrium for the growth of large protrusions.

The next point requiring explanation is how fine polishing suppresses the formation of protrusions in the coating with 0.5 wt.% Y. Recent studies by Gil *et al.* [4] have shown that Y-rich oxide protrusions form preferentially in the concave regions of rough NiCoCrAlY coatings upon initial exposure to oxidising conditions. They found that the flux of Y to the concave regions is greater both because of the larger

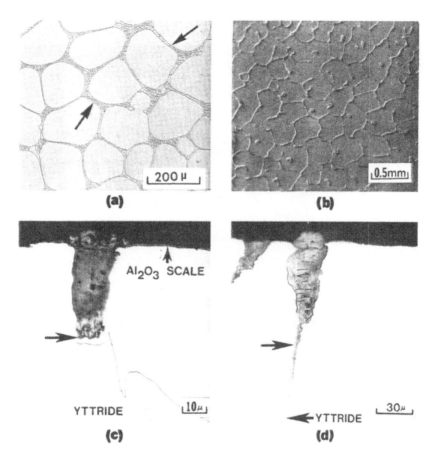

*18.17* Features developed on a Ni–15Cr–6Al–3Y alloy (plane-front solidified) after oxidation at 1200 °C in 1 atm of air. (a) Microstructure of alloy before oxidation (arrows indicate Ni–Al yttrides); (b) surface after 30 min of oxidation; (c) and (d) transverse sections through specimens after 1 h and 16 h of oxidation, respectively, showing features of preferential yttride oxidation. The arrow in (c) shows that the yttride is not in contact with the oxidation product

surface area of that region and because the diffusion distance from the yttrium reservoir is shorter to the concave regions than to the convex regions. The results of the current study are consistent with this model. The finely polished surface does not have concave regions of a size to cause significant preferential oxidation of yttrium and the yttrium-rich oxides form uniformly along the specimen surface.

The final point is the explanation of how the oxide protrusions propagate through the coating. Figure 18.17(d) indicates that the protrusions can grow to depths comparable to those observed for the NiCoCrAlY coatings (Fig. 18.10) simply by preferential oxidation if the supply of yttrium is sufficient. (The temperature for Fig. 18.17 was 1200 °C, whereas the coatings were exposed at 1100 °C, but the times at which the protrusions were examined in the coatings were much longer.) Furthermore, accelerated oxygen diffusion along Ni/Al$_2$O$_3$ interfaces has been observed in the internal oxidation of Ni–Al alloys [5].

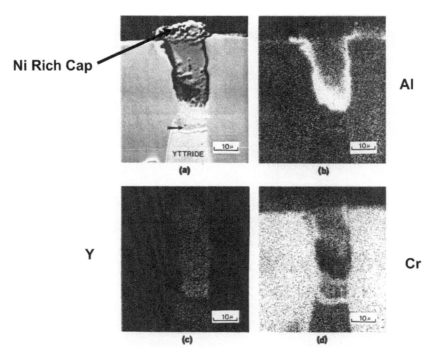

*18.18* Electron backscatter photomicrograph and X-ray maps of a protrusion in the alloy in Fig. 18.17 showing features of preferential yttride oxidation

On the other hand, the spacing and morphology of the protrusions in the coatings (e.g. Figs. 18.10 and 18.11) raise the possibility of cracking as a propagation mechanism. Features similar to those in Figs. 18.10 and 18.11 have been presented for a CoNiCrAlY coating on a directionally-solidified superalloy substrate after long-term cycling at 1010 °C [6]. The features were described as 'thermal fatigue cracks' but their origin was not described. Cracking has also been suggested to be the cause of rapid internal oxidation and nitridation of bulk Ni–Cr–Al alloys [7,8]. The linear coefficient of thermal expansion (CTE) of NiCoCrAlY coatings varies with composition [9] and for the composition studied here has an average value of approximately $17 \times 10^{-6}$ K$^{-1}$ and that for the N5 substrate is reported as approximately $14 \times 10^{-6}$ K$^{-1}$ [10]. The average CTE for α-alumina is approximately $9 \times 10^{-6}$ K$^{-1}$ [10] over the same temperature range. Therefore, on cooling, an alumina protrusion would cause a tensile stress in the surrounding coating. If this stress resulted in cracking of the coating during thermal cycling, the protrusion could possibly propagate by oxidation at the crack tip. Examination of the tips of the protrusions has not revealed cracking in the coating. It is possible that, if cracking occurred in the coating, it did not occur on every cycle and the specimens were not sectioned at a time when cracks were evident. Cracks were evident in the oxides within the protrusions (Figs. 18.9 and 18.11). Based on the relative values of CTE, these presumably formed in the heating portion of the thermal cycles.

The data presently available do not allow a definitive propagation mechanism to be proposed. Further experiments involving shorter exposure times and the comparison of isothermal and cyclic exposures are planned to resolve the mechanism by which the protrusions propagate.

## 18.4    Summary and conclusions

The cyclic oxidation behaviour of a NiCoCrAlY coating on a single crystal superalloy substrate has been found to be affected by yttrium content and surface preparation. In particular:

- The formation of deep oxide protrusions occurs in the high Y-content coatings with rough surfaces but does not occur in the low Y-content coatings for any surface preparation
- Fine polishing of the high Y-content coating suppresses the formation of deep oxide protrusions.

It is proposed that the protrusions are initiated by preferential oxidation of yttrium in concave regions of the coating with high Y-content and that fine polishing removes these initiation sites.

The determination of the mechanism by which the protrusions propagate will require further experimentation.

## Acknowledgments

Financial support of this research by ONR (Contract No. N00014-03-0030) and specimen preparation by Praxair Surface Technologies, Inc. are gratefully acknowledged.

## References

1. N. M. Yanar, F. S. Pettit and G. H. Meier, *Metall. Mater. Trans. A*, 37A (2006), 1563.
2. D. R. Mumm and A. G. Evans, *Acta Mater.*, 48 (2000), 1815.
3. U. Schulz and K. Fritscher, 'Behaviour of sub-surface modified EB-PVD processed thermal barrier coatings on cyclic tests', 163–172, Ceramic Coatings, MD-Vol. 44, ASME, 1993.
4. A. Gil, V. Shemet, R. Vassen, M. Subanovic, J. Toscano, D. Naumenko, L. Singheiser and W. J. Quadakkers, *Surf. Coat. Technol.*, 201 (2006), 3824.
5. F. H. Stott and G. C. Wood, *Mater. Sci. Technol.*, 4 (1988), 1072.
6. K. S. Chan and N. S. Cheruvu, 'Degradation mechanism characterisation and remaining life prediction for NiCoCrAlY coatings', in Proc. ASME Turbo Expo 2004, 14–17 June 2004, Paper GT2004-53383.
7. W. Cheung, B. Gleeson and D. J. Young, *Corros. Sci.*, 35 (1993), 923.
8. S. Han and D. J. Young, *Oxid. Met.*, 55 (2001), 225.
9. T. A. Taylor and P. N. Walsh, *Surf. Coat. Technol.*, 177–178 (2004), 24.
10. J. A. Haynes, B. A. Pint, W. D. Porter and I. G. Wright, *Mater. High Temp.*, 21 (2004), 87.

# 19

# Parameters affecting the growth rate of thermally grown oxides and the lifetime of thermal barrier coating systems with MCrAlY bond coats*

## Juan Toscano, Dmitry Naumenko, Lorenz Singheiser and W. Joe Quadakkers

*Forschungszentrum Juelich GmbH, IEF 2, D-52425, Juelich, Germany*
*j.toscano@fz-juelich.de*

## Aleksander Gil

*AGH University of Science and Technology, Faculty of Materials Science and Ceramics, Kraków, Poland*

## 19.1 Introduction

MCrAlY-type coatings (M = Ni,Co) have commonly been used as overlay coatings and bond coats for Thermal Barrier Coatings (TBCs) to protect the components in industrial gas turbines against high-temperature oxidation and corrosion. During exposure to the gas at high temperature, an oxide scale will form on the MCrAlY surface, becoming part of the original system composed of the superalloy, bond coat and TBC. This Thermally Grown Oxide (TGO), mainly consisting of alumina, that builds up between the TBC and the bond coat, often represents the starting point for failure which ends up with TBC delamination [1–4]. The growth rate and adhesion of the oxide scale and the aluminium depletion in the bond coat [5,6] are, therefore, among the determining factors for the lifetime of a TBC and they have been the subject of extensive studies in recent years. Controlled and predictable oxidation of the MCrAlY coating, which also involves knowledge of the TGO characteristics at failure, is certainly to be desired in such a coating system in connection with the design and modelling of turbine blades. Several authors [7–9] have suggested that a critical oxide scale thickness to TBC failure (or alternatively, a critical thickness for the onset of oxide spallation) can be defined for TBCs produced by electron-beam physical vapour deposition (EB-PVD). Various values of the scale thickness to failure have been reported in the literature for different EB-PVD TBC systems [10]. Moreover, indications have been found [11,12] that, for a given system, the critical oxide scale thickness might depend on the pre-treatment procedures of the MCrAlY bond coat before deposition of the thermal barrier coating.

---

\* Reprinted from J. Toscano et al.: Parameters affecting TGO growth rate and the lifetime of TBC systems with MCrAlY bondcoats. *Materials and Corrosion.* 2008. Volume 59. pp. 501–7. Copyright Wiley-VCH Verlag GmbH & Co. KGaA.

If the assumption of a critical TGO thickness for TBC failure is correct, then variation of the TGO growth rate should result in variation of the time to TBC failure. Therefore, in the present paper, the effects of several parameters on the oxide growth rates on MCrAlYs have been studied. These parameters are:

- Exposure temperature
- Surface roughness of the MCrAlY coating
- Yttrium reservoir of the MCrAlY coating
- Oxygen partial pressure in the oxidising atmosphere.

Specifically in this study, an attempt has been made to establish experimental parameters, which can influence the growth rate without significantly affecting the composition and morphology of the scale. It is well known that an increase in exposure temperature will cause faster scale growth rate and, in the present study, the particular parameter studied with respect to the temperature change has been the yttrium distribution in the coating and its effect on scale growth. Regarding the second parameter, the surface roughness of the bond coat has been shown to affect scale adherence during cyclic oxidation [10,13]; however, the effect of roughness on the oxide scale morphology and microstructure has not been extensively studied.

Rather than yttrium concentration, the yttrium reservoir has been considered in this study. This is defined as the effective amount of yttrium available for incorporation in the TGO. More precisely, the yttrium reservoir is determined by the yttrium concentration and the volume to surface ratio of the coating (to a first approximation, coating thickness) [14]. The reason for including this parameter was the observation of a relatively large scatter of the bond coat thicknesses both in test specimens and in turbine blades after service.

Finally, variation of the oxygen partial pressure of the atmosphere was used to influence the oxidation rate based on the results of previous studies by the present authors with FeCrAl-base, alumina-forming alloys [15] and on observations by other authors [16].

## 19.2    Experimental

Three different types of specimens were prepared for the experiments, having MCrAlYs (M = Ni,Co) as bond coats with compositions typical of those used in gas turbine applications; their yttrium content was 0.5 wt.%. The first specimen type (Type 1) consisted of a 3.5 mm thick IN738LC plate coated by vacuum plasma spraying (VPS) with a 200 μm layer of a CoNiCrAlY coating; the specimens were subsequently subjected to a conventional two-stage vacuum annealing (2 h at 1100 °C and 24 h at 850 °C). Following annealing, the rough bond coat surfaces were smoothed using a cut wire peening technique and subsequently were coated with a 300 μm thick ceramic top layer of Yttria Stabilised Zirconia (YSZ) thermal barrier coating using EB-PVD. Finally, 20 mm × 20 mm specimens were cut from the plate with a low speed saw.

The second type of specimen (Type 2) was free-standing CoNiCrAlY coatings. To produce them, a 3 mm thick CoNiCrAlY layer was sprayed by VPS onto a steel plate, and then removed from the substrate by spark erosion. The eroded plate was cut into 20 mm × 10 mm specimens which were ground down to 2 mm thickness on the bottom side to remove contamination that originated from the erosion process; the surface finish was obtained by abrasion with 1200 grit SiC, so that the specimens ended up with a rough, as-sprayed surface on the top and a ground surface on the

*Table 19.1* Description of tests and specimen types

| Test | Specimen type | Atmosphere | Exposure temp. | Exposure time(s) |
|---|---|---|---|---|
| Temperature variation (TG) | 2 (2 mm thick) | Lab. air | 1000 °C; 1100 °C | 72 h |
| Roughness variation | 2 (2 mm thick) | Lab. air | 1100 °C | 16 h, 25 h, 111 h, 406 h |
| Y reservoir variation | 1, 2 | Lab. air | 1100 °C | 16 h, 111 h, 406 h |
| Oxygen partial pressure variation | 3 | Lab. air ($pO_2 = 0.2$ bar) Ar–4% $H_2$–2% $H_2O$ ($pO_2 = 10^{-14}$ bar) | 1100 °C Cyclic (4 h heat/1 h cool) | 100 h, 300 h, TBC failure |

Specimen type 1 = flat IN738/CoNiCrAlY/TBC.
Specimen type 2 = flat free-standing CoNiCrAlY.
Specimen type 3 = cylindrical IN738/NiCoCrAlY/TBC.

bottom. Finally, the specimens were subjected to the same vacuum heat treatment as described above.

The third type of specimen (Type 3) consisted of cylindrical rods of IN738LC with 10 mm diameter which were coated with a 200 µm NiCoCrAlY bond coat by means of VPS (as described for specimens of Type 1). After the respective heat treatment and surface smoothing with the cut wire peening technique, a 300 µm YSZ layer was deposited by EB-PVD.

Various tests were carried out with the three different specimen types, focusing on one parameter affecting the TGO growth in each test, as presented in Table 19.1. In addition to thermogravimetric measurements, the specimens were analysed using optical metallography, SEM/EDX and cathodoluminescence.

## 19.3    Results and discussion

### 19.3.1    Influence of temperature on TGO growth

The mass change ($\Delta m$) data from thermogravimetric (TG) experiments with the 2 mm thick, free-standing CoNiCrAlY specimens (Type 2) during oxidation in air at 1000 °C and 1100 °C for 72 h are presented in Fig. 19.1a. It can be seen, as would be expected from the Arrhenius temperature dependence of the oxidation rate, that the mass change is higher at 1100 °C than at 1000 °C. The instantaneous oxidation rate, $k_P(t) = d(\Delta m^2)/dt$, was calculated from the TG data by local linear fitting of the data set $\Delta m^2 = f(t)$, with each $k_P(t)$ point derived from a time interval of approximately 14 min (Fig. 19.1b). The plot in Fig. 19.1b indicates that, at 1000 °C, the oxidation rate exhibits non-monotonic behaviour, with a kink at approximately 10 h exposure, where the $k_P(t)$ value is higher than that at 1100 °C. This effect is attributable to the formation of metastable theta alumina at 1000 °C whereas the decrease in $k_P(t)$ at longer times is related to the transformation of metastable theta into alpha alumina as shown in Ref. 17. In cross-sections of the oxidised samples, it was found that the oxide grown on the air-exposed specimen at 1000 °C was 2.1 µm thick (Fig. 19.2a), whereas that grown on the specimen exposed at 1100 °C was 3.8 µm thick (Fig. 19.2b), in agreement with the above TG data. However, the change in temperature not

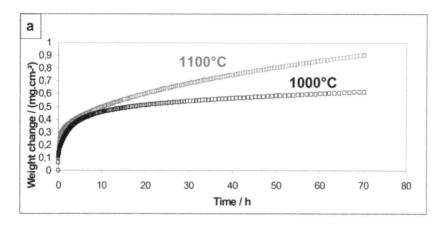

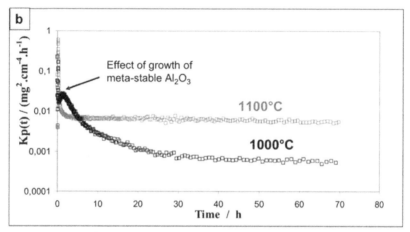

*19.1* (a) Mass change data for free-standing, 2 mm thick CoNiCrAlY coatings during thermogravimetric experiments in air at 1000 °C and 1100 °C for up to 72 h. (b) Instantaneous $k_P$ from thermogravimetric data for free-standing, 2 mm-thick CoNiCrAlY specimens during isothermal oxidation in air at 1000 °C and 1100 °C

only altered the oxidation rate of the free-standing coatings, but also the oxide composition and morphology. After oxidation at 1000 °C, yttrium was found to be present within the alumina scale and in the sub-scale regions as tiny, 0.5 to 1 μm Y/Al mixed oxide precipitates. In contrast, at 1100 °C, much more yttrium was present in the scale and sub-scale regions in the form of coarse yttrium aluminates up to 3 μm in size (compare Figs. 19.2a and 19.2b). Therefore, it can be argued that the higher oxidation rate at 1100 °C compared to that at 1000 °C is to a large extent related to more extensive incorporation of yttrium into the scale and internal oxidation of yttrium at the former temperature, as has been shown by other authors [18]. These observations indicate that the oxidation of a MCrAlY coating at different temperatures leads to the growth of oxide scales which are significantly different with respect to morphology and composition. Consequently, the TGO critical thickness for spallation is likely to be different at 1100 °C compared to that at 1000 °C and the time to scale failure cannot be extrapolated from one temperature to another simply by considering the Arrhenius temperature dependence of the oxidation rate.

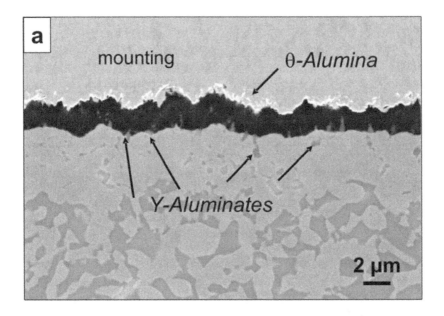

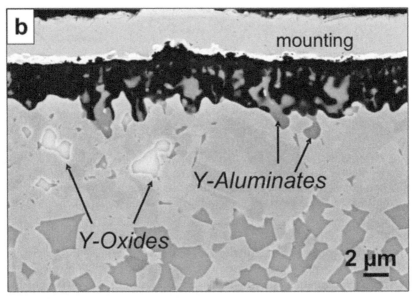

*19.2* SEM micrographs of free-standing 2 mm-thick CoNiCrAlY specimens exposed isothermally for 72 h in air at (a) 1000 °C; (b) 1100 °C

### 19.3.2   Influence of bond coat surface roughness on TGO growth

On the rough side of the free-standing, 2 mm thick CoNiCrAlY specimens (Type 2), inhomogeneous TGO growth was observed. As can be seen in Fig. 19.3a, after 25 h oxidation at 1100 °C, the (internal) yttrium aluminates and oxides are mainly concentrated in the concave parts of the coating surface, whereas in the convex parts of the coating, spinel formation can be observed. The yttrium-rich oxides could be

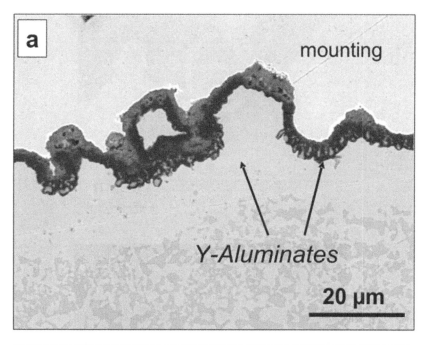

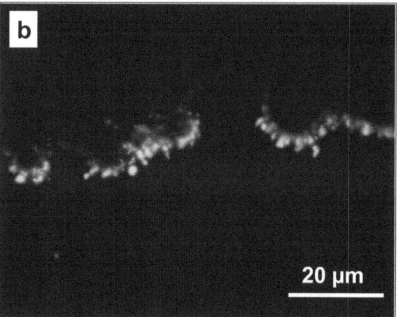

*19.3* Free-standing, 2 mm-thick CoNiCrAlY specimen after isothermal exposure for 25 h in air at 1100 °C; (a) SEM micrographs; (b) cathodoluminescence image

unequivocally identified and located using cathodoluminescence imaging (Fig. 19.3b). In contrast, the scale formation and the distribution of yttrium-rich oxides are much more uniform on the ground side of the same coating specimen. The EDX mapping in Fig. 19.4 indicates that, in the concave parts, the internal oxide precipitates are

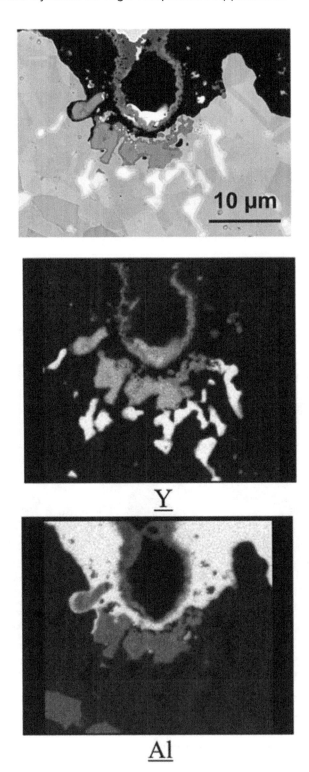

*19.4* EDX Maps for Y and Al of a 'valley' on the rough surface of a free-standing, 2 mm thick CoNiCrAlY specimen after 111 h of exposure in air at 1100 °C

yttrium oxide ($Y_2O_3$), which transforms into yttrium aluminates in the vicinity of and within the alumina scale. After 406 h of exposure in air at 1100 °C, the oxidation on the rough side tends to be more accelerated in the concave coating areas compared to the convex areas (Fig. 19.5a). This effect is related to the more extensive incorporation of yttrium into the scale on the concave areas of the coating. In contrast, the scale morphology and thickness are uniform on the flat surface after 406 h exposure

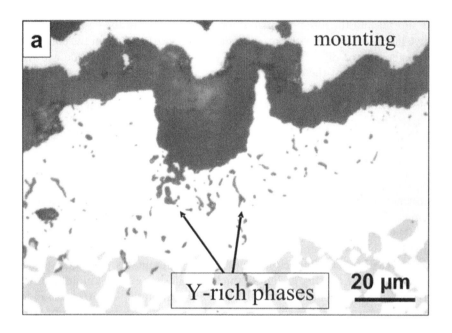

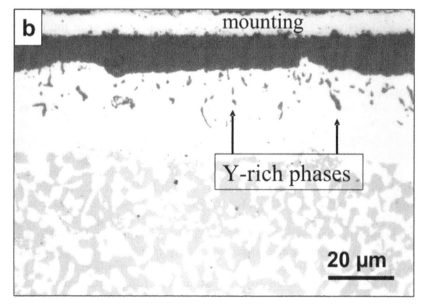

*19.5* Optical micrographs of free-standing, 2 mm-thick CoNiCrAlY specimens exposed for 406 h in air at 1100 °C; (a) rough surface; (b) ground surface

(Fig. 19.5b). The mechanistic explanation for the observed effect of coating rough-ness was given in Ref. 19. It was proposed that, during vacuum heat treatment and the initial stages of oxidation, yttrium as well as aluminium are depleted from the convex parts of the coating. In these areas of the coating with high surface to volume ratio, the depletion of aluminium results in local formation of non-protective nickel-rich spinel oxides, under which a protective alumina layer is soon established. At later stages of exposure, yttrium is consumed from the bulk of the coating for the growth of yttrium-rich precipitates (pegs) in the concave areas, where the diffusion path is shorter and the aluminium activity is higher than in the convex areas. More extensive incorporation of yttrium-aluminates in the oxide scale in the concave areas allows faster inward diffusion of oxygen, causing locally faster oxide growth [19]. The local variations of the oxide composition and morphology on rough MCrAlY surfaces indicate that definition of a critical TGO thickness in such systems is not possible. In addition, the stress in the oxide, which has been shown to be the driving force for failure, has a different distribution on the rough compared to the flat surfaces; in the latter, the out-of-plane stress component prevails [20–22].

### 19.3.3    Influence of the coating yttrium reservoir

The influence of the yttrium reservoir was observed after exposing two different types of specimens in air: 2 mm thick free-standing CoNiCrAlY (Type 2) and IN738 + 0.2 mm thick CoNiCrAlY + TBC (Type 1) at 1100 °C. It can be seen (Fig. 19.6a) that after exposure for 111 h, the alumina scale formed in the TBC system is uniform in thick-ness with occasional small precipitates of yttrium-aluminates. In contrast, the scale on the free-standing coating is slightly thicker, containing large amounts of yttrium aluminate pegs (Fig. 19.6b). After 111 h, the aluminium depletion zone underneath the TGO/BC interface is very similar in both cases, having a depth of approximately 8 µm. Cathodoluminescence studies of the free-standing coating revealed significant formation of internal yttrium-rich oxides (Fig. 19.6c), which were not observed in the TBC system. Between 111 h and 406 h of oxidation, the TGO in the TBC system thickened only slightly (Fig. 19.7a), whereas the oxide scale on the free-standing coating (Fig. 19.7b) grew significantly compared with that on the 111 h specimen and was thicker by a factor of three than that in the TBC system after the same exposure time (compare Figs. 19.7a and 19.7b). It is important to note that, in the TBC coated specimen after 406 h exposure, the aluminium-depleted zone due to interdiffusion with the substrate was only about 40 µm thick. Therefore, this depletion is not expected to have any significant effect on the bond coat oxidation behaviour. The significant scale thickening on the free-standing coating between 111 and 406 h is apparently related to the incorporation of yttrium-rich internal oxide precipitates providing short-circuit paths for oxygen diffusion. Since the CoNiCrAlY free-standing coating and the bond coat for the TBC had the same chemical composition and were produced by the same VPS process, the above difference in oxidation behaviour must be related to the amount of yttrium available for incorporation into the scale, i.e. to the yttrium reservoir, as was verified in Ref. 14 by testing free-standing coatings with different thicknesses. The measurements of the scale thickness as a function of time derived from SEM images for all studied specimens are presented in Fig. 19.8.

The failure of the TBC with the CoNiCrAlY coating occurred after 1200 h of discontinuous exposure at 1100 °C while no macroscopic oxide spallation was

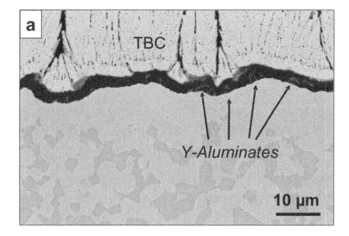

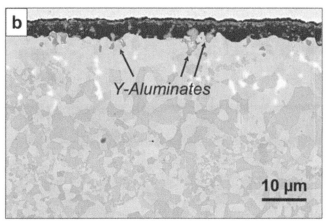

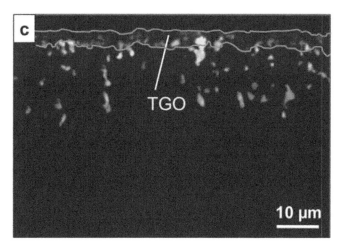

*19.6* (a) TBC system with CoNiCrAlY bond coat after 111 h exposure in air at 1100 °C; (b) SEM image of a free-standing, 2 mm-thick CoNiCrAlY specimens after 111 h exposure in air at 1100 °C; (c) Cathodoluminescence image of the same area as in (b), above, showing the accumulation of Y in the oxide/alloy interface area

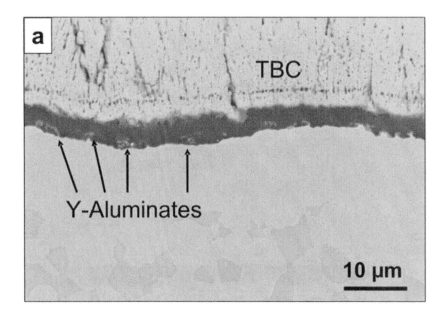

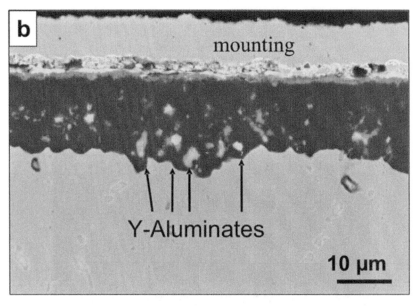

*19.7* SEM micrographs of specimens with CoNiCrAlY coatings exposed for 406 h in air at 1100 °C; (a) IN738/CoNiCrAlY (0.2 mm thick)/TBC; (b) Free-standing, 2 mm thick CoNiCrAlY; ground surface

observed on the 2 mm thick free-standing coating after exposure for up to 1600 h, in spite of the fact that a very thick scale formed. Although the early spallation in the TBC-coated specimen can be partly related to the strain energy stored in the ceramic coating, it is believed that the excellent scale adherence on the free-standing coatings

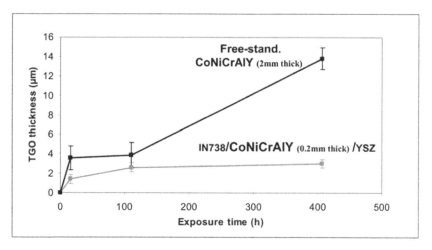

*19.8* TGO thickness as a function of time on specimens of Types 1 and 2 with a CoNiCrAlY Bond Coat after exposure in air at 1100 °C

(in spite of a higher growth rate) is related to a larger yttrium reservoir, as shown recently for two TBC systems [14]. Consequently, for the same coating composition, the critical oxide thickness for spallation is expected to increase with increasing coating thickness.

### 19.3.4    Altering TGO growth by changing the oxygen partial pressure of the test atmosphere

In the preceding sections, it was shown that the oxidation rate can be influenced by variation of the temperature and coating thickness, and that the major reason for both effects was a change in the amount and distribution of yttrium-rich internal oxide precipitates incorporated into the scale. The latter mechanism, however, implies that the scale composition and adherence, which are known to be affected by the presence of yttrium, also vary with the scale growth rate. In order to establish whether a change in the scale growth rate in MCrAlY bond coats quantitatively affects the TBC lifetime, one should, ideally, be able to change only the scale growth rate without changing other parameters such as scale morphology and adherence. In previous studies with alumina forming FeCrAl-alloys [15], it has been shown that the alumina scale growth rate can be influenced by a change in oxygen partial pressure ($pO_2$) in the atmosphere. The main reason for the latter effect was shown to be the change in the chemical potential gradient across the scale. Hence, the oxidation rate was predicted to decrease with decreasing $pO_2$ and good agreement was observed between the calculated rates and thermogravimetric measurements.

In the present study, TBC specimens of Type 3 (see Section 19.2) were exposed at 1100 °C in two different atmospheres: a high $pO_2$ atmosphere (air); and a low $pO_2$ atmosphere (Ar–4% $H_2$–2% $H_2O$; $pO_2$~$10^{-14}$ bar). A variation of the TGO growth rate was achieved; after 300 h, the oxide scale grown on the specimen exposed in air was 4.7 μm thick while that formed on the specimen exposed in Ar–4% $H_2$–2% $H_2O$ was 3.2 μm thick (see Fig. 19.9). An important observation was the fact that the oxide scales were almost identical with respect to microstructure and morphology.

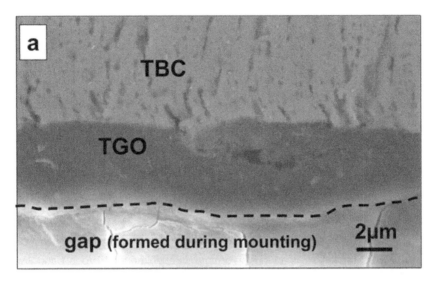

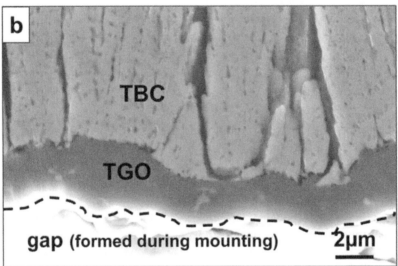

*19.9* SEM micrographs of specimens IN738/NiCoCrAlY/TBC exposed for 300 h; (a) in air at 1100 °C; (b) in Ar–4% $H_2$–2% $H_2O$ at 1100 °C

Subsequently, new identical specimens were exposed in the two atmospheres until TBC failure; the lifetime of the ceramic coating was 560 h when exposed in air and 1060 h when exposed in Ar–4% $H_2$–2% $H_2O$. The TGO thickness at failure could not be measured since the oxide had completely spalled off together with the ceramic coating, as is frequently found for EB-PVD TBC coatings. Therefore, the available TGO thickness data after 100 h and 300 h in both atmospheres was extrapolated to estimate the TGO thickness at failure assuming a power law time dependence for the thickening rate [5]. As can be seen in Fig. 19.10, the estimated TGO thickness at TBC failure is ca. 6 μm for both atmospheres, which is apparently the critical TGO thickness for the studied EB-PVD TBC system under the temperature cycling conditions used.

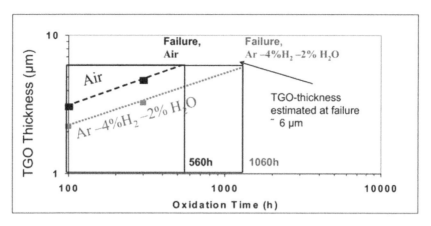

*19.10* Double logarithmic plot of TGO thickness as a function of time for specimens IN738/NiCoCrAlY/TBC exposed in air and in Ar–4% $H_2$–2% $H_2O$ at 1100 °C

## 19.4 Summary and conclusions

This investigation showed that the composition, morphology and, therefore, the growth rate of the oxide scales on MCrAlY coatings can be significantly affected by variations of the exposure temperature, coating thickness (yttrium reservoir) and surface roughness. In particular, these factors were found to have an effect on the amount and distribution of the internal yttrium-rich oxides in the sub-surface coating regions. Consequently, the variations in oxide morphology and growth rate could be explained in terms of differences in the incorporation of the yttrium-rich oxides into the scale.

The exposure in atmospheres with considerably different oxygen partial pressures enabled the formation of oxide scales at different growth rates which were identical in morphology and composition. The outcome of the latter experiment indicated that a critical TGO thickness to TBC failure exists for a given EB-PVD TBC system with a MCrAlY coating, provided that other parameters such as exposure temperature, surface roughness, coating manufacturing process and yttrium reservoir are kept constant.

## Acknowledgements

One of the authors, D. Naumenko, would like to acknowledge the Deutsche Forschungsgemeinschaft (DFG) for funding his work under Grant. No. NA 615-1-1. The authors would like to thank Professor G. H. Meier from the University of Pittsburgh for his comments on the manuscript and valuable discussions, and E. Wessel and A. Everwand for their contribution with the SEM micrographs and cathodoluminescence images. The authors also acknowledge R. Vaßen for supplying the MCrAlY-coated specimens and free-standing coatings for this study.

## References

1. M. Gell, E. Jordan, K. Vaidyanathan, K. McCarron, B. Barber, Y.-H. Sohn and V. Tolpygo, *Surf. Coat. Technol.*, 120–121 (1999), 53.

2. B. A. Pint, I. G. Wright, W. Y. Lee, Y. Zhang, K. Prüßner and K. B. Alexander, *Mater. Sci. Eng.*, A245 (1998), 201.
3. E. P. Busso, J. Lin, S. Sakurai and M. Nakayama, *Acta Mater.*, 49 (2001), 1515.
4. N. M. Yanar, M. J. Stiger, M. Maris-Sida, F. S. Pettit and G. H. Meier, *Key Eng. Mater.*, 197 (2001), 145.
5. H. Echsler, D. Renusch and M. Schütze, *Mater. Sci. Technol.*, 20 (2004), 307.
6. D. Renusch, H. Echsler and M. Schütze, *Mater. Sci. Forum*, 461–464 (2004), 72.
7. V. K. Tolpygo, D. R. Clarke and K. S. Murphy, *Surf. Coat. Technol.*, 146–147 (2001), 124.
8. G. M. Kim, N. M. Yanar, E. N. Hewitt, F. S. Pettit and G. H. Meier, *Scrip. Mater.*, 46 (2002), 489.
9. V. K. Tolpygo and D. R. Clarke, *Surf. Coat. Technol.*, 163–164 (2003), 81.
10. N. M. Yanar, G. Kim, S. Hamano, F. S. Pettit and G. H. Meier, *Mater. High Temp.*, 20 (2003), 495.
11. N. M. Yanar, F. S. Pettit and G. H. Meier, *Mater. Sci. Forum*, 426–432 (2003), 2453.
12. T. J. Nijdam, L. P. H. Jeurgens, J. H. Chen and W. G. Sloof, *Oxid. Met.*, 64 (2005), 355.
13. N. M. Yanar, F. S. Pettit and G. H. Meier, *Met. Mater. Trans.*, 37A (2006), 1563.
14. J. Toscano, R. Vaßen, A. Gil, M. Subanovic, D. Naumenko, L. Singheiser and W. J. Quadakkers, *Surf. Coat. Technol.*, 201 (2006), 3906.
15. A. Kolb-Telieps, D. Naumenko, W. J. Quadakkers, G. Strehl and R. Newton, 'High temperature corrosion of FeCrAlY/Aluchrom YHf in environments relevant to exhaust gas systems', in *Materials Aspects in Automotive Catalytic Converters*, Wiley-VCH Verlag GmbH, Weinheim, Germany, 2001.
16. T. J. Nijdam, L. P. H. Jeurgens and W. G. Sloof, *Acta Mater.*, 53 (2005), 1643.
17. F. H. Stott, *Mater. Sci. Forum*, 251 (1997), 19.
18. D. R. Mumm and A. G. Evans, *Acta Mater.*, 48 (2000), 1815.
19. A. Gil, V. Shemet, R. Vassen, M. Subanovic, J. Toscano, D. Naumenko, L. Singheiser and W. J. Quadakkers, *Surf. Coat. Technol.*, 201 (2006), 3824.
20. M. Y. He, A. G. Evans and J. W. Hutchinson, *Mater. Sci. Eng.*, A245 (1998), 168.
21. V. K. Tolpygo and D. R. Clarke, *Acta Mater.*, 46 (1998), 5167.
22. D. R. Mumm and A. G. Evans, *Key Eng. Mater.*, 197 (2001), 199.

# 20

## Influence of bond coat surface roughness on chemical failure and delamination in thermal barrier coating systems*

### M. P. Taylor, W. M. Pragnell and H. E. Evans

*Department of Metallurgy and Materials, School of Engineering,*
*The University of Birmingham, Edgbaston, Birmingham, UK*
*M.P.Taylor@bham.ac.uk*

### 20.1 Introduction

Thermal Barrier Coating (TBC) systems consist of a thermally-insulating yttria stabilised zirconia (YSZ) top coat attached to the superalloy substrate by means of an oxidation-resistant bond coat. TBCs have the potential of offering an increase in the efficiency of both aero-engine and land-based gas turbine engines by increasing the inlet temperature and reducing the amount of cooling air required in high-temperature components. However, the commercial potential of these coating systems has not been realised fully due to the lack of understanding and thus control of the failure mechanisms, e.g. spallation of the outer ceramic layer. Failure of the TBC system through spallation of the ceramic under these circumstances would lead to excessively rapid oxidation of the bond coat and underlying component [1–4].

Bond coats are designed to provide oxidation protection through the formation of a continuous alumina layer and to establish a sufficient aluminium reservoir so that this protective layer can be maintained. Current bond coats are either of overlay MCrAlY type (where M is Ni, Co, Fe or a combination of these), often deposited by low pressure plasma spraying (LPPS), HVOF or electroplating, or they may be diffusion coatings produced by platinising and/or aluminising the outer surface of the nickel-based superalloy substrate. These bond coats have initially high aluminium contents to counter the depletion that occurs as a result of selective oxidation and the continued growth with time at temperature of the surface alumina layer. Even so, it is possible for local regions of the bond coat to become diffusionally isolated from the bulk of the reservoir and to experience substantial depletion of aluminium after short exposure times [5]. In sprayed coatings, such regions may be individual splats which have been oxidised during spraying to form a continuous surface layer of oxide [5] or, in denser coatings, may be asperities, with large surface area to volume ratios [4,6], associated with a rough bond coat surface.

There is increasing experimental evidence [7,8] that reasonably flat bond coats can prolong TBC lifetimes. Related modelling activity has also confirmed that out-of-plane stresses can increase with increasing surface roughness, either as-manufactured [9–11] or induced during exposure [12]. This role of roughness in the

---

\* Reprinted from M. P. Taylor et al.: The influence of bond coat surface roughness on chemical failure and delamination in TBC systems. *Materials and Corrosion.* 2008. Volume 59. pp. 508–13. Copyright Wiley-VCH Verlag GmbH & Co. KGaA.

modelling studies, to date, has centred solely on its influence on stress development within the TBC system but, as indicated above, bond coat asperities are also regions where aluminium supply can be limited and excessive depletion can result.

A consequence of severe aluminium depletion is that *chemical failure* may occur. This could be manifested by the reduction of alumina by chromium or the formation of Ni-based spinels under the alumina layer (processes termed Intrinsic Chemical Failure, InCF). It might also occur by the inability to re-heal the alumina should it crack (termed Mechanically Induced Chemical Failure, MICF). A description of the general concept is given in Ref. 13. InCF is a thermodynamic criterion which, in this context, requires aluminium activities in contact with the alumina layer to be extremely low. Spinel oxides underlying an alumina layer have been reported [14] but the more likely chemical failure route, especially at surface asperities where out-of-plane stresses can readily develop, is via MICF. As a result of chemical failure, fast-growing oxides (containing nickel, cobalt and chromium) form at the localised regions where the initial severe aluminium depletion occurred. These oxides grow into a constrained space because of the presence of the YSZ top coat and will generate out-of-plane tensile stresses within the top coat, at temperature, at locations between the regions undergoing chemical failure [5,6]. Recent finite-element analyses [15] of this process, using the approach described in Ref. 11, show that these out-of-plane tensile stresses can be very large, ~1 GPa, and increase quickly once localised fast-growing oxides have initiated. Qualitative evidence exists for top coat cracking associated with asperities undergoing chemical failure [10] and for failure lifetimes to be associated with the formation of non-protective oxides [16].

The purpose of this present work is to examine the behaviour of a TBC system in which the surface of the MCrAlY bond coat consists of small asperities of high aspect ratio and for which early susceptibility to chemical failure is expected. The influence of localised non-protective oxidation on crack development is of particular interest.

## 20.2    Experimental procedure

The substrate alloy used in this study was Haynes 230 (30Cr/6W/6Co/4Fe/2Mo/ balance Ni in at.%) supplied as 150 mm × 10 mm × 1 mm bars recessed at 20 mm intervals along the length, for later sectioning. The NiCoCrAlY bond coat (35Ni/ 32Co/22Cr/10Al/0.4Y in at.%) was electroplated onto all surfaces of the bar and this, in turn, was coated with an air plasma sprayed yttria partially stabilised zirconia (YSZ) ceramic top coat. The bars were sectioned to produce coupons of approximate dimensions 20 mm by 10 mm by 1 mm, and exposed metal surfaces were coated with alumina paste. The specimens were oxidised isothermally at 1000 °C or 1100 °C in laboratory air for periods up to 1000 h.

At the end of each period of isothermal exposure, the specimens were cooled to room temperature and metallographically sectioned. The sections were examined using scanning electron microscopy (SEM), including Energy Dispersive Spectroscopy (EDS) to identify the thermally grown oxide (TGO) phases and produce compositional profiles through selected specimens.

## 20.3    Microstructural development

The electroplating process produced NiCoCrAlY bond coats with highly roughened outer surfaces onto which the YSZ top coat was plasma-sprayed. The extent of the roughness can be easily seen in the section through an as-received specimen (Fig. 20.1)

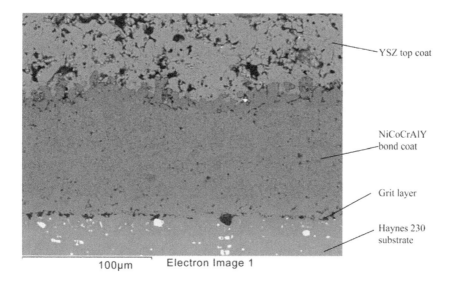

*20.1* Back-scattered SEM image of a section through an as-received specimen showing the Haynes 230 substrate, electroplated NiCoCrAlY bond coat and plasma-sprayed YSZ top coat

and consists of asperities approximately 10 μm wide and 10 μm high. There is intimate contact between the outer surface of the bond coat and the APS YSZ top coat with no detectable cracking at this interface. Such surfaces are desirable for good bonding of APS coatings to underlying alloys or coatings. The bond coat is dense and approximately 100 μm thick. The interface between this layer and the substrate is indicated by the alumina particles remaining from the grit blasting procedure used to clean the substrate surface prior to coating.

During oxidation testing, a TGO grows at the bond coat outer surface and in the early stages its dark contrast highlights the interface between the bond coat and top coat (Fig. 20.2). The TGO formed has been identified as alumina by EDS analysis.

With longer exposure at temperature, the TGO grows but has an irregular thickness and the layer contains pores, as shown in the section taken through the specimens held at 1100 °C for 600 h (Fig. 20.3(a)) and 400 h (Fig. 20.3(b)). In addition, the roughness of the bond coat/TGO interface has decreased whereas that of the TGO/top coat interface has not. Localised regions of breakaway oxides, identified using EDS, are now found dispersed above the surface of the continuous alumina layer (Fig. 20.4). Extensive cracking in inter-asperity regions of the top coat and within the TGO, indicated as sites A and B, respectively in Figs. 20.2 and 20.3 is also present. Some of the latter appear to be shear-type cracks extending through the TGO and connected to cracks at the TGO/bond coat interface. Spallation of the top coat has not occurred in these specimens.

## 20.4    Discussion

### 20.4.1    Origin of breakaway oxides

Chemical failure arises when the aluminium content in the bond coat depletes to levels where a protective alumina TGO can no longer be sustained. For this reason,

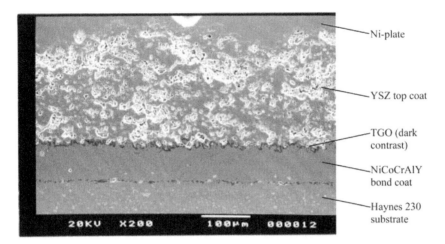

*20.2* Scanning electron micrograph of a section through the TBC consisting of the Haynes 230 alloy substrate, an electroplated bond coat and a plasma-sprayed YSZ ceramic top coat; the bright contrast in the top coat layer is the result of charging. The specimen has been exposed at 1000 °C for 50 h

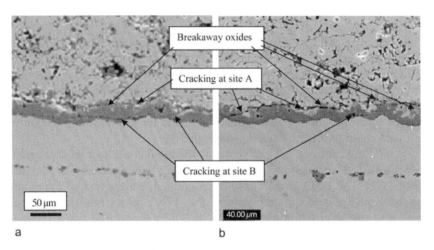

*20.3* SEM image of a specimen held at (a) 1100 °C for 600 h and (b) 1100 °C for 400 h

bond coats whose microstructures do not contain barriers to elemental diffusion are preferred. In the TBC studied here, the bond coat appears homogeneous and free of internal oxidation (Figs. 20.3(a) and 20.3(b)) and a continuous alumina layer is present adjacent to the bond coat, demonstrating an adequate supply of aluminium to the surface. However, at longer times at 1100 °C, regions of breakaway oxides have been identified above the alumina layer. Close examination of the bond coat/ YSZ top coat interfacial region, including the TGO, using SEM techniques, has led to the model described schematically in Fig. 20.5. Here it is proposed that breakaway oxidation arises as a consequence of aluminium depletion localised to the original asperities on the bond coat surface. The underlying aluminium concentration is determined by the balance between the flux of aluminium entering the oxide layer

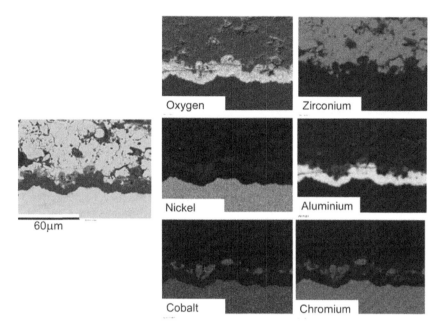

*20.4* Backscattered SEM micrograph and EDS elemental maps of a section
through a specimen held for 400 h at 1100 °C showing a dispersion of mixed Ni,
Co, Cr oxides above a continuous alumina layer

and that arriving at the asperity from the bulk of the coating. The depletion rate of
aluminium in the asperities is then dependent on the geometry and volume of each
asperity. As the volume of these spheroidal structures decreases there is an increase
in surface area to volume ratio and thus an acceleration of depletion. Over any area
of the electroplated bond coat surface, there will be a range of depletion rates each
specific to individual asperity dimensions. As can also be seen in the micrographs,
replenishment of these regions is restricted to a narrow section or neck attached
to the bulk of the coating. This reduces the section area over which diffusion of
aluminium from the bulk into the asperity can occur.

In an initial attempt to demonstrate this process, the two-dimensional ODIN
finite-difference (FD) model [17,18] was used to calculate aluminium depletion in an
asperity with a height and width of 10 μm, similar to the dimensions found in practice
(Fig. 20.1). Because of the two-dimensional nature of the FD model, however, the
asperity is a linear rather than a spheroidal feature with infinite length into and out
of the paper, i.e. no end effects are present. This means that in the real case, depletion
rates within the asperity will be greater than calculated by ODIN but, as will be
shown, the results from the two-dimensional model readily demonstrate the enhanced
depletion depicted in Fig. 20.5.

The calculations were performed for a 100 μm thick coating oxidised on one
surface only. Relatively short times at 1100 °C were used to ensure that the depletion
profile within the bulk of the coating did not extend more than 50 μm (the mid-plane)
and that only alumina formation needed to be considered. The ternary interdiffusion
coefficients obtained by Nesbitt [19] for NiCrAlY at 1100 °C were used in the com-
putations. The alumina growth kinetics were sub-parabolic, as found previously for
a LPPS MCrAlY coating [20]:

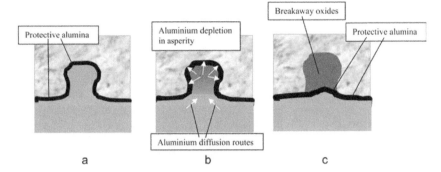

*20.5* Schematic diagrams describing the proposed model, (a) sufficient aluminium present in asperity to maintain a protective alumina layer, (b) rapid depletion and restricted replenishment reducing aluminium levels within the asperity, and (c) aluminium content in the asperity reaches levels where protective alumina can no longer be maintained, the remaining bond coat oxidises rapidly but re-healing occurs at the base of the asperity where aluminium levels are sufficiently high

$$\xi = 2.77 \times 10^{-8} \, t^{0.4} \qquad\qquad [20.1]$$

where $\xi$ is the oxide thickness, m, and t is exposure time, s.

The computational results shown in Fig. 20.6 refer to an initial aluminium concentration of 10 at.% throughout the bulk of the coating and within the asperity. It can be appreciated that selective oxidation to form the alumina layer at this short time (1 h at 1100 °C) leads to a significant reduction in aluminium content at the surface of the bulk coating and a consequent gradient of aluminium away from this interface.

The other notable feature of these results is that the initial high rates of alumina growth produce significantly more depletion in the asperity than in the bulk of the coating. Figure 20.6 shows that the aluminium concentration within this particular asperity geometry is essentially everywhere less than 1 at.% after exposure for 1 h at 1100 °C. Although the asperity shown in Fig. 20.6 has an extremely low aluminium concentration throughout, this situation represents a relatively steady state and the composition does not change significantly except after much longer times when the bulk coating has become depleted in aluminium. This means that over an extended period of time, asperities may be susceptible to chemical failure but the formation of breakaway oxides by MICF requires an initiating event. This may be cracking of the TGO at temperature and the exposure of the depleted coating to the oxidising species. At aluminium concentrations of around 1 at.%, it is unlikely that the alumina layer would readily repair. This sort of mechanical damage within the TGO can be induced by tensile stresses developed within the oxide layer at temperature simply as a result of the growth of *protective* alumina [15]. Figure 20.7 shows a set of interfacial features after 1000 h at 1000 °C, and it can be seen that although at least one asperity has reached breakaway, others remain intact, indicating that the triggering event does not occur simultaneously at all locations. Computational results for this lower temperature are not available due to the scarcity of interdiffusion data and oxidation kinetics, both of which are parameters vital for the model.

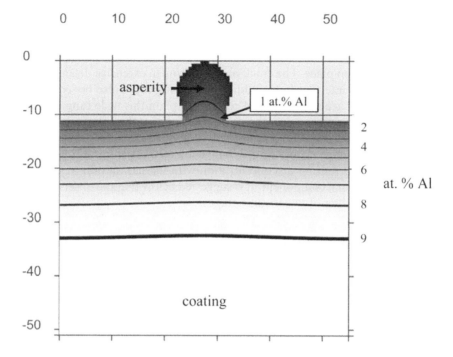

*20.6* ODIN predictions of aluminium concentration in and around a two-dimensional asperity on a 100 μm thick MCrAlY coating after oxidation for 1 h at 1100 °C. The solid lines show the contours for the indicated aluminium concentrations. The dimensions of the model are shown in micrometres

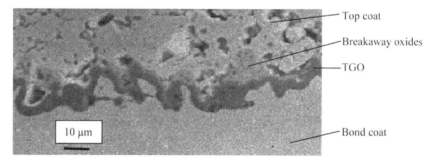

*20.7* Micrograph of a section through a specimen held at 1000 °C for 1000 h showing the various stages in the oxidation process, i.e. some asperities still intact and others have gone into breakaway oxidation

Unfortunately, experimental confirmation of the aluminium levels in the asperities using EDS was not possible due to the limitations of the technique. The probe size and depth of penetration compared with the dimensions of the asperities meant that the aluminium levels recorded included both the surrounding TGO and the sub-surface alumina.

### 20.4.2    Cracking mechanisms

After a few hundred hours exposure (Figs. 20.3 and 20.4), extensive cracking had occurred within the TGO and the adjacent top coat region, although spallation of the top coat had not taken place. The crack path resulting in extensive delamination was at the TGO/top coat interface but in addition, cracking could also be seen within the TGO and at its interface with the bond coat. To explain this wide range of cracking locations it is necessary to consider the stress generation at this interface both at temperature, i.e. isothermally, and on cooling. It is important to note that the thermal history of these specimens contains only one cooling transient and thus systematic changes in crack distribution or concentration with increasing time at temperature must have an isothermal component. It is suggested below that the observed cracking morphology is a product of two distinct stages.

### 20.4.3    Stress generation during isothermal exposure

At temperature, the alumina TGO grows on the bond coat surface with an associated volume increase. As noted above, this change in volume takes place in a mechanically constrained environment. As a consequence, the oxidation process will generate stresses, isothermally at the oxidation temperature, particularly if the bond coat surface is not planar. In this case, the upward displacement due to oxide growth will vary along the uneven surface profile of the bond coat and will produce a complementary variation in imposed strains within the top coat required to maintain continuity [6]. It can be appreciated, at least qualitatively, that these strains will result in the development of out-of-plane tensile stresses in the top coat which are expected to maximise between the peaks and troughs of the surface of the bond coat.

A consequence of chemical failure of the asperities is that fast-growing, non-alumina oxides will develop at the asperity tips [10,14], as shown schematically in Fig. 20.5. It is expected that their faster growth rate will lead to larger out-of-plane tensile stresses located between the asperities. Estimates of the magnitude of these stresses cannot be made reliably without the use of finite element (FE) procedures that fully account for the volume change associated with oxide formation and for relaxation effects due to creep within the TBC system. Detailed FE analysis of this situation is under way [15] but initial results show that out-of-plane tensile stresses of order 1 GPa can form within top coat valleys at 1100 °C only 100 h after breakaway oxidation has commenced. These stresses tend to be localised to the flanks of bond coat asperities and increase with aspect ratio. In the present experiments, the situation is even more extreme since the entire bond coat asperity is converted into oxide (Figs. 20.3 and 20.4, for example). The associated large volume increase is likely to produce substantial out-of-plane tensile stresses between these, now oxide, asperities as shown schematically in Fig. 20.8. It is postulated that these will be sufficient to nucleate sub-critical cracks within the top coat at the oxidation temperature. They will be discrete and isolated between the oxide asperities.

### 20.4.4    Cracking during cooling

Examination of the specimens after cooling shows extensive cracking within the top coat, within the TGO and along the TGO/bond coat interface (Figs. 20.3 and 20.4). There are examples where the latter are associated with cracks within the TGO and

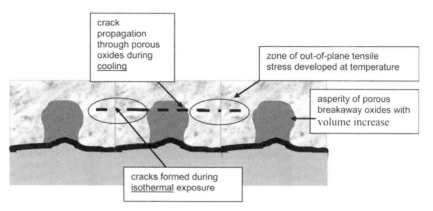

crack propagation through porous oxides during <u>cooling</u>

zone of out-of-plane tensile stress developed at temperature

asperity of porous breakaway oxides with volume increase

cracks formed during <u>isothermal</u> exposure

*20.8* Schematic diagram showing how the volume change associated with the oxidation bond coat asperities may lead to the formation of tensile cracks within the top coat between the asperities at the oxidation temperature. Subsequent cooling can result in the link-up of these cracks through the porous breakaway oxides formed when the asperities oxidised

may arise from wedge cracking. However, the dominant degradation process is top coat cracking at or near the TGO/top coat interface.

During isothermal exposure, the cracks that formed between the asperities could not extend since, at temperature, there was no driving force for crack growth into the oxide asperities because these were under out-of-plane compression. However, during cooling, the stress pattern will change because of the mismatch between the thermal coefficient of expansion of the metallic alloy and that of the ceramic top coat. It is important to recognise that the thermal coefficient of expansion of the TGO is probably not significant in this context since the complete oxidation of the asperities tends to reduce the roughness of the bond coat/TGO interface. Indeed, extensive delamination cracking is seen in specimens where few if any asperities remain. Nevertheless, out-of-plane tensile stresses develop across the TGO during cooling because the metallic alloy core contracts to a greater extent than the ceramic top coat shell (Fig. 20.9). This then provides a driving force both for the initiation of new cracks within the porous breakaway oxides [14] and for the extension of those cracks produced during the isothermal oxidation. The crack path is shown schematically in Fig. 20.8 and the result is the extensive decohesion seen at the TGO/top coat interface (Figs. 20.3 and 20.4). The overall effect of chemical failure of the asperities at the bond coat surface is to produce an effective weakening of the bond coat/top coat interface.

## 20.5    Conclusions

TBCs are used in high-temperature environments to provide a thermally insulating ceramic top coat with oxidation protection provided by aluminium sacrifice from the bond coat to form a protective oxide. Aluminium loss from the bond coat is a critical factor determining the lifetime of TBCs and to maximise the available aluminium content of the bond coat, a production route is required which produces a dense structure with no internal barriers to diffusion. It has been shown in this study that breakaway oxides, signifying chemical failure, can be formed due to localised

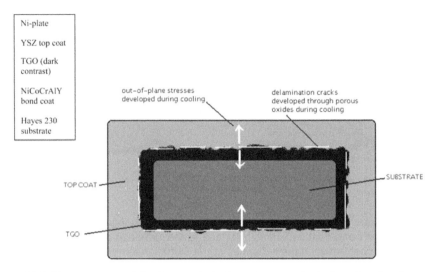

*20.9* The presence of the top coat shell produces out-of-plane tensile stresses across the TGO during cooling. These provide a driving force for delamination cracking

depletion at bond coat asperities and this is linked to asperity geometry. The application of the FD model, ODIN, predicted the rapid depletion of aluminium from the asperities and showed that the aluminium content of the bulk of the bond coat is sufficient to enable the formation of a continuous protective alumina underneath these regions, as observed experimentally. It is suggested that the volume increase associated with the conversion of the bond coat asperities into breakaway oxide results in out-of-plane tensile stress development at the oxidation temperature. These stresses are anticipated to be at a maximum between the oxide asperities and are thought to nucleate small, sub-critical cracks at these locations *at temperature*. Further delamination will occur during cooling both by the extension of these cracks and by the growth of crack-like defects within the porous breakaway oxides.

## References

1. D. J. Wortman, B. A. Nagaraj and E. C. Dunderstadt, *Mater. Sci. Eng.*, A121 (1989), 433.
2. J. T. Demasi-Marcin, K. D. Scheffler and S. Bose, *J. Eng. Gas Turbines Power*, 112 (1990), 521.
3. A. H. Bartlett and R. D. Maschio, *J. Am. Ceram. Soc.*, 78 (1995), 1018.
4. J. A. Haynes, E. D. Rigney, M. K. Ferber and W. D. Porter, *Surf. Coat. Technol.*, 86–87 (1996), 102.
5. H. E. Evans and M. P. Taylor, *Oxid. Met.*, 55 (2001), 17.
6. H. E. Evans and M. P. Taylor, *J. Corros. Sci. Eng.*, 6 (2003), H10.
7. B. A. Pint, I. G. Wright, W. Y. Lee, Y. Zhang, K. Prüßner and K. B. Alexander, *Mater. Sci. Eng.*, A245 (1998), 201–211.
8. N. M. Yanar, F. S. Pettit and G. H. Meier, *Metall. Mater. Trans.*, 37A (2006), 1563–1580.
9. A. M. Freborg, B. L. Ferguson, W. J. Brindley and G. J. Petrus, *Mater. Sci. Eng.*, A245 (1998), 182–190.

10. C-H. Hsueh, J. A. Haynes, M. J. Lance, P. F. Becher, M. K. Ferber, E. R. Fuller Jr, S. A. Langer, W. C. Carter and W. R. Cannon, *J. Am. Ceram. Soc.*, 82 (1999), 1073.
11. E. P. Busso, L. Wright, H. E. Evans, L. N. McCartney, S. R. J. Saunders, S. Osgerby and J. Nunn, *Acta Mater.*, 55 (2007), 1491–1503.
12. A. W. Davis and A. G. Evans, *Metall. Mater. Trans.*, 37A (2006), 2085–2095.
13. H. E. Evans, A. T. Donaldson and T. C. Gilmour, *Oxid. Met.*, 52 (1999), 379.
14. P. Niranatlumpong, C. B. Ponton and H. E. Evans, *Oxid. Met.*, 53 (2000), 241.
15. E. P. Busso, Z. Q. Qian, M. P. Taylor and H. E. Evans, in preparation.
16. E. A. G. Shillington and D. R. Clarke, *Acta Mater.*, 47 (1999), 1297.
17. W. M. Pragnell and H. E. Evans, *Modelling Simul. Mater. Sci. Eng.*, 14 (2006), 1.
18. W. M. Pragnell and H. E. Evans, *Oxid. Met.*, 66 (2006), 209.
19. J. A. Nesbitt, in *Lifetime Modelling of High Temperature Corrosion Processes*, 359, ed. M. Schütze, W. J. Quadakkers and J. R. Nicholls, Maney Publishing, London, 2001.
20. S. Gray, M. P. Taylor, E. Chau and H. E. Evans, *J. Corros. Sci. Eng.*, 6 (2003), H16.

# 21

## Investigation on the oxidation behaviour of gamma titanium aluminides coated with thermal barrier coatings*

Reinhold Braun[1], Maik Fröhlich[1], Andrea Ebach[1] and
Christoph Leyens[1,2]

*¹DLR – German Aerospace Center, Institute of Materials Research,
D-51170 Köln, Germany*
*reinhold.braun@dlr.de; maik.froehlich@dlr.de; andrea.ebach@dlr.de;
leyens@tu-cottbus.de*
*²Technical University of Brandenburg at Cottbus, Physical Metallurgy and
Materials Technology, D-03046 Cottbus, Germany*

### 21.1 Introduction

Gamma titanium aluminide-based alloys are considered to be suitable as structural materials for high-temperature applications due to their attractive properties, including low density, high specific stiffness and tensile strength, and good creep resistance [1–5]. Numerous engineering $\gamma$-TiAl alloys have been developed in the last two decades for structural applications in aerospace vehicles and automotive engines [6]. Possessing a favourable combination of low density and attractive high-temperature capability, these alloys, however, exhibit insufficient oxidation resistance when operated at temperatures above 800 °C for long periods [7]. Compared to binary TiAl alloys, the oxidation resistance can be increased by alloying ternary elements, in particular, niobium which has proven to be the most effective alloying element to reduce the oxidation rate [6]. Furthermore, W and Mo additions can improve the oxidation behaviour [8,9]. Ag-containing TiAl-based alloys have also been reported to display improved oxidation behaviour due to the formation of a continuous alumina scale [10]. Considerable efforts have been made to develop coatings providing efficient oxidation protection to increase the service temperature range of $\gamma$-TiAl alloys [11–14]. Metallic and ceramic coatings have been studied. However, the development of oxidation-resistant coatings with long-term protection capability is still a major issue for high-temperature structural applications of gamma titanium aluminides [13]. A very promising technique, reported recently, to improve the oxidation resistance is the utilisation of the halogen microalloying effect [15]. The addition of small amounts of halogens introduced into the surface of TiAl-based alloys can improve the oxidation resistance by a selective oxidation mechanism of aluminium, resulting in the formation of a protective alumina scale [16,17].

Thermal barrier coatings (TBCs) are widely used in aeroengines and land-based gas turbines [18–22]. Typical TBC systems consist of nickel-based superalloy substrates with bond coats and ceramic top coats [23,24]. Bond coats used are typically

---

* Reprinted from R. Braun et al.: Investigation on the oxidation behaviour of gamma titanium aluminides coated with thermal barrier coatings. *Materials and Corrosion.* 2008. Volume 59. pp. 539–46. Copyright Wiley-VCH Verlag GmbH & Co. KGaA.

MCrAlY (M = Ni, Co) or PtAl layers, which form a slow-growing alumina scale. On top of this thermally grown oxide, the ceramic top layer with low thermal conductivity is deposited, usually being the current standard yttria stabilised zirconia. Through the use of TBCs and internal cooling, the components can operate at substantially lower temperatures than the surface temperature of the ceramic coating. The temperature of the metal surface can be lower by between 100 and 150 °C. Recent studies have demonstrated that thermal barrier coatings can also be applied successfully on $\gamma$-TiAl alloys [13,25–28]. Samples with and without oxidation-resistant coatings have been investigated. The aim of the present work was to study the performance of thermal barrier coatings deposited on $\gamma$-TiAl-based alloys with particular emphasis on microstructural examination of the oxide scales formed during thermal exposure. Pre-oxidation treatments should ensure the formation of outer scales rich in alumina to serve as bond coats to provide good adherence of the ceramic top coat.

## 21.2    Experimental

The materials used were two $\gamma$-TiAl-based alloys with chemical compositions (in atomic percent) of Ti–45Al–8Nb–0.2B–0.15C (alloy 98G) and Ti–45Al–6Nb–1Cr–0.4W–0.2B–0.5C–0.2Si (alloy 02G). The forgings provided by UES – Materials & Processes (Dayton, OH, USA) exhibited a two-phase microstructure consisting of $\gamma$-TiAl and $\alpha_2$-Ti$_3$Al phases. Alloy 02G contained a significant amount of intermetallic phases enriched with niobium, chromium and tungsten. From these forgings, disc-shaped specimens of 15 mm diameter and 1 mm thickness were machined by spark erosion. The specimens were ground to a 2500-grit surface finish using SiC emery paper. The specimens were pre-oxidised in air or exposed to oxygen at low partial pressure. The exposure experiments in oxygen at low partial pressure were conducted using a tube furnace connected to a mechanical pump and an oxygen tank. Temperature, exposure time and oxygen flow rate were optimised by examining the grown oxide scale using scanning electron microscopy and X-ray analysis (the pre-oxidation tests in low partial pressure oxygen atmosphere were performed at UES, Dayton, OH, USA). Both pre-treatments were carried out to form compact oxide scales and avoid rapid initial oxide growth during TBC deposition at high temperatures.

Thermal barrier coatings (TBCs) of 7 wt.% yttria partially stabilised zirconia (YPSZ) were deposited on the pre-oxidised specimens using electron-beam physical vapour deposition (EB-PVD). The thickness of the TBC was between 150 and 160 μm. Specimen holders made of titanium alloy Ti–6Al–4V were electron-beam welded to the samples for handling during TBC deposition. Before high-temperature exposure, the specimen holder was removed, but remnants of the holder material remained on the sample, being spots of preferential oxidation.

Cyclic oxidation tests were carried out at 850 and 900 °C in air using automated rigs. One cycle consisted of 1 h at temperature and 10 min cooling down to about 60 °C. The specimens were periodically inspected and weighed. The maximum exposure was 1100 cycles. Cross-sections of oxidised specimens were examined using scanning electron microscopy (SEM) and energy-dispersive X-ray spectroscopy (EDS).

## 21.3    Results and discussion

Figure 21.1 shows SEM micrographs of the oxide scales formed on pre-oxidised alloy 98G specimens. When exposed to air at 900 °C for 10 h, the specimens developed an outer scale consisting predominantly of alumina followed by a porous titania layer and a nitride inner layer interrupted by alumina particles (Fig. 21.1a). Beneath the nitride zone, Nb-rich precipitates were observed (bright phase). Exposure to a low partial pressure oxygen atmosphere at 950 °C for 2 h resulted in an outer mixed scale of titania and alumina and an inner $\alpha_2$-Ti$_3$Al phase layer containing embedded

a

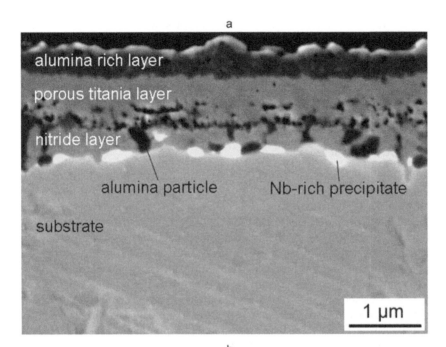

b

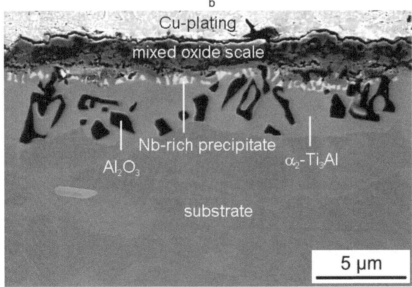

*21.1* Oxide scales formed on alloy 98G specimens which were pre-oxidised (a) in air at 900 °C for 10 h and (b) in low partial pressure oxygen atmosphere at 950 °C for 2 h

alumina particles (Fig. 21.1b). A niobium enriched phase formed at the outer/inner scale interface. As reported in the literature, pre-oxidation under low partial pressure oxygen (in a vacuum of $10^{-3}$ Pa) improved the oxidation behaviour of TiAl, being attributed to the formation of an adherent very rich alumina scale [29,30]. Similarly, annealing of $\gamma$-TiAl in low partial pressures of oxygen at temperatures between 550 and 900 °C produced an oxide scale consisting of an outer $Al_2O_3$ layer and an inner $TiO_2$ layer [31]. The pre-treatment at 950 °C for 2 h used in this work did not result in the formation of a protective alumina scale, probably associated with the alloying element Nb. Niobium-containing $\gamma$-TiAl oxidised in argon + 20% oxygen exhibited an Al-depleted subsurface layer beneath a heterogeneous alumina/titania scale and the occurrence of internal oxidation [32]. Such a depleted layer was not found during oxidation in air. For longer exposure times, the oxidation rate was higher in $Ar/O_2$ than in air due to the formation of a stable near-continuous nitride layer in the latter atmosphere [33,34].

Results of cyclic oxidation tests at 850 °C in air are presented in Fig. 21.2 for both alloys, showing plots of mass change versus number of cycles. Before exposure, the specimens were pre-oxidised in air and coated with TBCs. For both $\gamma$-TiAl alloys, no spallation of the thermal barrier coating occurred during the maximum exposure of 1100 cycles. The drop in mass change of the alloy 02G sample after 620 cycles was associated with flaked off oxides of remnants of specimen holder material, but not with spallation of the oxide scale grown on the $\gamma$-TiAl sample. The weight loss during the initial stages of exposure was related to the evaporation of water vapour adsorbed by the ceramic TBC during storage before testing. The mass gain of specimens coated with TBC was slightly higher for alloy 02G than for alloy 98G.

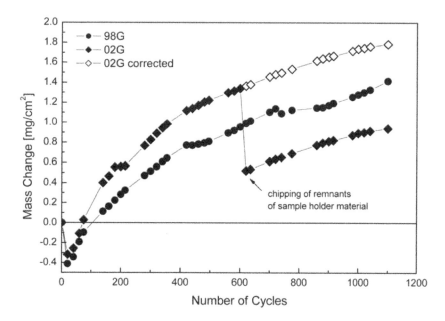

*21.2* Mass change versus number of cycles of specimens of the $\gamma$-TiAl alloys 98G and 02G exposed to air at 850 °C (mass change of the alloy 02G sample was corrected regarding spalled off oxides of remnants of sample holder material). Specimens were pre-oxidised in air and coated with TBC

Figure 21.3 shows SEM micrographs of the cross-section of a TBC coated sample of alloy 98G which was oxidised at 850 °C for 1100 cycles. A thick thermally grown oxide scale was observed with protrusions of an outer oxide mixture exhibiting a columnar structure (Fig. 21.3a). A large amount of porosity was found in this outer

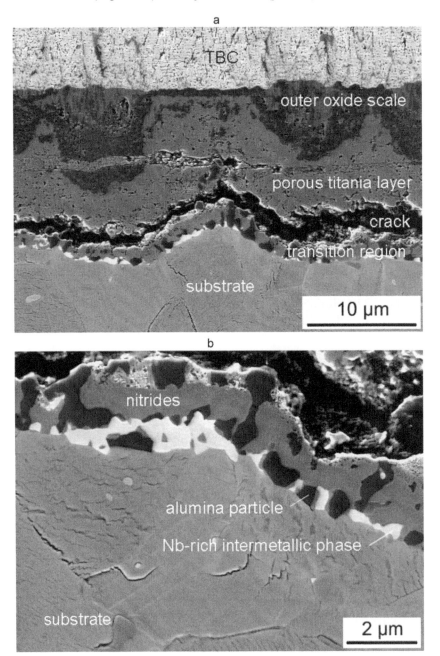

21.3 Scanning electron micrographs of an alloy 98G sample pre-oxidised in air and coated with TBC which was exposed to air at 850 °C for 1100 cycles, showing (a) the thermally grown oxide scale and (b) the transition region between oxide scale and substrate

oxide scale, in particular, near a surrounding alumina rich zone which separated the protrusions from an inner porous titania layer. In the transition region between the oxide scale and the substrate, a discontinuous nitride layer interspersed with alumina particles was observed (Fig. 21.3b, dark phase). At the nitride layer/substrate interface, niobium-rich particles precipitated (bright phase) with the chemical composition of 28.4Ti–38.9Al–32.7Nb (at.%) as measured by EDS analysis. The nitride layer probably consisted of two nitrides, $Ti_2AlN$ and TiN adjacent to the substrate and the oxide scale, respectively, as suggested by EDS analysis. However, the two nitride phases could not clearly be detected because of the thinness of the nitride layer (~1 μm). Using transmission electron microscopy, these two kinds of nitrides were identified at the oxide/substrate interface in a TiAl sample oxidised at 900 °C in air [35]. With increasing partial pressure of oxygen, TiN was oxidised to titanium oxide, thus limiting the presence of the nitride layer to a narrow zone adjacent to the substrate [36]. The niobium-rich intermetallic phase was oxidised above the nitride layer, probably resulting in a porous initial mixture of Al-, Ti-, and Nb-oxides.

The outer oxide scale beneath the TBC contained isolated zirconia particles (Fig. 21.4a). As revealed by EDS elemental profiles across TBC and oxide scale (Fig. 21.4b), titanium and aluminium were detected in the root area of the TBC. In the outer oxide protrusion consisting of titania and alumina, zirconium was found with a concentration of about 0.5 at.%. Niobium was not detected in the outer oxide scale but was present in the porous titania layer beneath.

Figure 21.5 shows an oxide protrusion beneath the TBC grown on an alloy 02G sample exposed to air at 850 °C for 1100 cycles. The outer oxide scale exhibited a columnar structure with strings of alumina grains embedded in rather dense titania (Fig. 21.5a). Large pores were observed in the lower region of the protrusion close to the inner scale. As shown in Fig. 21.5b, at the bottom of the protrusions near the bordering alumina rich zone, zirconia particles were observed, predominantly at the alumina/titania interface. Zirconia particles were also found in the outer oxide scale at different distances from the TBC (Fig. 21.5c). The intercolumnar gaps and pores of the root area of the TBC were filled with titania and alumina, whereas the smaller closed intracolumnar pores were free of oxides (Fig. 21.5d).

The presence of zirconia particles within the oxide, serving as markers to determine the diffusion direction, indicated that this outer oxide scale grew by the outward diffusion of titanium and aluminium cations to the TBC/oxide interface where they formed titania and alumina. During outward growth of the scale, zirconia particles detached from the TBC base region were buried in the oxides and moved inwards due to the outward mass flow. As supported by a quite straight front of the TBC on the oxide scale (Fig. 21.3a), there was an enhanced outward diffusion of titanium and aluminium in the area around the protrusions with concomitant formation of pores near the transition between the oxide scale with the columnar structure and the adjoining alumina-rich layer. This porosity might result from a Kirkendall mechanism of interdiffusion. The rapid outward diffusion of Ti and Al atoms can lead to a vacancy build up resulting in void formation. The porosity might also be associated with the dissolution of $Al_2O_3$ particles at the bottom of the protrusions which re-precipitated in the outer oxide scale near the TBC/oxide interface leaving pores behind. Such a dissolution mechanism of an intermediate $Al_2O_3$ barrier layer was observed during the oxidation of γ-TiAl in air when a critical scale thickness was exceeded [37,38]. The dissolution of alumina in the inner and its re-precipitation in the outer part of the scale was attributed to a decrease in the solubility of $Al_2O_3$ in

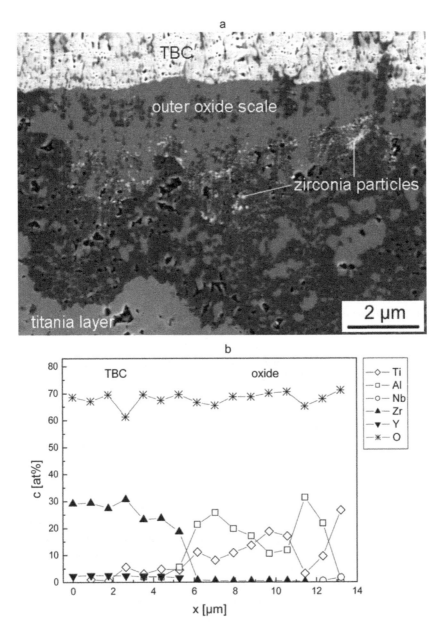

*21.4* Scanning electron micrograph (a) and EDS analysis (b) of a cross-section of an alloy 98G sample pre-oxidised in air and coated with TBC which was exposed to air at 850 °C for 1100 cycles

$TiO_2$ with increasing oxygen pressure [37]. A continuous network of pores along grain boundaries can cause titania grains to drop out during cross-section preparation, resulting in very large pores as observed in Fig. 21.5a. The presence of titania and alumina in the intercolumnar gaps and pores in the TBC root area indicated that outward diffusing Ti and Al cations formed oxides penetrating into interconnected cavities of the TBC.

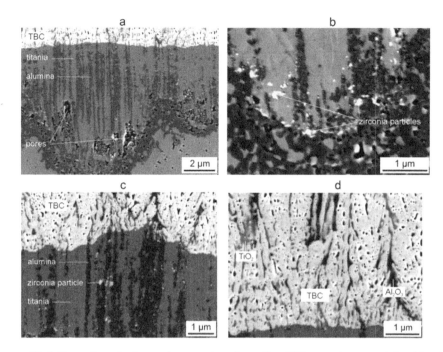

*21.5* Scanning electron micrographs of an alloy 02G sample pre-oxidised in air and coated with TBC which was exposed to air at 850 °C for 1100 cycles, showing (a) the outer oxide scale with columnar structure, (b) zirconia particles at the bottom of the bulge and (c) in the inner part of the outer oxide scale, (d) the root area of the TBC above the outer oxide scale

Figure 21.6 shows the mass gain versus number of cycles for alloy 98G specimens pre-oxidised in oxygen at low partial pressure which were exposed to air at 850 °C. Samples both bare and coated with TBC did not fail by spallation during the maximum exposure length of 1100 cycles. Taking into account the initial mass loss of specimens coated with TBC which was caused by evaporation of water vapour, the mass gain was rather similar for the specimens without and with thermal barrier coating (see corrected data). Figure 21.7 shows a micrograph and EDS analysis of the cross-section of the sample without TBC after oxidation at 850 °C for 1100 cycles. The grown oxide scale exhibited a layered structure with outer dense nodules of titania and alumina, followed by a layer consisting predominantly of alumina, a porous titania layer, and a discontinuous nitride layer interrupted by alumina particles. The scale thickness was about 12.5 μm. As revealed by the EDS elemental profiles, niobium was found throughout the oxide scale except in the outer nodules (Fig. 21.7b). Nitrogen was detected at the oxide/substrate interface, and the sub-surface region of the substrate was slightly depleted in titanium and enriched in aluminium. Micrographs of the cross-section of the sample with TBC again revealed protrusions of outer oxides with columnar structure (Fig. 21.8a). EDS analysis across the oxide scale (as marked by the arrow) indicated titania in the root area of the TBC and zirconia in the oxide protrusion below (Fig. 21.8b). In the adjacent porous $TiO_2$ layer, small amounts of niobium (~3.0 at.%) and aluminium (~0.3 at.%) were detected. The transition region again consisted of a discontinuous nitride layer (due

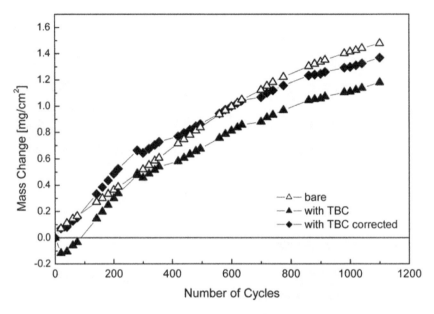

*21.6* Mass change versus number of cycles of alloy 98G specimens with and without thermal barrier coating exposed to air at 850 °C. Specimens were pre-oxidised in oxygen at low partial pressure (mass change of the sample with TBC was corrected for evaporation of water vapour during initial specimen heating)

to the stepwise recording of the elemental profiles, nitrogen was not indicated in the line scan presented, but it was detected in other profiles including an EDS point analysis of the thin transition region).

On the cross-section of the sample pre-oxidised in oxygen at low partial pressure and coated with TBC, areas were sometimes observed where no significant oxide growth occurred during the high-temperature exposure (Fig. 21.9). Intercolumnar gaps and pores in the TBC above these areas were free of oxides. The oxide scale was similar to that found after pre-oxidation in oxygen at low partial pressure (Fig. 21.1b). Metallographic examination of the cross-section revealed a typical width of these areas of about 20 μm. Beneath an outer, thin, mixed scale of alumina and titania, a niobium-containing $\alpha_2$-Ti$_3$Al phase layer formed which was interrupted by alumina particles and a niobium-rich intermetallic phase (Fig. 21.9). EDS analysis of this thin layer, inaccurate due to the approximately 1 μm wide spatial resolution of the electron beam, revealed a chemical composition of the outer part of the layer (dark grey) of 58.7Ti–20.6Al–3.0Nb–17.6O (at.%), whereas the inner part had a composition of 60.4Ti–30.8Al–8.9Nb with probably no oxygen. Nitrogen was not detected in the transition region between the oxide scale and substrate. Therefore, the upper part of the layer might be $\alpha_2$ phase with dissolved oxygen or the Z-phase. The cubic Z-phase with a chemical composition Ti$_5$Al$_3$O$_2$ was reported to form on $\gamma$-TiAl-based alloys when exposed to an oxygen–argon mixture or pure oxygen at high temperature [39–41]. The stability of a protective Al$_2$O$_3$-based scale on some surface regions of $\gamma$-TiAl alloys during oxidation has been attributed to the presence of a single Z-phase layer under the scale [39,42]. Thus, pre-oxidation in oxygen at low partial pressure

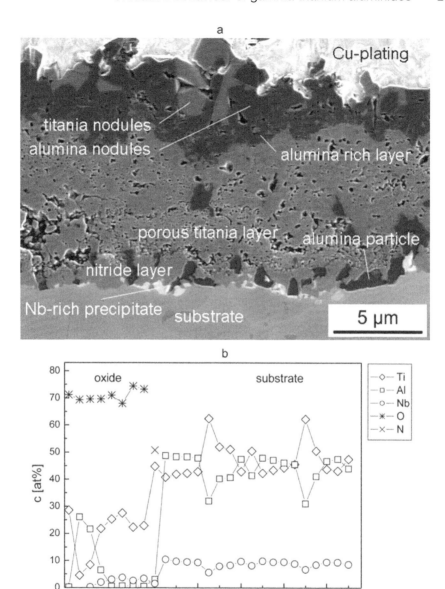

*21.7* Scanning electron micrograph (a) and EDS analysis (b) of a cross-section of an alloy 98G sample without TBC which was pre-oxidised in oxygen at low partial pressure and exposed to air at 850 °C for 1100 cycles

might result in a local improvement of the oxidation resistance due to the formation of the Z-phase; however, the treatment applied did not provide effective protection of the whole sample or even a component of it.

Figure 21.10 shows the mass change versus number of cycles of alloy 98G specimens, bare and with TBC, which were exposed to air at 900 °C. The specimens were

a

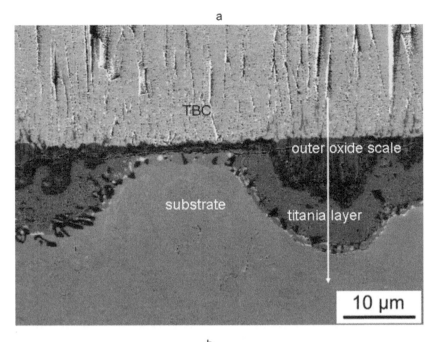

b

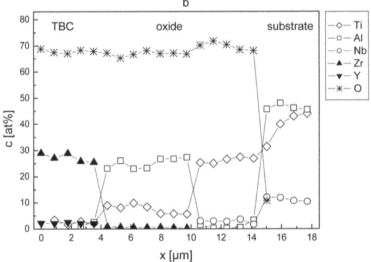

*21.8* Scanning electron micrograph (a) and EDS analysis (b) of a cross-section of an alloy 98G sample pre-oxidised in oxygen at low pressure and coated with TBC which was exposed to air at 850 °C for 1100 cycles (line scan position indicated by arrow)

pre-oxidised in oxygen at low partial pressure. For the uncoated sample, severe spallation of the oxide scale was observed after 540 cycles. The sample with TBC failed after 810 cycles. Mass gains as a function of exposure time were rather similar for both bare and coated specimens. However, failure of the sample with the TBC occurred at a higher number of cycles with a higher mass gain.

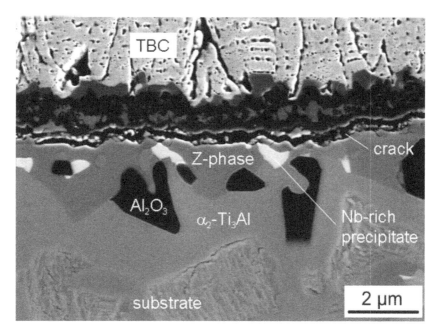

*21.9* Scanning electron micrograph of an alloy 98G sample pre-oxidised in oxygen at partial low pressure and coated with TBC which was exposed to air at 850 °C for 1100 cycles, showing area of thin oxide scale formation

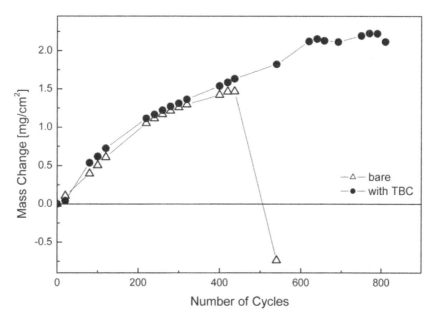

*21.10* Mass change versus number of cycles for alloy 98G specimens with and without a thermal barrier coating exposed to air at 900 °C. Specimens were pre-oxidised in oxygen at low partial pressure

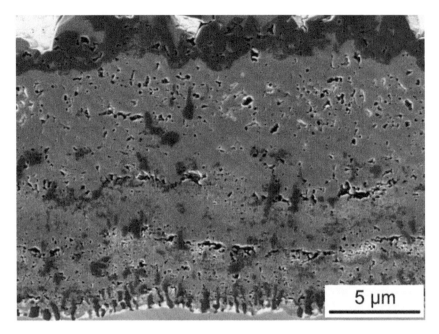

*21.11* Scanning electron micrograph of the oxide scale grown on an alloy 98G sample without TBC which was pre-oxidised in oxygen at low partial pressure and exposed to air at 900 °C for 540 cycles

Figure 21.11 shows the oxide scale on the bare sample after 540 cycles of exposure. In this area of the cross-section, the scale did not spall off, and the entire oxide scale still adhered to the substrate. The scale thickness was about 16 μm. This seemed to be the critical thickness for spallation at 900 °C, whereas the thickness of 12.5 μm observed on the uncoated sample exposed at 850 °C for 1100 cycles (Fig. 21.7a) was below the critical value for the latter exposure temperature. In the transition region between oxide scale and substrate, SEM examination of the cross-section of the sample oxidised at 900 °C revealed a partly discontinuous nitride layer similar to that observed on specimens oxidised at 850 °C (Figs. 21.3 and 21.7a). On the other hand, the transition region was also found to consist of alumina particles extending into a thin zone of $\alpha_2$-Ti$_3$Al phase in which niobium rich precipitates were also embedded (Fig. 21.12a). As confirmed by elemental profiles across the transition region, the $\alpha_2$-phase contained oxygen (Fig. 21.12b). Performing detailed EDS analysis, an average concentration of about 15 at.% was determined. Above the $\alpha_2$-phase zone, nitrogen was also detected.

Failure of the thermal barrier coating observed on the sample thermally cycled at 900 °C occurred in the porous titania layer, predominantly near the titania layer/nitride layer interface. The adherence of the TBC on the thermally grown oxide was excellent (Fig. 21.13a). This might be associated with the formation of an outer oxide beneath the TBC and the growth of oxides into the intercolumnar gaps and pores of the TBC root area. EDS analysis again confirmed the presence of titania in the TBC and zirconia in the oxide scale with the columnar structure (Fig. 21.13b). The measured concentration of zirconium was about 0.5 at.%. Niobium was not detected in the outer oxide, but in the underlying porous titania layer. The absence of niobium as

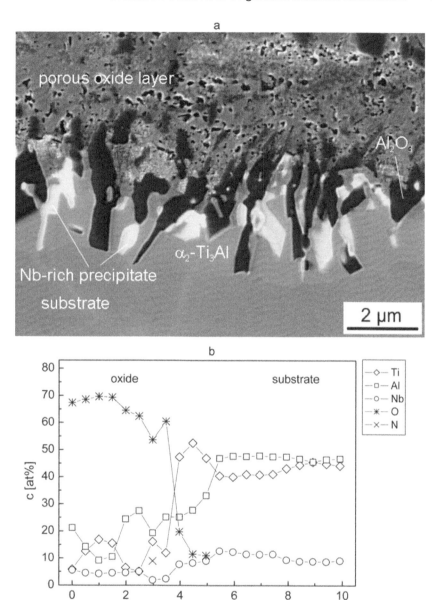

*21.12* Scanning electron micrograph (a) and EDS analysis (b) of the transition region between oxide scale and substrate of an alloy 98G sample without TBC which was pre-oxidised in oxygen at low partial pressure and exposed to air at 900 °C for 540 cycles

well as the presence of zirconia particles indicated that the outer oxide scale formed by outward diffusion of titanium and aluminium cations. The zirconium concentration measured in the outer oxide scale was quite constant. Thus, detection of zirconium was not associated with the zirconia particles but with a low amount of zirconia dissolved in titania and, maybe, alumina. The measured concentration of zirconium,

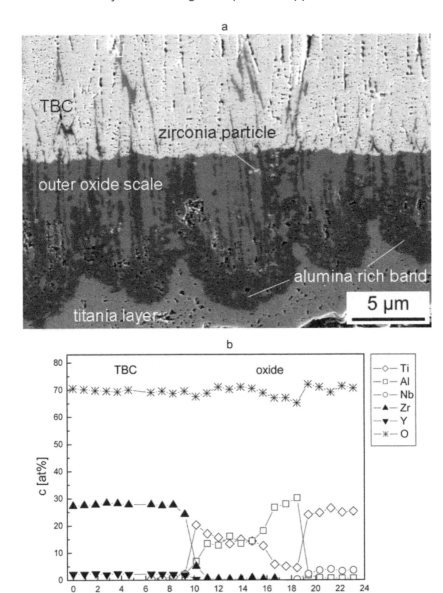

*21.13* Scanning electron micrograph (a) and EDS analysis (b) of a cross-section of an alloy 98G sample pre-oxidised in oxygen at low partial pressure and coated with TBC which was exposed to air at 900 °C for 810 cycles

however, seemed to be too low to be connected with a solid state reaction between the oxide scale and the zirconia top coat, such as the formation of the phase $ZrTiO_4$, which might occur according to the $TiO_2$–$ZrO_2$ phase diagram [43]. Transmission electron microscopy work is ongoing to elucidate this interrelationship.

The presence of thermal barrier coatings on γ-TiAl was found to have a significant effect on the oxide growth at least for long exposure time periods, resulting in a different morphology of the outer oxide layer compared to that on bare material. This change in scale morphology seems to reduce the propensity to spallation and, thus, increases the life-time of the TBC system. However, the mechanism causing the titania/alumina protrusions with columnar structure is not yet understood.

## 21.4    Conclusions

Thermal barrier coatings of yttria partially stabilised zirconia were deposited on $\gamma$-TiAl alloys using the EB-PVD technique. The TBCs were well adherent to the thermally grown oxide scales. No spallation was observed during cyclic oxidation in air at 850 °C for up to 1100 cycles.

For specimens with YPSZ top coats, an outer oxide scale with a columnar structure was observed underneath the thermal barrier coatings, formed by outward diffusion of Ti and Al cations. This scale also contained small amounts of zirconia. In the root area of the TBC, the intercolumnar gaps and pores were filled with titania and alumina.

Thermal cycling at 900 °C caused spallation of oxide scales and thermal barrier coatings with shorter life-time of the scales on the bare material. Failure of the sample with TBC occurred in the porous titania layer, while adhesion of the thermal barrier coating on the grown oxide scale was excellent.

Whereas the concept of thermal barrier coatings was successfully applied on titanium aluminides, effective oxidation protection, e.g. provided by coatings, remains a major issue for high-temperature application of $\gamma$-TiAl alloys. Pre-exposure in a low partial pressure oxygen atmosphere was insufficient to improve the oxidation resistance due to absence of a continuous alumina scale.

## Acknowledgment

The authors would like to thank Dr Y.-W. Kim from UES – Materials & Processes (Dayton, OH, USA) for providing the material and carrying out pre-oxidation in oxygen at low partial pressure.

## References

1. D. M. Dimiduk, *Mater. Sci. Eng.*, A263 (1999), 281.
2. H. Clemens and H. Kestler, *Adv. Eng. Mater.*, 2 (2000), 551.
3. H. Clemens, F. Appel, A. Bartels, H. Baur, R. Gerling, V. Güther and H. Kestler, *Ti-2003, Science and Technology*, Volume IV, 2123, ed. G. Lütjering and J. Albrecht, Wiley-VCH Verlag, Weinheim, 2004.
4. F. Appel and M. Oehring, *Titanium and Titanium Alloys*, 89, ed. C. Leyens and M. Peters, Wiley-VCH Verlag, Weinheim, 2003.
5. X. Wu, *Intermetallics*, 14 (2006), 1114.
6. W.-Y. Kim, *Niobium for High Temperature Applications*, 125, ed. W.-Y. Kim and T. Carneiro, The Minerals, Metals and Materials Society, Warrendale, PA, 2004.
7. M. Yoshihara and Y.-W. Kim, *Intermetallics*, 13 (2005), 952.
8. T. Carneiro and Y.-W. Kim, *Intermetallics*, 13 (2005), 1000.
9. Y. Shida and H. Anada, *Oxid. Met.*, 45 (1996), 197.
10. L. Niewolak, V. Shemet, C. Thomas, P. Lersch, L. Singheiser and W. J. Quadakkers, *Intermetallics*, 12 (2004), 1387.
11. C. Leyens, *Titanium and Titanium Alloys*, 187, ed. C. Leyens and M. Peters, Wiley-VCH Verlag, Weinheim, 2003.

12. L. Singheiser, L. Niewolak, U. Flesch, V. Shemet and W. J. Quadakkers, *Metall. Mater. Trans. A*, 34A (2003), 2247.
13. C. Leyens, R. Braun, M. Fröhlich and P. E. Hovsepian, *J. Met.*, 92 (2006), 17.
14. G. S. Fox-Rabinovich, D. S. Wilkinson, S. C. Veldhuis, G. K. Dosbaeva and G.C. Weatherly, *Intermetallics*, 14 (2006), 189.
15. M. Schütze, G. Schumacher, F. Dettenwanger, U. Hornauer, E. Richter, E. Wieser and W. Möller, *Corros. Sci.*, 44 (2002), 303.
16. A. Donchev, B. Gleeson and M. Schütze, *Intermetallics*, 11 (2003), 387.
17. A. Donchev, E. Richter, M. Schütze and R. Yankov, *Intermetallics*, 14 (2006), 1168.
18. D. J. Wortman, B. A. Nagaraj and E. C. Duderstadt, *Mater. Sci. Eng.*, A121 (1989), 433.
19. S. M. Meier and D. K. Gupta, *Trans. ASME J. Eng. Gas Turbines Power*, 116 (1994), 250.
20. W. Beele, G. Marijnissen and A. van Lieshout, *Surf. Coat. Technol.*, 120–121 (1999), 61.
21. P. K. Wright and A. G. Evans, *Curr. Opin. Solid State Mater. Sci.*, 4 (1999), 255.
22. N. P. Padture, M. Gell and E. H. Jordan, *Science*, 296 (2002), 280.
23. M. J. Stiger, N. M. Yanar, M. G. Topping, F. S. Pettit and G. H. Meier, *Z. Metallkd.*, 90 (1999), 1069.
24. C. Leyens, U. Schulz, K. Fritscher, M. Bartsch, M. Peters and W. A. Kaysser, *Z. Metallkd.*, 92 (2001), 762.
25. V. Gauthier, F. Dettenwanger and M. Schütze, *Intermetallics*, 10 (2002), 667.
26. C. Leyens, R. Braun, P. E. Hovsepian and W.-D. Münz, *Gamma Titanium Aluminides 2003*, 551, ed. Y.-W. Kim, H. Clemens and A. H. Rosenberger, The Minerals, Metals & Materials Society, Warrendale, PA, 2003.
27. R. Braun, C. Leyens and M. Fröhlich, *Mater. Corros.*, 56 (2005), 930.
28. M. Fröhlich, R. Braun and C. Leyens, *Surf. Coat. Technol.*, 201 (2006), 3911.
29. T. Suzuki, M. Goto, M. Yoshihara and R. Tanaka, *Mater. Trans. JIM*, 32 (1991), 1017.
30. S. Taniguchi, T. Shibata, A. Murakami and K. Chihara, *Oxid. Met.*, 42 (1994), 17.
31. D. Legzdina, I. M. Robertson and H. K. Birnbaum, *Acta Mater.*, 53 (2005), 601.
32. H. Nickel, N. Zheng, A. Elschner and W. J. Quaddakkers, *Mikrochim. Acta*, 119 (1995), 23.
33. N. Zheng, W. J. Quaddakkers, A. Gil and H. Nickel, *Oxid. Met.*, 44 (1995), 477.
34. W. J. Quaddakkers, P. Schaaf, N. Zheng, A. Gil and E. Wallura, *Mater. Corros.*, 48 (1997), 28.
35. C. Lang and M. Schütze, *Oxid. Met.*, 46 (1996), 255.
36. F. Dettenwanger, E. Schumann, M. Rühle, J. Rakowski and G. H. Meier, *Oxid. Met.*, 50 (1998), 269.
37. S. Becker, A. Rahmel, M. Schnorr and M. Schütze, *Oxid. Met.*, 38 (1992), 425.
38. A. Rahmel, W. J. Quadakkers and M. Schütze, *Mater. Corros.*, 46 (1995), 271.
39. N. Zheng, W. Fischer, H. Grübmeier, V. Shemet and W. J. Quadakkers, *Scripta Metall. Mater.*, 33 (1995), 47.
40. V. Shemet, H. Hoven and W. J. Quadakkers, *Intermetallics*, 5 (1997), 331.
41. E. H. Copland, B. Gleeson and D. J. Young, *Acta Mater.*, 47 (1999), 2937.
42. V. Shemet, A. K. Tyagi, J. S. Becker, P. Lersch, L. Singheiser and W. J. Quadakkers, *Oxid. Met.*, 54 (2000), 211.
43. H. M. Ondik and H. F. McMurdie (eds.), *Phase Diagrams for Zirconium and Zirconia Systems*, 136. The American Ceramic Society, Westerville, OH, 1998.

# 22

# Role of bond coat depletion of aluminium on the lifetime of air plasma sprayed thermal barrier coatings under oxidising conditions*

D. Renusch, M. Schorr and M. Schütze

*Karl-Winnacker-Institut der DECHEMA e.V., D-60486 Frankfurt am Main, Germany*

*renusch@DECHEMA.de*

## 22.1    Introduction

Failure by spallation of Air Plasma Sprayed – Thermal Barrier Coating (APS-TBC) top coats can be classified into three cases:

Case 1 – thermal fatigue failure
Case 2 – thermal ageing failure
Case 3 – Al depletion failure.

These are illustrated in Fig. 22.1 [1,2]. Cases 1 and 2 are considered to be more or less 'mechanical' failures, while case 3 is considered to be a type of 'chemical' failure. This classification produces a conceptual framework that separates 'mechanical' and 'chemical' spallation mechanisms, which can in turn be modelled separately when the appropriate measurement data are available.

### 22.1.1    Case 1 – Thermal fatigue failure

The sample on the left of Fig. 22.1 failed during thermal cyclic oxidation. The oxidation test was at 1100 °C and during each cycle, the sample experienced 1 h at temperature; also during each cycle, the sample was cooled to room temperature and remained at room temperature for at least 30 min. This type of cycle will produce the maximum thermal expansion mismatch strain (i.e. $\varepsilon = \Delta\alpha.\Delta T$). The macroscopic delamination crack seen in the figure, which plays a critical role with regard to top coat spallation, lies mostly within the yttria stabilised zirconia (YSZ) top coat with some of the cracking extending into the TGO. The sample spalled after 277 cycles. Here it is proposed that this sample failed due to a thermal cyclic fatigue mechanism. Additionally, it must be borne in mind that the TGO thickness is about 6.7 μm and is mostly $Al_2O_3$, which indicates that the bond coat is not critically depleted of Al.

### 22.1.2    Case 2 – Thermal ageing failure

The sample in the middle of Fig. 22.1 failed after isothermal oxidation. It was tested at 1100 °C for 1000 h and experienced only one cooling step. The macroscopic

---

* Reprinted from D. Renusch et al.: Role of bond coat depletion of aluminium on the lifetime of APS-TBC under oxidising conditions. *Materials and Corrosion.* 2008. Volume 59. pp. 547–55. Copyright Wiley-VCH Verlag GmbH & Co. KGaA.

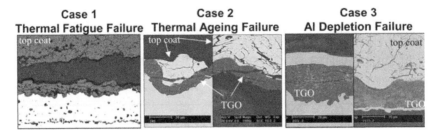

22.1 Spallation cases: Case 1, thermal fatigue failure; Case 2, thermal ageing failure; Case 3, Al depletion failure

delamination crack seen in this photo again passes through the YSZ top coat and with a significant part of the cracking extending into the thermally grown oxide. In contrast to Case 1, the Case 2 sample survived three times longer. The lifetime of cycled TBCs in many cases is shorter than that of isothermally exposed specimens, which is attributed to a thermal cyclic fatigue mechanism caused by thermal cycling that can dominate the lifetime of the sample. The degradation of the top coat under isothermal conditions is likely to be caused by sintering of the YSZ and TGO growth. The TGO in the Case 2 sample is about 9.3 µm thick, which is about 38% thicker than the Case 1 sample. The composition of the TGO is again mostly $Al_2O_3$, which is an indication that the bond coat is not critically depleted in Al.

### 22.1.3   Case 3 – Al depletion failure

The samples of Case 1 and Case 2 show the crack path passing through the top coat. Consequently, the mechanical properties of the YSZ and how the mechanical reliability of the top coat degrades with the respective thermal exposure are suspected to be of crucial importance. The sample on the right side of Fig. 22.1 (spallation Case 3) shows something completely different when compared to Cases 1 and 2. The Case 3 sample was isothermally oxidised at 1050 °C for 5000 h, when it spalled after cooling. The crack face passes only through the TGO. The TGO is >30 µm thick and is composed of $Al_2O_3$ with a large amount of Ni(Co,Cr)-spinels. The formation of the spinels and the TGO thickness, which should be about 13 µm thick when calculated from the $Al_2O_3$ oxidation kinetics data, is a strong indication that the bond coat is critically depleted in Al. Consequently, Case 3 is classified as an Al depletion failure. This type of failure occurs when the Al content within the bond coat falls to a critical value, thus causing the TGO to switch from slow growing $Al_2O_3$ to the fast growing Ni(Co,Cr)-spinels. This thick spinel-rich TGO then spalls during the next cooling period.

The ramifications that the observed results shown above may have on top coat lifetime are schematically plotted in Fig. 22.2. What is considered here is the temperature dependence of the lifetime for the three spallation classes depicted in Fig. 22.1. Here the 'maximum potential top coat lifetime' is expected to be limited by the bond coat depletion of Al. For this spallation case, the crack path is only in the TGO and the TGO contains large amounts of Ni, Cr and Co. For this type of failure, it is expected that the lifetime is limited by the bond coat properties, such as the temperature-dependent oxidation kinetics, the bond coat thickness, the bond coat Al content, and the inter-diffusion of Al into the substrate. For Case 2, thermal ageing failure, after isothermal oxidation, the macroscopic delamination crack passes through the YSZ top coat with some of the cracking extending into the TGO. The

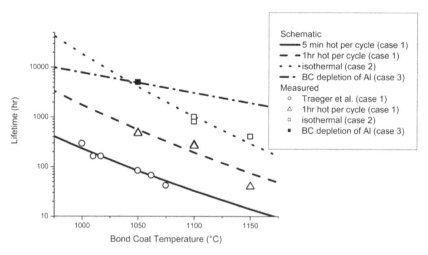

*22.2* Schematic of the temperature dependence of the spallation lifetimes of APS TBC top coats for different spallation mechanisms

degradation of the top coat under isothermal conditions is likely to be caused by sintering of the YSZ and TGO growth. As depicted in Fig. 22.2, this failure mode has a different temperature dependence than that of the bond coat depletion failure mode. This is to be expected because the degradation of the top coat due to sintering, TGO growth and bond coat depletion of Al should have different temperature-dependent kinetics. For spallation Case 1, thermal fatigue failure, acoustic emission measurement data show that the TGO and top coat suffer from cracking every time the sample is cooled during thermal cycling [3–6]. The consequence of the repeated cracking is thermal fatigue and accelerated damage accumulation, which in turn causes the top coat lifetime to shorten. The driving force behind the thermal fatigue is mostly the thermal expansion mismatch stresses/strains, which will have a different temperature dependence than the degradation mechanisms for Case 2 and Case 3 spallation. The solid line in Fig. 22.2 is for spallation caused by burner rig testing where the data points are taken from Traeger *et al.* [7]. It should be pointed out that the samples tested in this project and by Traeger *et al.* are produced by the same TBC manufacturer. It is expected that the rapid cycle frequency of the burner rig, where the sample is cooled every 5 min, causes much more thermal fatigue than cyclic oxidation, which in turn produces a considerably shorter top coat lifetime. The consequence of Fig. 22.2 is that the TBC lifetime is dependent on the exposure parameters, such as temperature and hot dwell time. Laboratory samples are tested under accelerated conditions (i.e. short lifetimes <1000 h), such as higher temperatures (>1050 °C) and rapid cycle frequencies (hot dwell times <2 h) and may show a thermal fatigue failure mechanism. However, the TBC in a land-based gas turbine with a lifetime of 25 000 h, an inspection interval of 1000 h and a bond coat temperature of 950 °C may suffer from spallation due to bond coat depletion of aluminium. Consequently, the failure mode observed in the laboratory samples may not be the same as the spallation cases observed in the turbine.

## 22.2     Measurements

The APS-TBC systems under investigation are listed in Table 22.1. The elemental compositions of the substrates and vacuum plasma sprayed (VPS) bond coats are

*Table 22.1*  TBC systems

| Short Name | Substrate | Bond coat |
|---|---|---|
| 12/56 SX | CMSX-4 | 150 µm of Abler Ni 192-8 |
| 12/56 DS 150 | CM247-DS | 150 µm of Abler Ni 192-8 |
| 12/56 DS 75 | CM247-DS | 75 µm of Abler Ni 192-8 |
| 12/34 | IN738 | 150 µm of Abler Ni 192-8 |
| 8/34 | IN738 | 150 µm of SC2231 |

*Table 22.2*  Elemental compositions

| Substrate or bond coat | Elemental composition (wt.%) |
|---|---|
| CMSX-4 | 61.7 Ni, 10 Co, 6.0 Cr, 5.6 Al, 6.0 W, 0.6 Mo, 1.0 Ti, 6.0 Ta, 3 Re, 1 Hf |
| CM247 | 61.9 Ni, 9.0 Co, 8.5 Cr, 5.6 Al, 9.0 W, 0.5 Mo, 0.7 Ti, 4.0 Ta, 0.07 C, 0.7 Si |
| IN738 | 61.7 Ni, 8.5 Co, 16 Cr, 3.4 Al, 1.7 W, 2.6 Mo, 3.5 Ti, 1.7 Ta, 0.9 Nb, 0.17 C |
| Abler Ni 192-8 | 47.25 Ni, 22 Co, 17 Cr, 12.5 Al, 0.6 Y, 0.25 Hf, 0.4 Si |
| SC2231 | 30 Ni, 38 Co, 27 Cr, 8.0 Al, 0.5 Y, 0.5 Si |

listed in Table 22.2. The TBC systems are given short names in order to expedite the text. As an example, the short name 12/56 SX is a TBC system with bond coat and substrate Al contents of 12.5 wt.% and 5.6 wt.%, respectively. The SX and DS stand for single crystal and directionally solidified, respectively. The bond coat thicknesses of the 12/56 DS 150 and 12/56 DS 75 are 150 µm and 75 µm, respectively. The top coats are APS yttria stabilised zirconia (YSZ) with ~12% cumulative porosity and with thicknesses of 150 µm and 300 µm.

For the prediction of spallation due to bond coat depletion of aluminium, the Al depletion kinetics have to be modelled and the Al content compared to a critical Al content. This requires a significant amount of measurement data. At this point in the project, 37 samples have been isothermally oxidised in lab air at 1100, 1050, 1000 and 950 °C for a variety of oxidation times up to 5000 h. For all 37 of these samples, the bond coat and substrate Al contents have been measured with the electron microprobe (EPMA) line scan. Additionally, another 36 samples are in the furnaces at 1050 and 950 °C for a variety of oxidation times up to 10 000 h, and the results will be published in a later paper.

An example of the EPMA line scan data is shown in Fig. 22.3 for the 8/34 TBC system. Three line scans are made per sample (Figs. 22.3a and 22.3c). The three line scans are averaged together (Figs. 22.3b and 22.3d). During the early stages of the Al depletion kinetics, this averaging has the effect of smoothing out the β-phase precipitates (compare Figs. 22.3a to 22.3b). At the end of the TBC lifetime, the averaging has a very minor effect (compare Figs. 22.3c to 22.3d).

The TGO thickness has been measured from 64 oxidised and cross-sectioned samples from all five systems in Table 22.1, where the thickness is measured at 15 to 30 different locations. The samples were isothermally and cyclically oxidised at 1100, 1050, 1000 and 950 °C for a variety of oxidation times up to 5000 h. The thickness

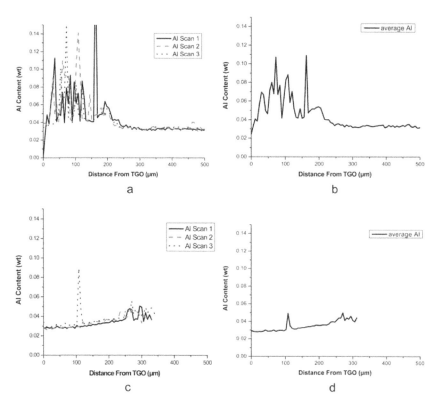

22.3 Example of electron microprobe line scan data for the 8/34 TBC system.
(a) Three line scans after oxidation for 24 h at 1100 °C. (b) The average of the
three scans in (a). (c) Three line scans after oxidation for 1000 h at 1100 °C.
(d) The average of the three scans in (c)

data from all 64 samples have been simultaneously fitted to the TGO kinetics
equation below:

$$d_{TGO}(t,T) = k_{o-TGO}\exp\left(\frac{-E_{A-TGO}}{R \cdot T}\right) \cdot \left(\frac{t}{t_o}\right)^n \qquad [22.1]$$

where $k_{o\text{-}TGO}$, $E_{A\text{-}TGO}$ and $n$ were found to be 6.5 µm, 55 030 J mol$^{-1}$, and 0.34,
respectively. In the equation, $T$, $t$ and R are the temperature (K), time (s) and the
Gas Constant, respectively, where $t_o$ is equal to 1 s. The systematic error of the fit
was found to be 0.2% with a standard deviation of 19%. Examples of the TGO data,
the best fit, and the Error Distribution are shown in Fig. 22.4.

Thirty-four of the samples investigated from the 12/56 SX system were suffering
from 'cauliflower' internal oxidation [2] (Fig. 22.5). The internal oxidation kinetics
were measured in the same way as the TGO growth. The data were again simultaneously fitted to the temperature dependent kinetics equation below:

$$d_{inox}(t,T) = k_{o-inox}\exp\left(\frac{-E_{A-inox}}{R \cdot T}\right) \cdot \left(\frac{t}{t_o}\right)^n \qquad [22.2]$$

where $k_{o\text{-}inox}$, $E_{A\text{-}inox}$ and $n$ were found to be 3710 µm, 155 800 J mol$^{-1}$, and 0.66, respectively. The systematic error of the fit was found to be 7.4% with a standard

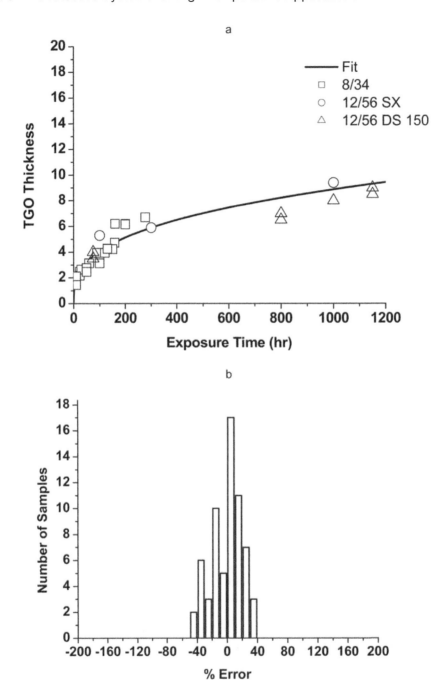

22.4 (a) TGO oxidation kinetics at 1100 °C, measured from cross-section samples. (b) Error distribution for the best fit of 64 TBC samples that were oxidised at 1100, 1050, 1000 and 950 °C for a variety of oxidation times up to 5000 h

a

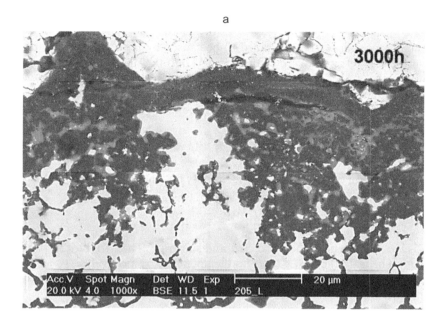

b

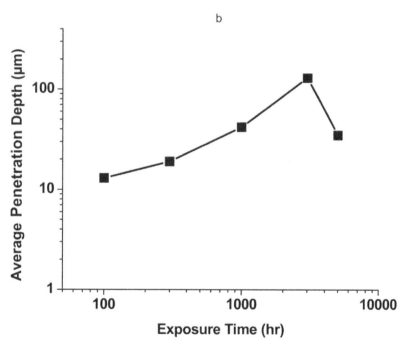

22.5  (a) 'Cauliflower' internal oxidation, measured from cross-section TBC system 12/56 SX samples, after oxidation at 1050 °C. (b) Average penetration depth of the internal oxidation

deviation of 31%, which reflects the non-systematic behaviour of this type of internal oxidation. This non-systematic behaviour is also shown in Fig. 22.5, where the mean penetration depth at 5000 h is less than the mean penetration depth at 3000 h.

## 22.3    Al depletion model

Bond coat depletion of Al is assumed to be caused by both the outward diffusion of Al into the TGO and inward diffusion of Al into the substrate. This behaviour is currently being modelled by considering the average bond coat Al content after an exposure time and at a position within the bond coat or substrate (i.e. binary diffusion). The model is a solution of Fick's Second Law, Eq. 22.3:

$$\frac{\partial C}{\partial t} = D \frac{\partial^2 C}{\partial z^2} + Q \qquad [22.3]$$

where the diffusion coefficient is dependent only on temperature, Eq. 22.4:

$$D(T) = D_o \exp\left(\frac{-E_{A-Al}}{R \cdot T}\right) \qquad [22.4]$$

and $C$ is the Al content of the bond coat and substrate. The variables $z$, $t$, and $T$ are the distance beneath the TGO, time, and temperature, respectively.

Equation 22.3 has been solved by separating diffusion of Al into the TGO ($C_{out}$) from inward diffusion of Al into the substrate ($C_{in}$), Eqs. 22.5–22.7:

$$C(Z,t,T) = C_{in}(Z,t,T) - C_{out}(Z,t,T) \qquad [22.5]$$

where

$$C_{in}(Z,t,T) = \frac{(C_{bc} - C_{sub})}{2}\left[erf\left(\frac{d_{bc} - z}{2\sqrt{tD}}\right) + erf\left(\frac{d_{bc} + z}{2\sqrt{tD}}\right)\right] + A \qquad [22.6]$$

and

$$C_{out}(Z,t,T) = \frac{(C_{tgo} + C_{inox})}{B}\left[1 - erf\left(\frac{z}{2\sqrt{tD}}\right)\right] \qquad [22.7]$$

The variables $C_{bc}$ and $C_{sub}$ are the Al contents in wt.% of the as-sprayed bond coat and substrate, respectively, while $d_{bc}$ and $d_{sub}$ are the thicknesses of the as-sprayed bond coat and substrate, respectively, and $C_{tgo}$ and $C_{inox}$ are the aluminium consumed by the TGO growth and internal oxidation, respectively. The parameters $A$ and $B$ have to be found by applying the mathematical boundary conditions. Plots of Eqs. 22.5, 22.6 and 22.7 are shown in Fig. 22.6. Here it is evident that inward and outward diffusion can be investigated separately, which is helpful for understanding the role that the substrate and TGO play in the consumption of bond coat Al.

Equations 22.5, 22.6 and 22.7 are not the complete solution to Fick's Second Law, because the boundary conditions have to be enforced. The first of the four boundary conditions is that as the time goes to zero ($t \rightarrow 0$ h), the modelled Al contents in the bond coat and substrate go to $C_{bc}$ and $C_{sub}$, respectively. The model does meet this condition, as shown in Fig. 22.6c, which is a calculation based on the 12/56 systems where $C_{bc}$ and $C_{sub}$ are 0.125 and 0.056 wt.%$_g$, respectively.

The second condition is that, in the absence of oxidation and only during inter-diffusion of Al between the bond coat and substrate, the overall Al reservoir content must remain constant, see Eq. 22.8 for $C_R$, the average Al reservoir content:

$$C_R = \frac{C_{bc}d_{bc} + C_{sub}d_{sub}}{d_{bc} + d_{sub}} \qquad [22.8]$$

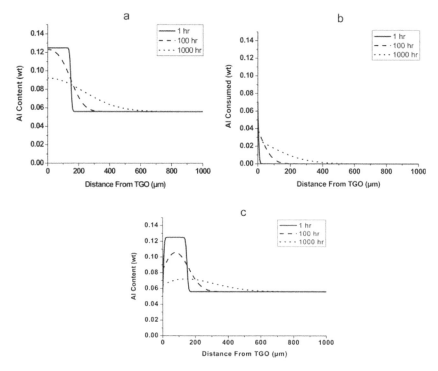

*22.6* Plots of Al diffusion equations 22.5, 22.6 and 22.7. (a) Al diffusion into the substrate. (b) Al consumed by TGO growth. (c) Al diffusion into substrate and consumed by TGO growth

Equation 22.8 gives a constant value for $C_R$ that is based only on the as-sprayed sample parameters. In order to meet this condition, Eq. 22.6 has to be integrated with respect to the distance below the TGO, as in Eq. 22.9:

$$C_R = \frac{\int_{0}^{d_{bc}+d_{sub}} C_{in}(Z,t,T)dz}{d_{bc}+d_{sub}} \quad [22.9]$$

After solving Eq. 22.9, the parameter $A$ was defined by Eq. 22.10:

$$A = C_R - \frac{(C_{bc}-C_{sub})}{2(d_{bc}+d_{sub})}\left[\begin{array}{l}(2d_{bc}+d_{sub})erf\left(\frac{2d_{bc}+d_{sub}}{2\sqrt{tD}}\right)-d_{sub}erf\left(\frac{d_{sub}}{2\sqrt{tD}}\right)\\ +2\sqrt{\frac{tD}{\pi}}\left(\exp\left(-\left(\frac{2d_{bc}+d_{sub}}{2\sqrt{tD}}\right)^2\right)-\exp\left(-\left(\frac{d_{sub}}{2\sqrt{tD}}\right)^2\right)\right)\end{array}\right]+$$

$$\frac{(C_{bc}-C_{sub})}{2} \quad [22.10]$$

The equation for parameter $A$ is long, which is unfortunate. However, Eq. 22.10 is necessary because it prevents Eq. 22.6 from 'creating' or 'destroying' Al during inter-diffusion of aluminium from the bond coat into the substrate. This conservation of Al has been verified by numerically integrating the mass balance of aluminium.

The third boundary condition is that, during oxidation, the outward-diffusing Al must be consumed by the growing TGO. This condition can be enforced again by integrating the following equation for the TGO growth:

$$d_{tgo} = \frac{\rho_{BC}}{\rho_{ox}} \frac{3}{2} \frac{M_O}{M_{Al}} \int_0^{d_{bc}+d_{sub}} C_{out}(Z,t,T)dz = k_n(T) \cdot t^n \qquad [22.11]$$

After solving Eq. 22.11, the parameter $B$ in Eq. 22.7 was found to be:

$$B = (d_{bc} + d_{sub})\left[1 - erf\left(\frac{d_{bc}+d_{sub}}{2\sqrt{tD}}\right)\right] + 2\sqrt{\frac{tD}{\pi}}\left(1 - exp\left(-\left(\frac{d_{bc}+d_{sub}}{2\sqrt{tD}}\right)^2\right)\right) + 1 \quad [22.12]$$

Again, Eq. 22.12 is necessary because it prevents Eq. 22.7 from 'creating' or 'destroying' Al during outward diffusion of aluminium into the TGO. This conservation of Al has also been verified by numerically integrating the mass balance of aluminium.

The fourth boundary condition is that, as the exposure time becomes extremely large (>$10^5$ h), the bond coat and substrate Al content should go to zero. For the above equations as the time goes to infinity, the Al content goes to negative infinity. This is not physically possible but mathematically necessary. The modelled Al content starts with a positive value and passes through zero before reaching negative infinity. This approach is commonly use for analytical solutions of Fick's Second Law.

The aluminium consumed by the TGO growth ($C_{tgo}$) is in Eq. 22.13, where $\rho_{bc}$ and $\rho_{ox}$ are the densities of the bond coat and the $Al_2O_3$ TGO, respectively, and values used are 7.1 and 3.8 g cm$^{-3}$, respectively. The terms $M_{O_2}$ and $M_{Al}$ are the molar masses of oxygen and Al, respectively:

$$C_{tgo} = \left(\frac{\rho_{ox}}{\rho_{bc}} \frac{2}{3} \frac{M_{Al}}{M_{O_2}}\right) d_{tgo}(t,T) \qquad [22.13]$$

The aluminium consumed by the internal oxidation is given by Eq. 22.14. The constant $C_{bc}/2$ is considered a *first approximation*. The amount of Al consumed by 'cauliflower' internal oxidation is not described in the literature. Consequently, the value of $C_{bc}/2$ was found by analysing the measured Al profiles for samples oxidised through 5000 h at four different temperatures.

$$C_{inox} = \frac{1}{2}C_{bc}d_{inox}(t,T) \qquad [22.14]$$

## 22.4    Results

Currently, the model is being used by curve fitting it to the measured Al concentration profiles. The fit parameter is only the temperature-dependent diffusion coefficient. The number of samples investigated, and under which conditions, are given in Table 22.3. The model has been curve fitted once for each temperature in Table 22.3, where the temperature-dependent diffusion coefficient is extracted for each of the four temperatures. For each temperature listed in Table 22.3, all of the samples for all of the systems listed are fitted simultaneously. At this point in the study, the model has been applied to about 2400 line scan data points from 37 samples from

*Table 22.3*   Statistics of measured and modelled results

| | Temperature (°C) | | | |
|---|---|---|---|---|
| | 1100 | 1050 | 1000 | 950 |
| Number of samples | 15 | 12 | 5 | 5 |
| Number of data points | 998 | 828 | 259 | 297 |
| Systems | 12/56 SX | 12/56 SX | 12/56 SX | 12/56 SX |
| | 12/34 | 12/34 | | |
| | 8/34 | | | |
| Systematic error (%) | 2.1 | 3.6 | 1.4 | 4.6 |
| Standard deviation (%) | 19 | 20 | 18 | 17 |

three different TBC systems that have been oxidised for up to 5000 h at four different temperatures. The error distributions are found in Fig. 22.7. The systematic errors and standard deviations are listed in Table 22.3.

An example of the measured and modelled Al concentration profiles for the 8/34 system after oxidation at 1100 °C can be seen in Fig. 22.8. The measured and modelled results agree quite nicely for the samples oxidised for 24, 300, and 1000 h. The sample oxidised for 100 h also shows some agreement between measured and modelled results, but not as good as the other three examples in this figure. More results are plotted in Fig. 22.9. This shows the measured and modelled Al concentration profiles for two samples oxidised for 5000 h at 1050 °C. The systematic error and

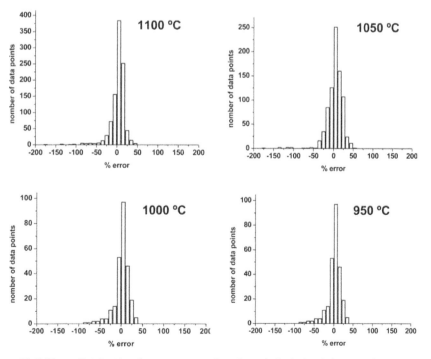

*22.7* Error distribution from measured and modelled aluminium profiles

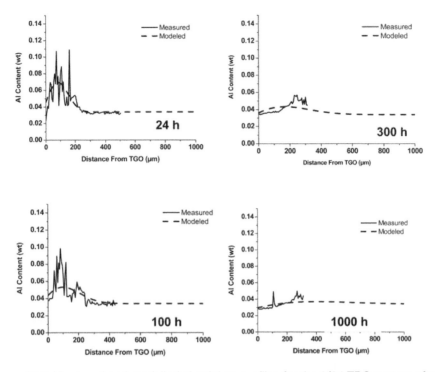

*22.8* Measured and modelled aluminium profiles for the 8/34 TBC system after oxidation at 1100 °C

standard deviation for the 12/56 SX system are 0.4% and 17%, respectively. For the 12/34 system, the systematic error and standard deviation are 7% and 14%. The agreement between the measured and modelled Al concentration profiles is shown in Table 22.3 and Figs. 22.7, 22.8 and 22.9 are considered to be quite good from the standpoint that the only fit parameter is the temperature-dependent diffusion coefficient and that the measured data set is huge.

The values found for the diffusion coefficients are plotted in Fig. 22.10. The above model was run twice, once by ignoring the 'cauliflower' internal oxidation and once by using Eq. 22.14. Modelling with Eq. 22.14 caused a net improvement with regard to the systematic error and the standard deviation. By including the internal oxidation, the values found for the diffusion coefficients behave more systematically, which is reflected in Fig. 22.10. The best fit to Eq. 22.5 found $D_o$ and $E_{A-Al}$ to be 2.38 m$^2$ s$^{-1}$ and 369 503 J mol$^{-1}$, respectively. Also plotted in Fig. 22.10 are values from the literature for the diffusion coefficients, from Evans and Taylor [8], and Nesbitt [9], where the agreement is nearly perfect.

The bond coat Al depletion kinetics are found by numerically integrating Eq. 22.5 over the bond coat thickness and dividing the result by the bond coat thickness, as shown in Eq. 22.15. This approach was found to produce the most stable description of bond coat aluminium depletion:

$$C_d(Z,t,T) = \frac{1}{d_{bc}} \int_0^{d_{bc}} C(Z,t,T)dz \qquad [22.15]$$

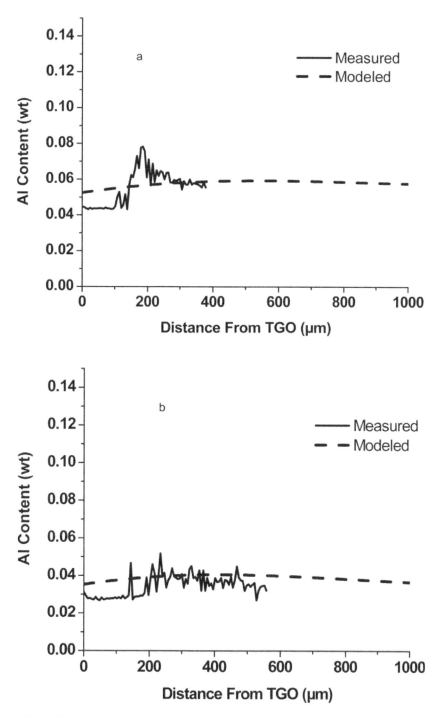

*22.9* Measured and modelled aluminium profiles after 5000 h of oxidation at 1050 °C for: (a) the 12/56 SX system and (b) the 12/34 system

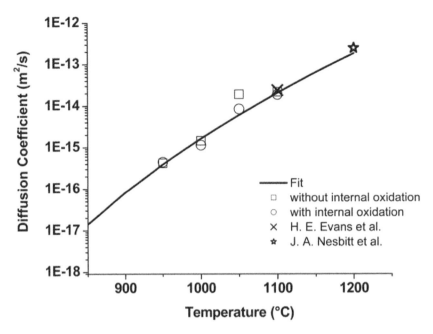

*22.10* Binary diffusion coefficient for bond coat aluminium diffusion

The measured and modelled bond coat Al depletion kinetics for exposures at 1050 °C for the 12/56 SX and 12/34 TBC systems are plotted in Fig. 22.11, where the measurement data extend out to 5000 h. The measured values are the average bond coat Al contents, where the error bars are the standard deviations of the measured values. As stated above, 12/56 SX suffered from 'cauliflower' internal oxidation. Consequently, the dashed modelled line plotted in Fig. 22.11a, which takes into account internal oxidation, fits through the measured data points. The 12/34 system did not suffer from as much internal oxidation. Consequently, the solid modelled line plotted in Fig. 22.11b nearly fits the measured values perfectly. Figure 22.11 also shows that 'cauliflower' internal oxidation consumes a large amount of aluminium to the extent that it can be considered a life limiting factor.

In order to use the above model to predict TBC lifetime, the Al depletion kinetics has to be compared to a critical Al content ($C_{Al,cr}$). The numerical value of the critical Al content was determined using the oxidation parameters that produce spallation Case 3 in Fig. 22.1, namely 1050 °C and 5000 h. The value calculated for $C_{Al,cr}$ is 3.14 wt.%, when internal oxidation was taken into the calculation. The lifetimes predicted by this approach are plotted in Fig. 22.12, while lifetimes with and without internal oxidation are shown in Figs. 22.12a and 22.12b, respectively. The model predicts that TBC systems with larger Al reservoirs have longer lifetimes. It also shows that internal oxidation is a major life limiting factor. For example, the 12/56 SX system has a minimum lifetime of less than 5000 h with internal oxidation and greater than 100 000 h without internal oxidation. The above model assumes that the substrate acts as part of the Al reservoir, thus supplying the bond coat with Al, which can be consumed by the growing TGO. At this point in the study, the measurement data seem to support this assumption. However, the samples tested have only been

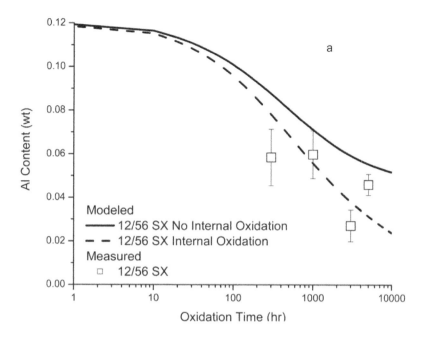

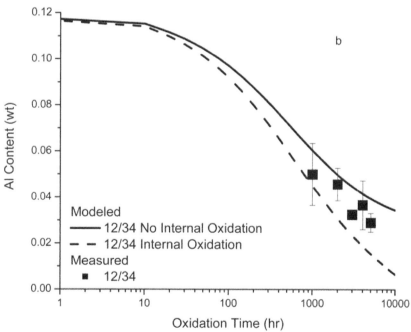

*22.11* Measured and modelled bond coat Al depletion kinetics for oxidation at 1050 °C for: (a) the 12/56 SX system and (b) the 12/34 system

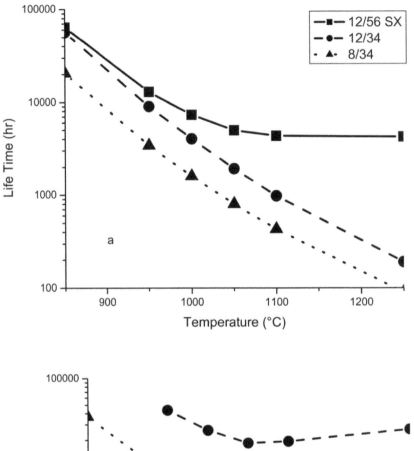

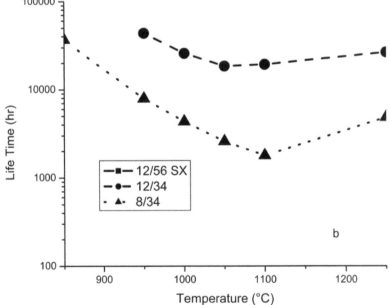

22.12 Spallation lifetime due to bond coat depletion of Al. (a) With internal oxidation and (b) without internal oxidation

oxidised for up to 5000 h. This assumption may have to be revaluated when the current exposure tests, which are up to 10 000 h, are completed. There is a possibility that the aluminium diffusion rate in the substrate is a little slower than that in the bond coat, because the two alloys have different grain structures. The bond coat diffusion rate may be higher because the bond coat is polycrystalline and grain boundary diffusion is possible. However, in the case of the substrate CMSX-4, which is single crystal, grain boundary diffusion is not possible. This can be modelled by using two diffusion coefficients, one for the bond coat and one for the substrate. This of course would change the prediction made in Fig. 22.12. At this point in the study, it is not believed to be necessary to add the additional complexity of two diffusion coefficients, but by the end of the current investigation, it may become necessary.

## 22.5    Summary and conclusions

In this paper, the spallation of the APS-TBC top coat is classified into three cases, namely: Case 1 – thermal fatigue failure, Case 2 – thermal ageing failure, and Case 3 – Al depletion failure. This classification produces a conceptual framework that separates 'mechanical' and 'chemical' spallation mechanisms, which can in turn be modelled separately when the appropriate measurement data are available. This paper focuses on the Al depletion failure and how it relates to top coat spallation.

The bond coat and substrate Al concentration profiles have been measured with the electron microprobe in line scan mode for 37 samples that have been isothermally oxidised in lab air at 1100, 1050, 1000 and 950 °C for a variety of oxidation times up to 5000 h. These measurement data have been modelled by considering Al diffusion into both the substrate and TGO. The measured and modelled results are in agreement to within a standard deviation of about 20%. This modelling approach produces values for the binary diffusion coefficient for Al in a vacuum plasma sprayed bond coat. Additionally, the bond coat aluminium depletion kinetics have been calculated and again compared to the measurement data. Finally, the Al depletion kinetics have been compared to a critical value of Al content for the purpose of making spallation lifetime predictions.

## References

1.  D. Renusch, H. Echsler and M. Schütze, *Mater. Sci. Forum*, 461–464 (2004), 729–736.
2.  H. Echsler, D. Renusch and M. Schütze, *Mater. Sci. Technol.*, 20 (2004), 307.
3.  D. Renusch, H. Echsler and M. Schütze, *Mater. High Temp.*, 21 (2004), 65–76.
4.  D. Renusch, H. Echsler and M. Schütze, *Mater. High Temp.*, 21 (2004), 77–89.
5.  D. Renusch, H. Echsler and M. Schütze, 'Life time modelling of APS-TBC by using acoustic emission analysis', in Conf. Proc. Turbomat-Symposium, 48–52, Bonn, 17–19 June 2002. DLR Köln, 2002.
6.  R. Herzog, O. Trunova, R.W. Steinbrech and L. Singheiser, 'Failure modes and damage evolution of plasma-sprayed thermal barrier coatings', in First Japanese-German Workshop on: Properties and Performance of Thermal Barrier Coating Systems and Factors Affecting it, 13–15 September 2006, Institute for Materials Technology, Darmstadt University of Technology.
7.  F. Traeger, M. Ahrens, R. Vaßen and D. Stöver, *Mater. Sci. Eng.*, A358 (2003), 255–265.
8.  H. E. Evans and M. E. Taylor, *Oxid. Met.*, 55 (2001), 17.
9.  J. A. Nesbitt, *J. Electrochem. Soc.*, 136 (1989), 1511.

# 23

## A software tool for lifetime prediction of thermal barrier coating systems*

### E. P. Busso

*Centre des Matériaux, Mines Paris, Paristech, UMR CNRS 7633, 91003 Evry, France*

*esteban.busso@ensmp.fr*

### H. E. Evans

*Department of Metallurgy and Materials, University of Birmingham, Birmingham, UK*

*H.E.Evans@bham.ac.uk*

### L. Wright, L. N. McCartney and J. Nunn

*National Physical Laboratory, Teddington, Middlesex TW11 0LW, UK*

*louise.wright@npl.co.uk; neil.mccartney@npl.co.uk; john.nunn@npl.co.uk*

### S. Osgerby

*Alstom Power, Rugby, UK*

*steve.osgerby@power.alstom.com*

### 23.1   Introduction

In modern gas turbines, thermal barrier coatings (TBCs) provide the essential technology controlling the performance and lifetime of key high-temperature components. It is therefore critical to be able to understand the mechanisms controlling the failure of TBCs and to be able to predict these events. A typical TBC system, particularly for land-based gas turbines, is made of a top layer of yttria stabilised zirconia (YSZ, the insulating layer) and a MCrAlY (where M is a combination of Ni and Co) low pressure plasma sprayed (LPPS) bond coat. During exposure to high temperatures, the metallic bond coat forms a thermally grown oxide (TGO) that consists predominantly of alumina. Most current approaches relate TBC failure to stress generation at or near the TGO (e.g. see Refs. 1–5).

Methods for predicting spallation and failure in TBCs have generally followed two main approaches. The first approach uses empirical fatigue life models that link the life of the TBC to the total amount of damage caused by oxidation and mechanical straining [6–8]. This approach has merit in that interpolation may be made within the data base but phenomenological and mechanistic processes are generally not

---

* Reprinted from E. P. Busso et al.: A software tool for lifetime prediction of thermal barrier coating systems. *Materials and Corrosion.* 2008. Volume 59. pp. 556–65. Copyright Wiley-VCH Verlag GmbH & Co. KGaA.

considered. The second approach uses delamination models to consider edge crack-ing and buckling [9]. This approach neglects the fact that there is generally a lengthy period of sub-critical crack formation and growth before the delamination stage which, largely, determines the life of the TBC system.

The work reported here investigates the local stresses that drive crack formation and growth within a TBC system, and the use of these stresses within a software tool for predicting lifetime of a coating system. The approach is based on linking models of TGO formation, continuum mechanics (including thermal stresses, viscoplasticity and sintering), and measurements of the properties of a coating system (material properties of the individual components, morphological properties of its internal microstructure, and damage behaviour before coating failure).

The models of TGO formation and continuum mechanics are solved using a finite element (FE) method. Such methods [10] have a well-established history of use in the field of continuum mechanics. The methods solve partial differential equations by:

- subdividing the region of interest into non-overlapping smaller regions (the 'elements' or 'mesh', with the element's corners being the 'nodes')
- approximating the variables of interest, in this case displacements, within each element using simple parametric functions such as linear variations interpolating the values at the nodes
- applying the governing partial differential equations at each point to generate a large set of simultaneous equations linking the unknown values at the nodes and
- solving the simultaneous equations, using an appropriate numerical algorithm, to calculate the values of the unknowns at the nodes and using the local approxima-tions and the nodal values to calculate results at any other points of interest.

As computational power has become more widespread, FE methods have become more widely used, and the models that can be solved using FE methods have become more complicated and sophisticated than the simple linear elasticity models that the techniques were originally developed to solve. In particular, time-dependent pro-cesses such as creep, sub-critical crack growth [11] and displacements due to oxide growth can now be incorporated [12–14].

The full details of the FE models used in this work are described later in the paper. They simulate an IN738/MCrAlY/EB-PVD system in the region of the top coat/bond coat interface and the TGO, predicting the stress distribution as it evolves during TGO growth and calculating the effects of the cooling cycle on the peak stress values and locations. Each model represents a single possible shape of the top coat/bond coat interface and a single exposure temperature, and so each model represents a unique pairing of coating system and operating conditions. The interface shape has been parameterised based on micrographs of the interface, as is explained in the next section, in order to quantify the roughness of the interface so that its effect on stresses and hence lifetime can be investigated.

The results of these FE models have been used in a software tool that interpolates between the values of temperature and the geometric parameter in order to calculate lifetime estimates for service conditions and coating system interface shapes that have not been explicitly modelled. This paper describes the principles of the software tool, and the development of the algorithms that are implemented therein. Validation of the approach is in progress through comparison of predictions with non-destructive measurements on the coating system.

The paper is divided into five further sections. Section 23.2 describes the measurements that are required in order to generate the input data for the software tool, including damage measurement and material properties. Section 23.3 outlines the finite element models used to generate data for the interpolation routines used in the software tool. Section 23.4 describes the three-step methodology underpinning the software tool. Section 23.5 explains the details of the software, the ways in which it can be used, and its strengths and limitations. Section 23.6 summarises the key points of the paper and details the planned future work that will lead to improved software and a better understanding of the assumptions made in the development of the software tool.

## 23.2    Software input data

The simulation of TGO formation and the resultant stress distributions within the TBC is a complex process, and requires a large amount of information about the various properties of the TBC system and its failure mechanisms.

Two main types of measured property are required for lifetime prediction. The first type consists of properties of the TBC system, including the material thermo-mechanical properties and the details of the morphology of the interface between the top coat and the bond coat. This set is required for creation of FE models that simulate the formation of the TGO. The second type is required for use of the lifetime prediction tool. This set consists of damage and possibly stress measurements, and a failure stress.

The finite element model, described more fully in the next section, requires various thermo-mechanical properties to be known for the various TBC system constituents, including the TGO. The system modelled in the work presented here consists of a low pressure plasma sprayed NiCoCrAlY bond coat and an electron beam physical vapour deposition (EB-PVD) YSZ top coat on a superalloy substrate. The bond coat and TGO have been modelled as elastic-viscoplastic isotropic materials, with all of their properties being temperature-dependent. The top coat has been modelled as a transversely isotropic elastic material with temperature-dependent properties, and incorporates a sintering model to account for the consolidation of the top coat as a function of exposure time. The superalloy is not modelled explicitly, and only its coefficient of thermal expansion is required for the model. Sources for the values used in the model and experimental methods for their determination are given in Refs. 12–16.

In addition to the thermo-mechanical properties, the FE model also requires an expression for the growth rate of the TGO layer and a measure of the expansion caused by oxidation. The growth rate expression is an experimentally determined function of time and temperature. The expression used in the model is dependent on the accumulated time spent above a critical temperature. The expansion caused by oxidation has been determined using the Pilling–Bedworth ratio and an experimentally determined ratio for the in-plane and out-of-plane expansions; see further details in Refs. 12, 13 and 16.

Finally, in order to create the geometry of the FE model, a description of the shape of the original interface between the top coat and the bond coat is required. A typical unit cell of the interface is characterised as shown in Fig. 23.1. The quantities $a$ (half-wavelength) and $b$ (amplitude) vary along the interface, since the interface is not a regular sinusoid. Analysis of micrographs from a scanning electron microscope makes it possible to calculate distributions for the quantities $a$ and $b$ [17]. The work

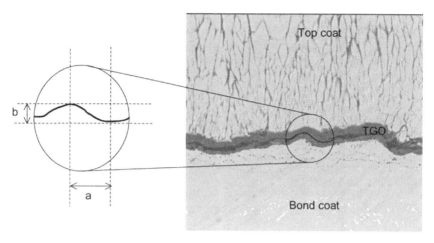

*23.1* Geometric parameters used to characterise the interface between the top coat and the bond coat

described here used FE models based on the 17th, 50th, and 83rd percentiles of these distributions.

The objective of the damage measurements is to estimate the percentage of the area of the TBC that has become debonded due to microcracks. As will be explained later, the methodology described in this paper links such damage measurements to critical stresses and to lifetime predictions. Two methods have been used for damage measurement: Raman spectroscopy and thermography.

Raman spectroscopy [18] links a relaxation of compressive stress to damage within the TBC. As well as indicating the damaged regions of the TBC, it measures the volume-averaged hydrostatic pressure in the TGO. The technique can be difficult to perform and the results can be difficult to interpret since the signals are sometimes of low power. The results can also suffer from spatial granularity because the area over which the measurement is taken is often much smaller than the gap between measurements.

Thermography [19] exploits the fact that the presence of cracks normal to the free surface inhibits heat transfer between the layers of the TBC and the substrate. The temperature distribution of the top surface of a sample is measured using a thermal imaging camera after either the top layer or the substrate of a sample has been heated for a short time. If the substrate has been heated (via conduction from a hot plate), then damaged regions appear cooler than undamaged ones because the heat cannot flow up from the substrate. Conversely, if the top surface has been heated (by radiation) then damaged areas appear hotter because the heat does not flow to the substrate.

The technique only requires a small amount of heat (successful results have been obtained after using the experimenter's hands to warm the sample), and the measurements only need to be relative temperatures. The granularity of the technique is less of a problem than it is for the Raman technique because the thermal image covers the whole sample and pixel size is quite small. The main drawbacks of the technique are that it may not pick up fine detail since the heat flow in the plane of the sample will blur small areas of damage, and it does not provide a measure of the stresses within the sample.

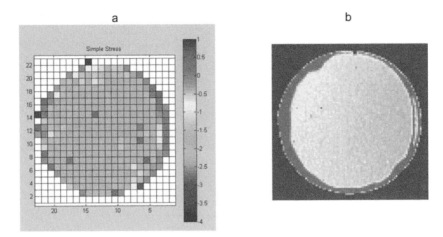

*23.2* Results of damage measurement methods: (a) results of Raman spectroscopy; (b) results of thermography. In both images, red areas indicate damage

Comparison of the techniques has shown that their results agree, as shown in
Fig. 23.2. The samples shown in Fig. 23.2 have experienced a total of seven 100-h
cycles in a Carbolite F1500 muffle furnace, with a dwell temperature of 1000 °C, a
cooling rate of –100 K per hour down to room temperature, and a heating rate of
+100 K per hour. The dwell time at room temperature varied between 2 days and
approximately 2 weeks.

The left-hand part of Fig. 23.2 shows the Raman spectroscopy measurements
taken after cooling the specimen to room temperature. Here, red squares indicate a
stress of approximately 0 (i.e. damaged regions) and blue squares indicate a compressive stress of about 4 MPa (i.e. undamaged regions). The right-hand figure shows a
thermal image of the same sample after surface heating, and the red (damaged) areas
are in the same place as those found by Raman spectroscopy.

The software tool requires a failure stress in order to predict coating lifetime. This
stress should be one of the four stress types interpolated by the software tool, namely
(i) volume-averaged hydrostatic pressure in the TGO (which is negative), (ii) peak
out-of-plane stress in the top coat, (iii) peak maximum principal stress in the TGO,
or (iv) peak traction on the TGO interfaces. If a suitable failure stress value cannot
be found experimentally, it is possible to obtain a suitable estimate by using the software tool on previously studied coatings that have failed. This approach will be
described in Section 23.5.1 of this paper.

As well as damage data, a large amount of lifetime data has been assembled. Figure
23.3 plots the time to spallation of the top coat, over at least 20% of the specimen
surface, against the exposure temperature. The solid line indicates the time at which
the TGO would reach a thickness of 3 μm for each temperature. The plot suggests
that the time to reach a thickness of 3 μm is a suitable lower bound for failure time,
and gives a reasonably accurate estimate of failure times at higher temperatures.

Whilst the 3 μm TGO thickness has been used in this work to define limits for the
FE model run time, it is a conservative limit for practical applications. In Fig. 23.3,
it can be seen that many of the specimens would have spalled at TGO thicknesses
larger than 3 μm. The spread in failure times is roughly a factor of 10 and it is clearly
important to understand the mechanistic factors which contribute to this spread. In

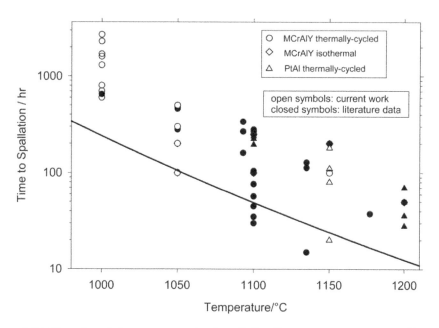

*23.3* Experimental measurements of spallation time versus temperature. The solid line shows time taken for the TGO to reach 3 µm thickness. The diagram is based on that given in Ref. 17

the absence of such understanding, it can be necessary in practice to use a conservative estimate of TGO thickness with a consequent loss of the potential that a TBC system offers.

The driving force for damage and spallation is not the TGO thickness itself, it is the local stresses caused by the TGO growth. These stresses are dependent on the local shape of the top coat/bond coat interface. For instance, if the interface is perfectly flat then the interfacial tractions will be zero for an adherent oxide and the out-of-plane YSZ stresses will be very small, so it is likely that the TGO would be able to reach a thickness significantly greater than 3 µm without damage occurring. Conversely, a more convoluted interface will fail at smaller TGO thicknesses because of large interfacial tractions. It is the present contention that variations in bondcoat surface roughness play a significant role in producing the scatter in TBC lifetimes shown in Fig. 23.3. This premise is supported by recent experimental work [20] which shows that hand polishing a MCrAlY bond coat substantially increased (by a factor of ∼10) the lifetime of a TBC system.

## 23.3    FE models

The aim of the FE models was to generate information about the peak stresses within the TBC system, given its interface morphology (i.e. values of *a* and *b*) and a thermal profile. Throughout the modelling activity, the thermal profile has been characterised by a prolonged period at a fixed dwell temperature followed by cooling to 25 °C in half an hour. The models simulate the oxidation and its effects as a quasi-static process. Crack nucleation and growth are not modelled explicitly: the aim of the models is to examine the stresses that lead to the damage, rather than to follow the progress of the damage as it develops.

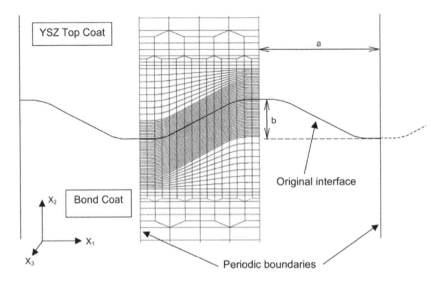

*23.4* Illustration of the use of periodic cells to simulate an infinite sinusoidal interface

The FE models used in this work have been developed as periodic cell models using generalised plane strain elements. The use of periodic cells, as illustrated in Fig. 23.4, means that the interface is treated as a single infinite sinusoid, equivalent to modelling a volume at the centre of a large object. Whilst the damage shown in Fig. 23.2 has occurred at the edges of the sample, this is not the only location at which damage has been seen during testing. In real turbine blades, damage and spallation occur at a range of locations on the blade, not just at the coating edges. The periodic cell illustrated in Fig. 23.4 has been used for the modelling as it is representative of damage to more general shapes of object than disc samples.

The use of generalised plane strain elements means that the model treats the strain in one direction as a spatially uniform value that varies in time, and so the model can be reduced to two dimensions. The combination of these assumptions means that the interface is modelled as the surface created by sweeping a sinusoid in the direction perpendicular to the plane it lies in, such as a corrugated sheet.

The FE models assume that the temperature distribution within the system is uniform. The only differential expansion that occurs at high temperature is caused by oxidation (the differential thermal shrinkages are taken into account during the cooling stage). Oxidation is simulated by changing the material properties of the elements around the top coat/bond coat interface to the properties of the TGO at the appropriate time, determined from the TGO growth rate expression mentioned in Section 23.2 and the distance of the element from the initial interface. The thermal expansion of the substrate is used to create the in-plane boundary conditions for the model; as the substrate is relatively massive compared to the coating, it will control the overall deformation of the coating before spallation.

All of the FE models have been run until the TGO thickness reaches 3 μm. As was stated in the previous section, this thickness provides a lower bound for an estimate of failure time and so was a convenient time to choose as a stop time. Since the time does not represent an upper bound, some samples may not have failed by this time, which makes extrapolation necessary within the software tool, as will be described in the next section. However, running the model for significantly longer simulation times would have been computationally expensive and so the need for extrapolation was accepted.

Models have been run with exposure temperatures of 1000 °C, 1050 °C, and 1100 °C, and with values of interface morphology parameters ($a$, $b$) of (4.02, 0.79) µm, (7.64, 2.47) µm, and (9.49, 4.54) µm. Every possible combination of temperature and interface morphology parameters was used, leading to nine runs in total. Further runs with intermediate values of $a$ and $b$ are planned. The values used were chosen by considering the distributions of $b/a$ (which is considered as characterising the interface) and $b$ to be perfectly correlated, choosing appropriate percentiles of these distributions, and finding $a$ from the chosen values of $b/a$ and $b$. The results of the FE models will not be presented here. For further information, see Refs. 16 and 17.

The results of the FE models used for the interpolation consist of a set of stresses (either peak values or volume-averaged values within some defined region of the TBC/TGO) corresponding to a known time, a known dwell temperature, and a known value of $b/a$. Hence if three of the four values {stress, time, temperature, $b/a$} are known, the fourth could be determined by interpolation between the results.

The use of finite element models to simulate the oxidation process has a number of drawbacks. The main drawback is that the models are very specific: each model simulates a single dwell temperature, a single interface morphology and a single set of material parameters. If the system and conditions of interest are not described by these quantities, the model results will not describe the system.

Another drawback is that the models are computationally expensive and require a long time to run. The time taken for the TGO to reach a thickness of 3 µm is inversely proportional to the temperature, so at the low dwell temperatures within the range considered the model time is longer than that at high temperatures. Since some of the material models involve visco-plasticity, and since oxidation leads to a sudden change in material properties, the simulations often require small time steps for convergence. The lowest temperature used in the work reported here was 1000 °C, which required approximately 250 h simulated time to reach 3 µm TGO thickness, with a computer run time of 4 days using an IRIX64 UNIX-based operating system.

These drawbacks mean that a more general methodology for life prediction in TBC systems with a shorter running time would be extremely useful. Finite element models give an excellent insight into the effects of the oxidation on the stress distribution within the TBC system, since they allow consideration of how the effects at a microscopic scale lead to an effect at the macroscopic scale. Their results provide a basis for the interpolations carried out during lifetime predictions which could not be obtained by any other means. The next section explains the approach developed by the authors, which uses the results of the FE models to predict coating lifetime under known operating conditions.

## 23.4    Lifetime prediction methodology

The lifetime prediction methodology combines a set of damage measurements of a TBC system with the results of a set of FE models of the same TBC system to predict the lifetime of the coating system without modelling the system and conditions explicitly, thus saving time and computational expense. The methodology is outlined in brief here and is described in more detail elsewhere [16,17]. The methodology is based on the following assumptions:

1.  The measured damage is caused by crack nucleation in the regions where the interface between the top coat and bond coat has its steepest local slope (i.e. largest value of $b/a$). These are regions where FE analyses [17] show that maximum stresses develop.
2.  Stress values for temperatures and interface geometries that have not been modelled explicitly can be obtained via interpolation of existing model results.
3.  Failure stress is a property of the TBC system and is not dependent on temperature nor interface morphology.

Figure 23.5 shows a flow chart of the methodology, which is a three-step process. The first step associates measured damage data with the roughness of the interface between the top coat and bond coat. As stated above, it is assumed that the damage occurs where the value of $b/a$ is largest. So for example, if the measurements show that 16% of the TBC is damaged, then the damage has occurred in the regions with the highest 16% of values of $b/a$. Since the distribution of the values of $b/a$ is known from micrographs, the value of $b/a$ (called $r_c$) corresponding to each damage measurement can be found. This process is illustrated in Fig. 23.6. Hence for a given time and temperature, a value of $b/a$ is associated with each damage measurement.

The second step involves interpolation of the FE model results to associate a stress with the known time and temperature, and the value of $b/a$ identified in the first step. As was stated in the previous section, the form of the FE results means that if three

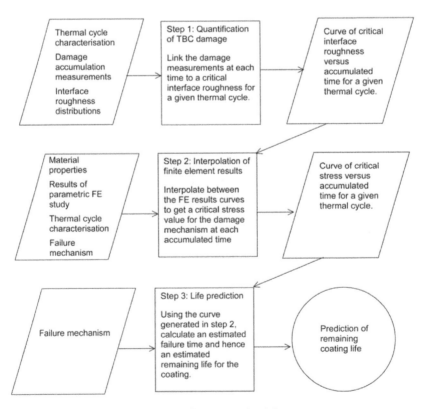

*23.5* Flowchart illustrating the software methodology

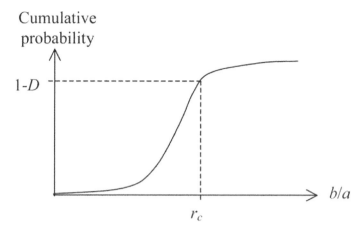

23.6 Sketch illustrating the calculation of $r_c$ for a measured fractional damage level $D$ $(0 \leqslant D \leqslant 1)$. The cumulative probability distribution for $b/a$ is determined from micrographs

of the values {stress, time, temperature, $b/a$} are known, the fourth value can be determined by interpolation. Hence the known temperature and time combined with the value of $r_c$ from step 1 lead to a critical stress value $\sigma_c$. This stress is the critical value of the local stress responsible for local microcrack initiation and growth within the TBC. The stress is regarded as critical because it is associated with the measured damage level via the value of $r_c$ identified in step 1. The output of this step is a curve of $\sigma_c$ versus accumulated time, corresponding to a known dwell temperature. The interpolation procedure implemented in the software will be described in more detail in the next section. The general procedure is illustrated in Fig. 23.7.

The final step, illustrated in Fig. 23.8, uses the results of step 2 and a known failure stress value, $\sigma_f$, to estimate the remaining life of the coating. The failure stress must be the same type of stress (e.g. maximum principal stress within the TGO, interfacial traction, etc.) as the interpolated values $\sigma_c$, and is assumed to be a property of the TBC system independent of time or interface morphology. The curve of $\sigma_c$ versus accumulated time produced in step 2 is interpolated with respect to stress to find the time at which the failure stress $\sigma_f$ is achieved. This time is the expected failure time for the coating system. The remaining coating life is found by subtracting the current time from the failure time.

## 23.5    Software implementation

The software methodology outlined above has been implemented using Fortran for the interpolation routines, with an Excel spreadsheet and associated Visual Basic macros as the front end. The user inputs all of the required information into the spreadsheet, and the macros run the interpolation routines and display the results. Figure 23.9 shows a screen shot of the input sheet of the tool. The FE model results used in the software were generated using Abaqus Standard, a finite element package using an implicit solver for static and quasi-static problems. The simplicity of the methodology used means that the tool takes a matter of seconds to carry out the calculations.

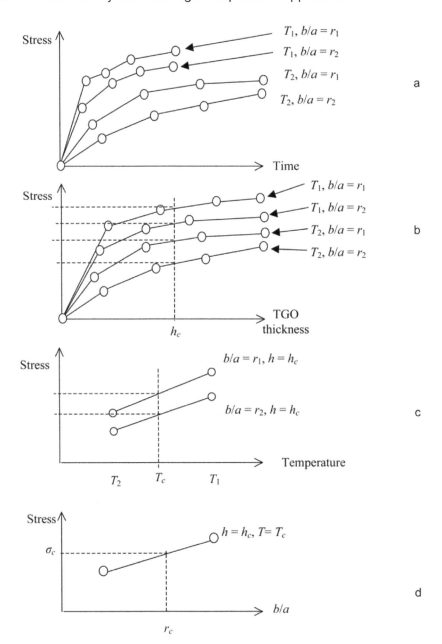

*23.7* Sketch illustrating the interpolation of FE model results to obtain a value of $\sigma_c$ relating to a time $t_c$, and oxide thickness $h_c$, a value of $b/a$ of $r_c$, and a dwell temperature $T_c$. (a) Original data with each curve being a different combination of $T$ and $b/a$. (b) Time axis converted to oxide thickness. (c) Single value $h_c$. (d) Single values of $h_c$ and $T_c$. Only four curves are shown for clarity: in general, there are more. Dotted lines indicate that stress is interpolated with respect to the quantity on the horizontal axis at the value shown on the horizontal axis

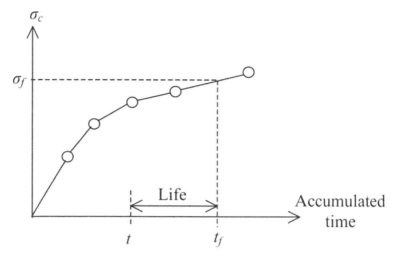

23.8 Sketch illustrating the interpolation of a curve of $\sigma_c$ versus time at a stress value $\sigma_f$ to obtain failure time $t_f$ and hence predicted life ($t$ is the time now)

The front end of the tool consists of three main spreadsheets. The 'Input and control' sheet, as shown in Fig. 23.9, allows the user to enter the input values (e.g. temperature, time, $b/a$, stress, etc.) and to choose the mode of operation (described in detail later). A set of push-button controls run the interpolation macros and allow the user to read in and plot the interpolated results. The results are displayed as numerical values on the 'Output' sheet. A number of different plots are offered via a tick-box menu, with the available plots depending on the mode of operation. The plots are created automatically on the Plots sheet. The front end file also contains the FE results that are interpolated but these sheets are hidden from the user as they are unnecessary for normal operation of the tool.

The tool allows four different peak stress measures to be used in the interpolation routines: volume-averaged hydrostatic pressure in the TGO; peak out-of-plane stress in the top coat; peak maximum principal stress in the TGO; and peak interfacial traction. Of these, the first is most likely to be the measured stress, particularly if Raman spectroscopy is used for the measurement, and the other three stresses are thought to drive crack formation. The out-of-plane stress component is responsible for the formation of in-plane microcracks, which are generally seen within the YSZ near the TGO interface. The maximum principal stress is generally identified as the main 'driving force' for the nucleation of cohesive microcracks within the TGO. The interfacial tractions, normal to either the top coat-TGO or the bond coat-TGO interfaces, drive the nucleation of interfacial microcracks and propagation of interfacial defects.

As stated above, if three of the four quantities {stress, temperature, time, $b/a$} are known, then the fourth can be determined via interpolation. In the software tool, the choices require the user to know either {stress, temperature, time} to obtain $b/a$ [Mode 1 operation], or {temperature, time, $b/a$} to obtain stress [Mode 2 operation]. If the latter option is chosen and a failure stress $\sigma_f$ is supplied, then the software also calculates an estimate of the remaining life.

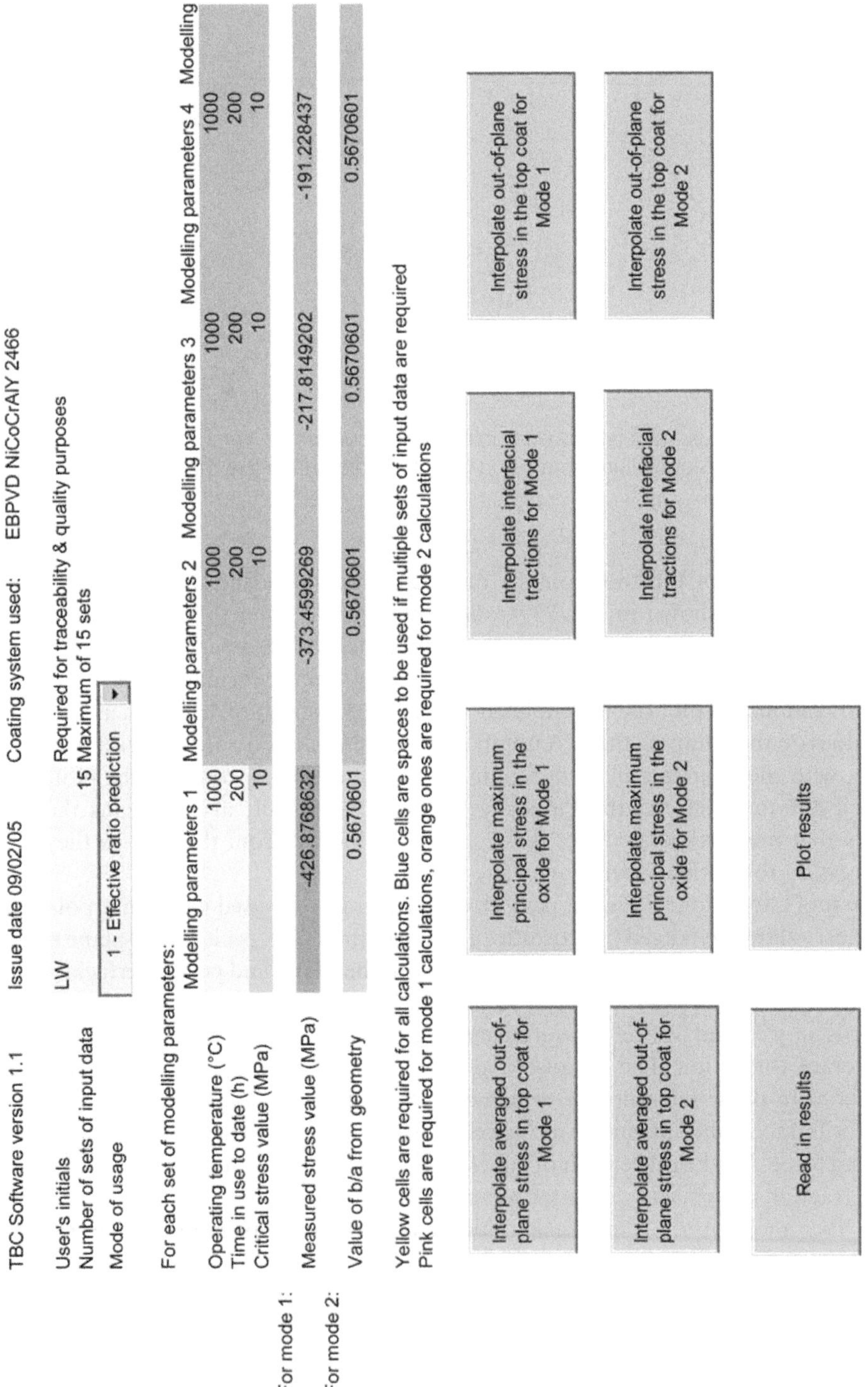

TBC Software version 1.1    Issue date 09/02/05    Coating system used:    EBPVD NiCoCrAlY 2466

User's initials    LW
Number of sets of input data    15 Maximum of 15 sets    Required for traceability & quality purposes
Mode of usage    1 - Effective ratio prediction

For each set of modelling parameters:

| | Modelling parameters 1 | Modelling parameters 2 | Modelling parameters 3 | Modelling parameters 4 | Modelling |
|---|---|---|---|---|---|
| Operating temperature (°C) | 1000 | 1000 | 1000 | 1000 | |
| Time in use to date (h) | 200 | 200 | 200 | 200 | |
| Critical stress value (MPa) | 10 | 10 | 10 | 10 | |
| For mode 1: Measured stress value (MPa) | -426.8768632 | -373.4599269 | -217.8149202 | -191.228437 | |
| For mode 2: Value of b/a from geometry | 0.5670601 | 0.5670601 | 0.5670601 | 0.5670601 | |

Yellow cells are required for all calculations. Blue cells are spaces to be used if multiple sets of input data are required
Pink cells are required for mode 1 calculations, orange ones are required for mode 2 calculations

Interpolate averaged out-of-plane stress in top coat for Mode 1

Interpolate maximum principal stress in the oxide for Mode 1

Interpolate interfacial tractions for Mode 1

Interpolate out-of-plane stress in the top coat for Mode 1

Interpolate averaged out-of-plane stress in top coat for Mode 2

Interpolate maximum principal stress in the oxide for Mode 2

Interpolate interfacial tractions for Mode 2

Interpolate out-of-plane stress in the top coat for Mode 2

Read in results

Plot results

23.9 Screen shot of the software tool

Within the software, the interpolations are not carried out with respect to time because the length of time simulated by the FE models depends on temperature. Instead, the times are converted to TGO thicknesses by using the empirical relationship between time, temperature and TGO thickness (called $h$). The assumption of linearity between result points is more reasonable with respect to $h$ than it is with respect to time. In addition, since all FE models were run until the same TGO thickness is reached (3 μm), interpolating between results of different models with respect to $h$ is simpler than interpolating with respect to time. When calculating the failure time, the thickness is converted back into an exposure time before reporting the result.

The FE models that provide the interpolation data for the tool have been run with every possible combination of three dwell temperatures and three values of $b/a$ (as stated in Section 23.3), giving nine different sets of results for each failure criterion. More sets of results will be added once the relevant FE model runs are completed. Before interpolation, the software identifies the values of temperature (and of $b/a$ if required) that lie on either side of the values specified by the user, and uses the stress versus $h$ curves corresponding to those temperatures for interpolation.

The majority of interpolations are linear, with one exception (see below). The steps in the interpolation procedures for each mode of usage are:

**Mode 1** (determination of $b/a$: stress, temperature and time (and hence $h$) known):

1. Linearly interpolate the chosen curves with respect to $h$ to obtain stress values for various values of $b/a$ and temperature at the known value of $h$.
2. Linearly interpolate the results of step 1 with respect to temperature to obtain stress values for various values of $b/a$ and the known values of $h$ and temperature.
3. Linearly interpolate the results of step 2 with respect to stress to obtain the appropriate value of $b/a$ for the known values of stress, $h$, and temperature.

**Mode 2** (determination of stress and life: failure stress, $b/a$, temperature, and time (and hence $h$) known):

1. Linearly interpolate all curves with respect to $h$ to obtain stresses for all values of $b/a$ and temperature at the known value of $h$.
2. Linearly interpolate the results of step 1 with respect to temperature to obtain stresses for all values of $b/a$ at the known values of $h$ and temperature.
3. Quadratically interpolate the results of step 2 with respect to $b/a$ to obtain a stress corresponding to the known values of $h$, temperature and $b/a$.
4. Linearly interpolate within all curves corresponding to the temperatures on either side of the known temperature so that they all have values of stress corresponding to the same values of $h$.
5. Linearly interpolate between results of step 4 that correspond to the same temperature with respect to $b/a$ to produce a curve of stress versus $h$ for the temperatures on either side of the known temperature and the known value of $b/a$.
6. Interpolate linearly between the two curves produced in step 5 to give a single curve of stress versus $h$ at the known temperature and known value of $b/a$.
7. Linearly interpolate the curve produced in step 6 with respect to stress to obtain the value of $h$ corresponding to the failure stress at the known values of temperature and $b/a$. This critical TGO thickness can then be used to calculate the failure time and hence the remaining coating life.

### 23.5.1   Obtaining the failure stress

Obtaining the failure stress requires damage measurements to full spallation from a TBC system whose interface morphology has been characterised. The last damage measurement before spallation can provide a bounding value for the failure stress. The damage measurement leads to a value of $b/a$ that can be used in the software to generate a value of critical stress that is a lower bound for the failure stress. The accuracy of the bound can be improved if additional measurements are available.

As an example, a set of measurement data obtained at NPL has been processed using the tool. The final measurement point before spallation, obtained after 700 h at 1000 °C, showed that 22% of the sample surface was damaged. Interpolation of the measured interface profile gave a value of $r_c = 0.411$ of the critical value of $b/a$. The values of $r_c$, time and temperature were then used in the software tool to determine limiting values of each possible failure stress. The results of these calculations are shown in Table 23.1.

### 23.5.2   Caveats and limitations

Whilst this software tool is much quicker to run than the full FE models and can allow for a wider range of operating conditions, it makes a number of assumptions that need to be noted. The first assumption, which may be key to the use of the tool for calculations on TBCs in service on turbine blades, is that thermal cycling has no effect over and above prolonged exposure to high temperature. In effect, the software assumes that exposing a TBC system to 1000 °C for 100 h is equivalent to exposing it to 1000 °C for four 25-h periods with cooling and reheating stages between the high-temperature periods. This assumption is currently being checked and conclusions will be reported at a later date.

As has been stated above, the software tool does not simulate crack nucleation and propagation directly. Instead, the tool relies on the assumption that the damage occurs wherever the ratio $b/a$ is largest, i.e. the steepest areas of the interface. In reality, it is possible that the cracks nucleate at these steep points, but then the damage increases via crack propagation rather than via further nucleation of cracks in areas with lower $b/a$ values. In many cases, crack growth causes damage to accelerate rapidly and leads to full failure, so it is likely that the damage measurements taken before full failure are caused by crack nucleation rather than growth.

All of the FE models were run until the TGO reached a thickness of 3 μm. If a system has not failed by this point, then the predictions for its stress levels are likely to involve extrapolation, and so may be less accurate than the interpolated values. Furthermore, the lifetime predictions may be negative since the system is expected to

Table 23.1   Possible values of each critical stress determined from experimental measurement data

| Failure stress criterion | Value from software tool, MPa |
| --- | --- |
| Volume-averaged hydrostatic pressure in the TGO | −812 |
| Peak out-of-plane stress in the top coat | 1190 |
| Peak maximum principal stress in the TGO | 1420 |
| Peak interfacial traction | 990 |

have failed by this point. Care must be taken when interpreting results under these circumstances.

Another issue related to the FE model results is that of existence and uniqueness. If, as is shown in Fig. 23.10(a), the curve of critical stress versus time passes through the nominal failure stress value more than once, the predicted failure time may be non-unique. Care must be taken when interpreting any non-unique results. Common sense suggests that the coating should fail the first time it passes through the failure value, but this may not be the case. Similarly, if the curve never reaches the failure stress (shown in Fig. 23.10(b)), the tool will predict that the coating never fails which is clearly not the case.

Another point to note about the tool is that it is still of comparatively limited applicability: the stresses used in the interpolation routine relate to a single set of material properties, and if the system of interest has different material properties then the FE model results are unlikely to be a suitable description for the system of interest. The methodology described in this paper could be applied to any set of results, so a new version of the tool could be produced to model a different system, but as it stands the tool only describes one system.

## 23.6    Conclusions and future work

A software tool has been created that uses the results of finite element models and measured damage values to predict the remaining lifetime of a thermal barrier coating system. The tool provides lifetime predictions in a matter of seconds, rather than the hundreds of hours required to run a full FE analysis, which makes it possible to investigate a number of potential failure scenarios in very little time. The tool can also be used to estimate failure stresses for subsequent lifetime predictions in similar systems.

The next phase of the work is to expand the range of FE model results used in the interpolation routine. More values of $b/a$ will be used in the new models in order to fill in the gaps in the results.

The effects of thermal cycling will be examined by comparing the results from various subdivisions of a 100-h dwell (e.g. ten 10-h cycles, two 50-h cycles, etc.) in order to test the assumption that thermal cycling can be neglected.

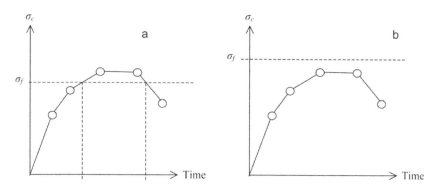

*23.10* Examples of (a) non-unique and (b) non-existent interpolation results

Finally, there is a need for validation of the tool by comparing damage measurements for a well understood TBC system to predictions of the tool. Experiments are under way that will lead to suitable validation data, but as yet the comparison has not taken place.

## References

1. V. K. Tolpygo, D. R. Clarke and K. S. Murphy, *Surf. Coat. Technol.*, 146–147 (2001), 124.
2. D. R. Mumm, A.G. Evans and I. T. Spitsberg, *Acta Mater.*, 49 (2001), 2329.
3. W. J. Brindley and J. D. Whittenberger, *Mater. Sci. Eng. A*, 163 (1993), 33.
4. S. Darzens and A. M. Karlsson, *Surf. Coat. Technol.*, 177–178 (2004), 108.
5. N. P. Padture, M. Gell and E. H. Jordan, *Science*, 296 (2002), 280.
6. R. A. Miller, *Trans. ASME*, 109 (1987), 448.
7. L. Singheiser, R. Steinbrech, W. J. Quadakkers and R. Herzog, *Mater. High Temp.*, 18 (2001), 249.
8. D. Renusch, H. Echsler and M. Schütze, in *Lifetime Modelling of High Temperature Corrosion Processes*, 324–336, ed. M. Schütze, W. J. Quadakkers and J. R. Nicholls, Maney Publishing, London, 2001.
9. M. Y. He, J. W. Hutchinson and A. G. Evans, *Mater. Sci. Eng. A*, 345 (2003), 172.
10. O. C. Zienkiewicz and R. L. Taylor, *The Finite Element Method, Volume 1: The Basis*, 5th edition, Butterworth-Heinemann, 2002.
11. H. E. Evans, G. P. Mitchell, R. C. Lobb and D. R. J. Owen, *Proc. Roy. Soc.*, A440 (1993), 1.
12. E. P. Busso, J. Lin, S. Sakurai and M. Nakayama, *Acta Mater.*, 49 (2001), 1515.
13. E. P. Busso, J. Lin and S. Sakurai, *Acta Mater.*, 49 (2001), 1529.
14. E. P. Busso and Z. Q. Qian, *Acta Mater.*, 52 (2006), 325.
15. S. Osgerby, J. W. Nunn and S. R. J. Saunders, presented at *Conf. Baltica VI: Life Management and Maintenance for Power Plants*, 345–354, Helsinki, Finland, and Stockholm, Sweden, 8–10 June 2004.
16. E. P. Busso, H. E. Evans, L. Wright, L. N. McCartney, S. R. J. Saunders and J. Nunn, *NPL Report DEPC MPE 003*, March 2005.
17. E. P. Busso, H. E. Evans, L. Wright, L. N. McCartney, S. R. J. Saunders, S. Osgerby and J. Nunn, *Acta Mater.*, 55 (2007), 1491.
18. A. Selcuk and A. Atkinson, *Mater. Sci. Eng.*, A335 (2002), 147.
19. J. Nunn, S. R. J. Saunders and J. Banks, presented at *6th Int. Conf. on Microscopy of Oxidation*, 219–226, Birmingham, UK, 4–6 April 2005.
20. N. M. Yanar, F. S. Pettit and G. H. Meier, *Metall. Mater. Trans.*, 37A (2006), 1563.

# 24

# Waste incineration corrosion processes: Oxidation mechanisms by electrochemical impedance spectroscopy*

F. J. Pérez, M. P. Hierro and J. Nieto

*Grupo de Investigación de Ingeniería de Superficies, Universidad Complutense de Madrid, Departamento de Ciencia de los Materiales, Facultad de Ciencias Químicas, 28040 Madrid, Spain*

## 24.1    Introduction

Incineration plants are recognised as a valid and efficient means of dealing with solid municipal wastes, particularly when they are part of integrated treatment plants. Waste incineration reduces the volume and weight of waste, thus saving space in landfill sites; also, the end products are sanitised by transforming organic matter into water and carbon dioxide ($CO_2$); it is also possible to recover steam and/or energy. The main environmental impact in an incineration plant is due to the residues of the process (solid, liquid and gaseous emissions). In addition, these residues cause the corrosion of boiler superheater tubes used in incinerators and biomass-fired plants. The gaseous species ($HCl$, $Cl_2$, $SO_2$, etc.) and solid impurities such as the alkali and heavy metals (K, Na, Zn, Pb, Sn, etc.) can form chlorides and sulphates with a low melting point; these eutectic salts promote accelerated attack of metallic materials in oxidising environments [1–5].

In this work, the corrosion tendencies of two Ni-base alloys, 625 and 617, were studied under the influence of a molten $KCl$ (70% molar)–$ZnCl_2$ (30% molar) salt mixture at 650 °C in air. This salt mixture has frequently been identified in the ash deposits of waste incinerators. Chlorine plays an important role in the degradation of these alloys due to the loss of the alloying constituents as volatile chlorides ($CrCl_3$, $NiCl_2$).

The corrosion process was monitored by electrochemical impedance spectroscopy. This technique allows real-time monitoring of the experiment without disturbing the thermodynamic system, enabling many measurements to be taken at different times to establish the corrosion mechanism.

## 24.2    Experimental procedure

The experimental set-up used in the present study is shown in Fig. 24.1. Test electrodes were placed in a crucible containing the salt mixture, $KCl$ (70% molar)–$ZnCl_2$ (30% molar). The crucible was contained in a stainless steel chamber that was heated

* Reprinted from F. J. Pérez et al.: Waste incineration corrosion processes: Oxidation mechanisms by electrochemical impedance spectroscopy. *Materials and Corrosion*. 2008. Volume 59. pp. 566–72. Copyright Wiley-VCH Verlag GmbH & Co. KGaA.

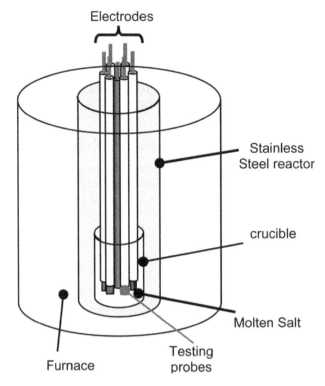

24.1 Experimental set-up

Table 24.1   Chemical composition of 625 and 617 alloys (wt.%)

| Alloy | Ni | Cr | Co | Mo | Fe | Mn | Al | Nb | Ti | S | C | O |
|-------|------|------|------|------|-----|-------|-----|--------|-----|--------|-------|--------|
| 625 | 58.0 | 23.0 | 1.0 | 10.0 | 5.0 | 0.5 | 0.4 | 3.15–5 | 0.4 | 0.015 | 0.1 | – |
| 617 | 54.8 | 22.4 | 11.1 | 9.0 | 1.2 | 0.051 | 1.1 | – | 0.4 | <0.09 | 0.053 | 0.0015 |

to 650 °C in a cylindrical furnace. A two-electrode system was used for the impedance measurements, the material under test (Ni-base alloy) being used for both the working electrode and the reference electrode. The alloy compositions are shown in Table 24.1.

The electrodes were prepared as follows. The sample under test was cut into a coupon with dimensions of 7 mm × 3 mm × 20 mm, and ground to a 600 grit finish on SiC paper. A Ni–Cr wire was spot welded to one end of the sample to provide the electrical connection.

Impedance measurements were carried out using a Voltalab 80 system. The amplitude of the input sine-wave was ±10 mV, and measurements were made over the frequency range from 50 kHz to 10 MHz. Software developed by B. A. Boukamp was used to fit the impedance spectra.

Scale morphology and composition were examined using scanning electron microscopy (SEM) and X-ray diffraction.

## 24.3    Results and discussion

In molten chlorides at 650 °C, the Ni-base alloys 625 and 617 can suffer the following corrosion mechanisms: protective scale formation; porous scale formation; and fast localised corrosion.

On initial exposure, the 625 and 617 alloys developed a protective oxide scale; the corrosion process was controlled by the transport of ions in this scale. The impedance spectra were composed of two capacitive loops, a small semi-circle at high frequency and a large semi-circle at low frequency, and could be described by the equivalent circuit of Fig. 24.2, where

$R_s$ represents the molten salt resistance
$R_t$ is the electrochemical transfer resistance
$C_{dl}$ is the double layer capacitance at the alloy–melt interface
$C_{ox}$ is the oxide capacitance; and
$R_{ox}$ is the transfer resistance of ions in the scale.

Taking into account the dispersion effect, a constant phase angle element $CPE = [Q(j\omega)^n]^{-1}$ was used to describe the elements Cdl and Cox in the fitting procedure, where

$j$ is the imaginary number
$Q$ is the frequency independent real constant
$\omega = 2\pi f$ is the angular frequency (rad s$^{-1}$)
$f$ is the frequency of the applied signal; and
$n$ is the CPE exponent [6–8].

Thus the total impedance of Fig. 24.2 can be expressed by Eq. 24.1 [6,9]

$$Z = R_s + \cfrac{1}{Y_{dl}(j\omega)^{n_{dl}} + \cfrac{1}{R_t}} + \cfrac{1}{Y_{ox}(j\omega)^{n_{ox}} + \cfrac{1}{R_{ox}}} \qquad [24.1]$$

where $Y_{dl}$, $n_{dl}$ and $Y_{ox}$, $n_{ox}$ are constants representing the elements $Q_{dl}$ and $Q_{ox}$, respectively.

Twenty-four hours after the start of the experiment on Alloy 617 in molten chlorides at 650 °C, the protective scale suffered partial failures through which the alloy suffered fast localised corrosion. This was characterised by the fast local growth of scale, not internal oxidation. Obviously, the reactions at sites of slow and fast corrosion occurred simultaneously. In the slow corrosion zone, the charge transfer resistance at the scale/melt interface may be ignored compared with the transportation resistance of ions in the scale. Thus, this kind of localised corrosion can be represented by the equivalent circuit of Fig. 24.3, where $C_{dl}$ and $R_t$, respectively, represent

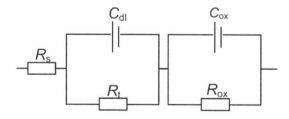

*24.2* Equivalent circuit of protective scale model

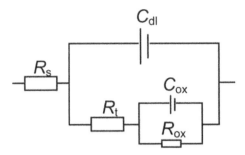

*24.3* Equivalent circuit of localised corrosion model

the double-layer capacitance and charge transfer resistance along the localised corro-sion zone, and $C_{ox}$ and $R_{ox}$ the oxide capacitance and transfer resistance of ions in the scale at sites of slow corrosion.

The Nyquist plot corresponding to the equivalent circuit of Fig. 24.3 also consists of two capacitance loops. The electrochemical impedance $Z$ can be expressed by Eq. 24.2

$$Z = R_s + \cfrac{1}{Y_{dl}(j\omega)^{n_{dl}} + \cfrac{1}{R_t} + Y_{ox}(j\omega)^{n_{ox}} + \cfrac{1}{R_{ox}}}$$ [24.2]

However, in the case of Alloy 625, the localised corrosion mechanism appeared after 48 h. Both materials are Ni-base alloys with high chromium contents, but the Alloy 617 suffered localised corrosion faster than the Alloy 625. The main difference between both alloys is the presence of niobium in Alloy 625, instead of the presence of cobalt in Alloy 617.

When the localised corrosion process was very advanced, the oxide scale on the metal surface had many failure sites and may be considered permeable to the molten chlorides. The Nyquist plot is composed of a semi-circle at high frequency and a line at low frequency, representing a diffusion-controlled Warburg impedance reaction. The diffusion-controlled reaction of Ni-base alloys after long periods of time in deep molten salt can be described by the equivalent circuit of Fig. 24.4 [8,9].

In Fig. 24.4, $R_s$ represents the molten salt resistance, $R_t$ the electrochemical transfer resistance, $C_{dl}$ the double layer capacitance at the alloy–melt interface, and $W$ the Warburg resistance.

The impedance can be expressed by the following equation [10]

$$Z = R_s + \cfrac{1}{Y_{dl}(j\omega)^{n_{dl}} + \cfrac{1}{R_t + W}}$$ [24.3]

The transport of ions through the scale and the diffusion of oxidants in the molten salt are the rate limiting processes. The impedance diagrams can be fitted to the equivalent circuit of Fig. 4 [9,11].

The fitting results at 24, 48, 72, 100, 150 and 200 h for Alloys 625 and 617 in molten KCl–ZnCl$_2$ at 650 °C in air are shown in Fig. 24.5.

The parameters in Eqs. 24.1, 24.2 and 24.3 were obtained by fitting the impedance spectra on the basis equivalent circuits of Figs. 24.2, 24.3 and 24.4, respectively, and are listed in Table 24.2 for Alloy 617 and in Table 24.3 for Alloy 625.

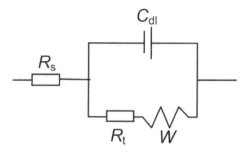

*24.4* Equivalent circuit of porous scale model

Based on the above discussion, it can be seen that the electrochemical impedance diagrams for metals undergoing either uniform corrosion with the formation of a protective scale or localised corrosion, all display the characteristics of double capacitance loops. Thus, analysis of the corrosion products is needed to establish a suitable impedance model.

To corroborate the impedance results, the corroded samples were analysed by SEM and XRD. The SEM analyses confirmed the presence of a protective $Cr_2O_3$ layer on the surfaces of the alloys. This scale was able to protect the material for 24 h in the case of Alloy 617, and for 48 h in the case of Alloy 625. Figure 24.6 shows the chromium oxide growth on the metal surfaces by the following reactions

$$2Cr + 3Cl_2 \rightarrow 2CrCl_3 \qquad [24.4]$$
$$4CrCl_3 + 3O_2 \rightarrow 2Cr_2O_3 + 6Cl_2 \qquad [24.5]$$

Alloy 617 was protected by the $Cr_2O_3$ layer for 24 h, after which molten phases such as the volatile chlorides $ZnCl_2$ and KCl and gas components (HCl, $Cl_2$, $O_2$, etc.) penetrated through defects in the scale and caused the loss of corrosion protection in some areas (Fig. 24.7(a)).

On the other hand, while chromium was forming an oxide layer, chlorine reacted with chromium oxide to form volatile $CrCl_3$; this process was favoured thermodynamically due to the evaporation of $CrCl_3$ at high temperature according to the following reaction

$$2Cr_2O_3 + 6Cl_{2(g)} \rightarrow 4CrCl_{3(g)}\uparrow + 3O_{2(g)} \qquad [24.6]$$

At the same time, during the corrosion process, $ZnCl_2$ was consumed continuously due to its strong evaporation and its oxidation to form ZnO, which reacted with $Cr_2O_3$ to form the $ZnCr_2O_4$ spinel (Fig. 24.7(b)) by the following reactions [1]

$$ZnCl_{2(l)} + 1/2O_{2(g)} \rightarrow ZnO_{(s)} + Cl_{2(g)} \qquad [24.7]$$
$$ZnO_{(s)} + Cr_2O_{3(s)} \rightarrow ZnCr_2O_{4(s)} \qquad [24.8]$$

Studies of the solubility of oxide scales in molten chlorides can also provide important information on the mechanism of dissolution of protective scales. Ishitsuka and Nose [12] measured the solubility of $Cr_2O_3$ in molten NaCl–KCl mixtures at 727 °C at different basicity levels, by changing the partial pressures of HCl and $H_2O$ and found that the $Cr_2O_3$ films can easily dissolve in the molten salt forming the hexavalent chromium ion according to the reaction

$$2Cr_2O_3 + 3O_2 + 4O^{2-} \rightarrow 4CrO_4^{2-} \qquad [24.9]$$

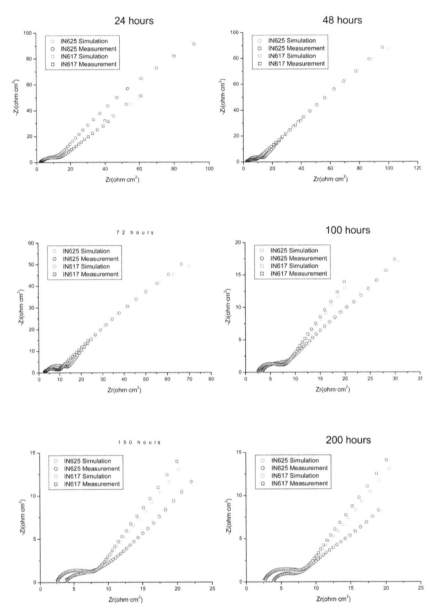

*24.5* Nyquist plots of Alloys 625 and 617 in molten KCl–ZnCl$_2$ at 650 °C in air

The chromium content decreased near the alloy–oxide interface due to evaporation as CrCl$_3$, dissolution as CrO$_4^{2-}$ [13] and the formation of ZnCr$_2$O$_4$ spinel. Thus, the Cr$_2$O$_3$ scale lost its protective properties. Subsequently, other alloying elements oxidised forming NiO and MoO$_3$. The oxide scales formed in this way are rather porous and can hardly provide any effective protection, so that the corrosion rate will be significantly enhanced (Fig. 24.8(a)). Figure 24.8(b) shows cross-section morphologies of Alloy 617 after 100 h in molten chlorides; the absence of chromium allowed salt penetration through the porous NiO scale.

*Table 24.2*   Adjusted results of the impedance spectra for Alloy 617 in molten KCl–ZnCl$_2$ at 650 °C in air

| Time (h) | $R_e$ (Ω) | $R_t$ (Ω) | $C_{dl}$ (Ω$^{-1}$ s$^n$) | $n_{dl}$ | $R_{ox}$ (Ω) | $C_{ox}$ (Ω$^{-1}$ s$^n$) | $n_{ox}$ | $W$ (Ω) |
|---|---|---|---|---|---|---|---|---|
| 2 | 1.88 | 304.93 | $3.45 \times 10^{-2}$ | $7.35 \times 10^{-1}$ | 20.37 | $1.36 \times 10^{-2}$ | $6.43 \times 10^{-1}$ | |
| 24 | 2.02 | 10.28 | $1.74 \times 10^{-3}$ | $6.89 \times 10^{-1}$ | | | | $5.68 \times 10^{-2}$ |
| 48 | 2.14 | 5.84 | $2.55 \times 10^{-3}$ | $6.41 \times 10^{-1}$ | | | | $9.19 \times 10^{-2}$ |
| 72 | 2.27 | 6.67 | $2.64 \times 10^{-3}$ | $6.19 \times 10^{-1}$ | | | | $1.92 \times 10^{-1}$ |
| 100 | 2.39 | 4.90 | $3.61 \times 10^{-3}$ | $6.15 \times 10^{-1}$ | | | | $2.17 \times 10^{-1}$ |
| 150 | 2.40 | 5.04 | $3.66 \times 10^{-3}$ | $6.03 \times 10^{-1}$ | | | | $2.15 \times 10^{-1}$ |
| 200 | 2.38 | 5.23 | $3.86 \times 10^{-3}$ | $5.84 \times 10^{-1}$ | | | | $2.13 \times 10^{-1}$ |

*Table 24.3*   Adjusted results of the impedance spectra for Alloy 625 in molten KCl–ZnCl$_2$ at 650 °C in air

| Time (h) | $R_e$ (Ω) | $R_t$ (Ω) | $C_{dl}$ (Ω$^{-1}$ s$^n$) | $n_{dl}$ | $R_{ox}$ (Ω) | $C_{ox}$ (Ω$^{-1}$ s$^n$) | $n_{ox}$ | $W$ (Ω) |
|---|---|---|---|---|---|---|---|---|
| 2 | 2.42 | 243.02 | $4.11 \times 10^{-2}$ | 0.52 | | | | |
| 24 | 2.55 | 9.65 | $2.45 \times 10^{-3}$ | 0.68 | 559.02 | $1.82 \times 10^{-2}$ | 0.67 | |
| 48 | 2.78 | 10.10 | $9.75 \times 10^{-4}$ | 0.73 | 636.12 | $3.73 \times 10^{-2}$ | 0.60 | |
| 72 | 2.92 | 10.30 | $1.39 \times 10^{-3}$ | 0.69 | 236.51 | $5.87 \times 10^{-2}$ | 0.62 | |
| 100 | 3.22 | 3.78 | $5.18 \times 10^{-3}$ | 0.61 | 298.17 | $1.05 \times 10^{-1}$ | 0.43 | |
| 150 | 3.22 | 9.15 | $3.39 \times 10^{-2}$ | 0.32 | | | | $1.59 \times 10^{-1}$ |
| 200 | 3.83 | 7.73 | $2.68 \times 10^{-3}$ | 0.66 | | | | $3.71 \times 10^{-1}$ |

*24.6* (a) Cr$_2$O$_3$ protective scale on Alloy 617 after 24 h in molten salts. (b) Cr$_2$O$_3$ protective scale on Alloy 625 surface after 24 h in molten salts

Alloy 625 suffered similar reactions to Alloy 617, but the corrosion process was slower. It was protected by the Cr$_2$O$_3$ layer for 48 h, after which molten phases such as the volatile chlorides, ZnCl$_2$ and KCl, and gas components (HCl, Cl$_2$, O$_2$, etc.) penetrated through scale defects causing the loss of corrosion protection in some places. Between 48 and 150 h, Alloy 625 alloy suffered localised corrosion through the cracks and fissures in the oxide scale.

24.7 SEM images of Alloy 617 after 48 h in molten salts

24.8 SEM images of Alloy 617 after 100 h in molten salts. (a) Surface appearance. (b) Cross-section image

24.9 SEM images of Alloy 625 after 200 h in molten salts. (a) Cross-section image. (b) Surface appearance

Following depletion of the chromium content, other alloying elements oxidised forming new oxides and spinels, for example, $MoO_3$, $NiO$ and $NbCrO_4$, by the following reactions

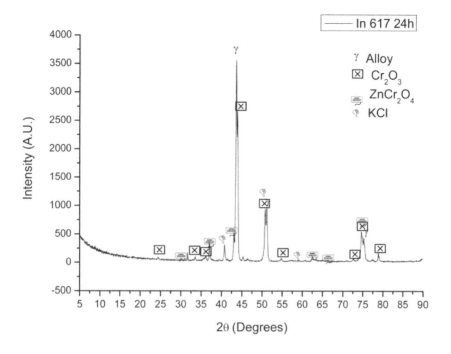

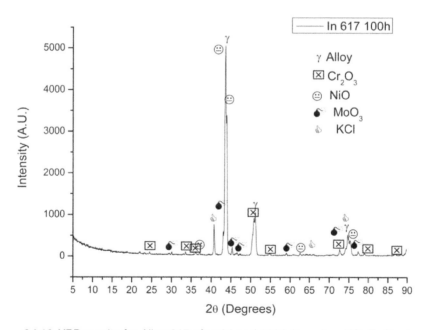

*24.10* XRD results for Alloy 617 after 24 and 100 h in molten KCl–ZnCl$_2$ at 650 °C in air.

$$Mo + 3/2O_2 \rightarrow MoO_3 \qquad [24.10]$$
$$2Ni + 1/2O_2 \rightarrow NiO \qquad [24.11]$$
$$Cr_2O_3 + Nb_2O_5 \rightarrow 2NbCrO_4 \qquad [24.12]$$

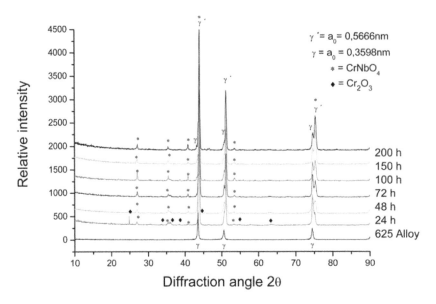

*24.11* XRD results for Alloy 625 after 24, 48, 72, 100, 150 and 200 h in molten KCl–ZnCl$_2$ at 650 °C in air

Figure 24.9(a) shows the cross-section morphologies of Alloy 625 after 200 h in molten chlorides. The image shows the Cr$_2$O$_3$, NiO and MoO$_3$ scales. Figure 24.9(b) shows the material surface after 200 h.

Kawahara [14] presented a corrosion model for Alloy 625 in molten chlorides, under high $pO_2$ and high $pCl_2$; the alloying elements reacted as shown in the following reactions:

$$2Cr + 3Cl_2 \rightarrow 2CrCl_3 \qquad\qquad [24.13]$$
$$Mo + O_2 (Cl_2) \rightarrow MoO_2 (MoCl_2) \qquad [24.14]$$
$$2Ni + O_2 (2Cl_2) \rightarrow 2NiO (2NiCl_2) \qquad [24.15]$$

The XRD results for Alloy 617 confirmed the presence of Cr$_2$O$_3$, ZnCr$_2$O$_4$ and NiO. The presence of MoO$_2$ was not established by this technique. Nevertheless, it was possible to attribute some peaks in the XRD spectrum at 100 h (Fig. 24.10) to another molybdenum oxide, MoO$_3$.

The XRD results for Alloy 625 confirmed the presence of Cr$_2$O$_3$ and NbCrO$_4$ (Fig. 24.11).

## 24.4    Conclusions

The corrosion behaviour of two Ni-base alloys, Alloy 625 and Alloy 617 in the presence of molten chlorides has been monitored by electrochemical impedance spectroscopy.

These alloys formed a protective oxide scale which became porous after a certain period of time leading to the oxidation and chlorination of alloying elements such as Cr, Ni, Mo, Nb and Co. The EIS profiles of Alloy 625 were fitted to a protective scale model for exposure times of up to 24 h. With longer exposures, penetration of the oxide layer by the molten salt occurred at local sites. Finally, the diffusion of corrosive species through the porous scale was the main corrosion process. Alloy 617 suffered the same corrosion processes as those observed for Alloy 625. However, the kinetics of the attack were faster in Alloy 617, suggesting that the presence of cobalt in the alloy reduces its resistance to corrosion at high temperature.

The high chromium contents of the Ni-base alloys failed to provide protection against corrosion in molten chlorides, due to evaporation and dissolution of Cr as $CrCl_3$ and $CrO_4^{2-}$. Nor were the oxide scales formed by Ni and Mo able to protect the alloy in this severely corrosive environment.

## References

1. Y. S. Li, Y. Niu and W. T. Wu, *Mater. Sci. Eng. Struct. Mater.*, 345(1–2) (2003), 64–71.
2. M. Spiegel, G. Schroer, and H. J. Grabke, 'Corrosion of high alloy steels and Fe–Cr-alloys beneath deposits from waste incinerator plants', in 4th Int Symp. on High Temperature Corrosion and Protection of Materials, 527–534, Les Embiez (Var), France, May 1996. Trans Tech Publications, 1997.
3. M. Spiegel, 'Corrosion mechanisms and failure cases in waste incineration plants', in 5th Int. Symp. on High Temperature Corrosion and Protection of Materials, 971–978, Les Embiez (Var), France, May 2000. Trans Tech Publications, 2001.
4. M. Spiegel, *Mater. Corros.*, 50(7) (1999), 373–393.
5. Y. S. Li and M. Spiegel, *Corros. Sci.*, 46(8) (2004), 2009–2023.
6. F. Mansfeld, H. Xiao and Y. Wang, *Mater. Corros.*, 46(1) (1995) 3–12.
7. F. Mansfeld, *Electrochim. Acta*, 38(14) (1993), 1891–1897.
8. F. Mansfeld, L. T. Han, C. C. Lee and G. Zhang, *Electrochim. Acta*, 43(19–20) (1998), 2933–2945.
9. C. L. Zeng, W. Wang and W. T. Wu, *Corros. Sci.*, 43(4) (2001), 787–801.
10. F. Mansfeld, *Electrochim. Acta*, 35(10) (1990), 1533–1544.
11. C. L. Zeng and J. Li, *Electrochim. Acta*, 50(28) (2005), 5533–5538.
12. T. Ishitsuka and K. Nose, *Corros. Sci.*, 44(2) (2002), 247–263.
13. R. A. Rapp, *Corros. Sci.*, 44(2) (2002), 209–221.
14. Y. Kawahara, *Corros. Sci.*, 44(2) (2002), 223–245.

# 25

## Behaviour of NiAl air plasma sprayed coatings in chlorine-containing atmospheres*

Hadj Latreche, Sébastien Doublet*, Patrick Masset*,
Till Weber and Michael Schütze

*Karl-Winnacker-Institut der Dechema e.V., Theodor-Heuss-Allee 25,
60486 Frankfurt am Main, Germany*

*masset@dechema.de*

*\*Present address: Department of Ferrous Metallurgy, RWTH Aachen University,
Intzestrasse 1, 52072 Aachen, Germany*

Guido Tegeder and Gerhard Wolf

*ATZ Entwicklungszentrum, Kropfersrichter Straße 6–10,
92237 Sulzbach-Rosenberg, Germany*

### 25.1 Introduction

Chlorine-containing atmospheres are encountered in several industrial processes (e.g. coal gasification, incineration of industrial and domestic waste, coal- and biomass-fired boilers), leading to severe corrosion phenomena. The partial pressures of oxygen and chlorine in industrial atmospheres can vary over several orders of magnitude, from $10^{-30}$ to $10^{-1}$ bar at temperatures from 300 to 1400 °C. Figure 25.1 depicts the combinations of oxygen and chlorine partial pressures and temperature encountered in industrial processes [1]. The application of thermally sprayed protective coatings is one of the solutions to improve the resistance of steels to chlorine corrosion at high temperature and to promote their long-term resistance [2,3]. Flame and arc sprayed coatings reveal comparatively low coating qualities, e.g. due to their low homogeneity and high porosity. Therefore, they are less suitable for high-temperature corrosion protection. Atmospheric plasma spraying (APS) and high-velocity-oxy-fuel-flame-spraying (HVOF) are thermal spraying processes that produce dense coatings [4]. Due to the high melting point and brittle character of the investigated NiAl alloy, APS was used in this work. Coating structure and properties can be well adjusted by parameter optimisation during APS [5–7]. High-temperature corrosion of pure metals by chlorine gas has been widely investigated (Ni [8], Fe [9], Cr [10]). This corrosion phenomenon is characterised by the formation and volatilisation of metal chlorides which can induce severe corrosion effects, especially at low oxygen partial pressures [11,12]. Thermodynamic calculations allow the establishment of the stability areas of metals as well as their corresponding oxides and chlorides as a function of the chlorine and oxygen partial pressures of the atmosphere.

---

\* Reprinted from H. Latreche et al.: Behaviour of NiAl APS-coatings in chlorine-containing atmospheres. *Materials and Corrosion*. 2008. Volume 59. pp. 573–83. Copyright Wiley-VCH Verlag GmbH & Co. KGaA.

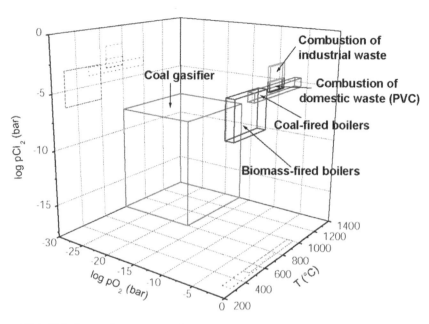

25.1 Chlorine and oxygen partial pressures and temperature conditions during operation for different industrial processes [1]

However, these diagrams do not include any gas phase, and consequently cannot be used to explain and predict the corrosion phenomena in chlorine-containing atmospheres. In order to achieve this purpose, Schwalm and Schütze [13–15] and later Bender and Schütze [16,17] calculated the so-called 'quasi-stability diagrams'. These diagrams combine values of the Gibbs free energy of formation and the vapour pressure of the metal chlorides. In addition, 'a borderline' characterised by the limit of $10^{-4}$ bar was fixed for the partial pressures of the volatile metal chlorides. This corresponds approximately to the annual corrosion rate of 0.1 mm year$^{-1}$. Somewhat arbitrarily, this criterion was taken as the boundary between protective and non-protective areas. In these diagrams, the critical metal chloride partial pressures are obtained in equilibrium with much lower chlorine partial pressures than by the boundary fixed between liquid or liquid metal chloride and solid metal or metal oxide in 'traditional' stability diagrams.

Recently, these diagrams were improved [18] by adding new parameters such as sample geometry, gas flow velocity and the mass transfer coefficients of the products and reactants during the corrosion process. They combine simultaneously thermo-dynamic and kinetic approaches. A quite good agreement was obtained between the results of the thermodynamic calculations and the experimental investigations under these conditions [19]. Nevertheless, discrepancies between experimental and thermodynamic results were observed when cracks formed in the oxide layer, which enhanced the chlorine transport from the process atmosphere through the oxide layer to the metal/oxide interface. In this case, the oxygen partial pressure at the tip of the crack was lower and the metal chloride partial pressure higher than in the surrounding environment, which increased the corrosion rate at the same time. In a recent study, on the basis of these advanced stability diagrams [20], Doublet et al. [19] presented and analysed results from experimental investigations at 800 °C, where nine commercial alloys were tested with regard to their corrosion behaviour in reducing chloridising atmospheres (<5 ppm oxygen). From these results, the corrosion resistance of each metal was analysed in order to provide a basis for the development of new coatings. The design of coatings should meet some basic requirements:

- chlorine resistance for the application
- compatible melting point with the application
- compatible thermal expansion coefficients (CTE) of the coating material and the substrate in order to minimise the residual thermal stresses
- ability to form a thick and uniform coating on the substrate for complex geometries.

Regarding the thermodynamic predictions, silicon would be the ideal candidate. However, the thermal expansion coefficient of $SiO_2$ ($0.8–1.2 \times 10^{-6}$ K$^{-1}$ [21] at 800 °C) is too low in comparison with austenitic (1.4948 X6CrNi18–11: $20.10^{-6}$ K$^{-1}$ at 723 °C) or ferritic (1.4762 X10CrAl24: $14.10^{-6}$ K$^{-1}$ at 723 °C) steels and oxide scale cracking or spallation can be expected during cooling. The calculated coefficient of thermal expansion of the stoichiometric phase NiAl [22] was found to be close to that of the steels and the high melting point of NiAl ($T_m = 1638$ °C, see Ni–Al phase diagram in Fig. 25.2 from Ref. 23) allows its use at up to more than 800 °C. For this reason, the stoichiometric phase NiAl (NiAl31.49 in wt.%) was chosen as a coating material and the present work reports the corrosion resistance of thermally sprayed NiAl coatings in reducing chloridising atmospheres, i.e. with a low oxygen partial pressure (3 ppm). In addition, NiAl was preferred to commercial alloys, e.g. Alloy 625, since the high chromium content would be detrimental in high chlorine-containing environments due to the formation of chromium oxychlorides [16].

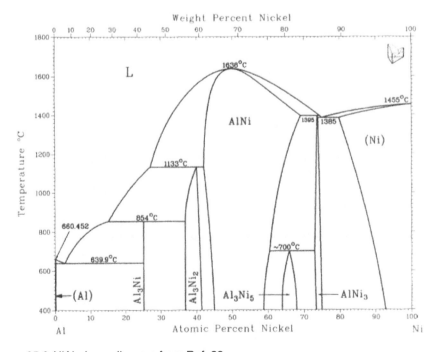

25.2 NiAl phase diagram from Ref. 23

*Table 25.1* Chemical compositions (mass%) of the steel substrate and the coating tested

|                     | Ni    | Al      | Mn    | C     | Cr    | Fe   | Si      |
|---------------------|-------|---------|-------|-------|-------|------|---------|
| NiAl APS-coating    | 68.51 | 31.49   | –     | –     | –     | –    | –       |
| Substrate X10CrAl18 | –     | 0.7–1.2 | <1.0  | 0.12  | 17–19 | Bal. | 0.7–1.4 |

## 25.2    Experimental

### 25.2.1    Materials and coatings

The starting material for the thermally sprayed coatings was a NiAl powder of the desired stoichiometric composition. This was produced by inert gas atomisation [24] and subsequently classified into the particle size fraction of 20 μm to 45 μm for the APS process. The samples to be coated for the corrosion tests were small cylinders of the commercial ferritic steel X10CrAl18 (see Table 25.1 for composition), approximately 15 mm in diameter, 20 mm in length and with rounded edges ($r \sim 1$ mm). High purity NiAl bulk material (see composition in Table 25.1) has been studied, too, for comparison.

Bulk samples with a disc form (diameter 15 mm and 2 mm thick) were machined by electro-erosion from the NiAl rod. The samples were ground to remove the outer surface zone influenced by the cutting procedure and mechanically polished down to 1200 grit with SiC abrasive paper. Subsequently, they were ultrasonically cleaned in acetone and ethanol for 10 min to degrease the surface. The NiAl-coated specimens were tested as received.

The coefficient of thermal expansion of a cylindrical bulk NiAl specimen (10 mm in diameter and 15 mm long) was measured using a L75/1550 dilatometer from Linseis. The measurements were carried out between 25 and 800 °C in Ar–5 vol.% $H_2$ to minimise the formation of oxides. Measurements of the thermal expansion coefficient of bulk NiAl specimens were repeated three times and the results averaged. Reproducible results were obtained with heating rates lower than or equal to 3 K min$^{-1}$. A blank measurement was carried out as a reference by using the same thermal cycle. The measurements were repeated three times.

Differential Thermal and Thermogravimetric Analysis (DTA/TG) was carried out using an STA 490 PC thermoanalyser from Netzsch Gerätebau GmbH (Germany) in Ar (gas flow 50 mL min$^{-1}$) at a heating rate of 20 K min$^{-1}$. The sample mass was close to 49.6 mg. It was contained in an alumina crucible.

The size of the NiAl particles was measured by laser granulometry (diffraction) using a HELOS granulometer from Sympatec GmbH equipped with a pneumatic dispersion unit. The measurement range covers particle sizes of between 1.8 and 350 μm.

The diffusion experiments were carried out in a tube furnace made of alumina under argon (3 ppm oxygen). The coated steel specimens were held in the furnace at 800 °C for 100 h in order to observe whether interdiffusion occurred across the coating/substrate interface. After the diffusion experiments, metallographic observations as well as elemental mapping of the cross-sections of the samples were recorded.

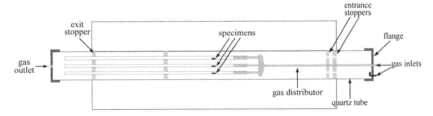

25.3 Schematic diagram of the apparatus used for the experimental investigations [18]

### 25.2.2  Corrosion experiments

The bulk materials as well as the coatings were tested in chlorine-containing atmospheres with a low level of oxygen. The composition of the gas mixture was 0.2 vol.% $Cl_2 + 3$ ppm $O_2 + Ar$, delivered as a gas mixture by Air Liquide.

A three-zone furnace (type TZT from Carbolite) was used for the exposure tests in chlorine-containing atmospheres. The test chamber of the furnace consisted of several parts (Fig. 25.3): a quartz tube of 150 mm diameter in which nine quartz tubes (15 mm in diameter) and a quartz gas separator were inserted. More details of the set-up can be found in Ref. 18. The quartz gas separator was necessary to avoid any de-mixing of the gas composition due to the low flow rate and to provide the same gas atmosphere for each specimen. The chlorine exposure tests were carried out at 800 °C for 280 h.

### 25.2.3  Analysis of as-applied and corroded coatings

X-Ray diffraction patterns of the NiAl powder and APS coating were recorded with a Siemens (now Bruker AXS) D500 diffractometer using a copper cathode ($K_\alpha$) in the $\Theta/2\Theta$ mode equipped with a position-sensitive detector (PSD).

Before the metallographic observations, the samples were gold sputtered, coated with a nickel foil and embedded in epoxy resin. Due to the presence of chloride species coming from the corrosion process, a water-free preparation method was used [25]. The cross-sections were polished with 4000 grit SiC abrasive paper using petroleum, followed by polishing with a silica suspension down to 0.1 μm. The coating cross-sections were examined before and after the corrosion tests with an optical microscope (Leica type DMRME) at several magnifications.

SEM images of the surface and cross-sections of the coatings were obtained with either a Zeiss DSM 950 or a Philips XL40 electron microscope coupled with an EDX probe from EDAX. Elemental maps with a spatial resolution close to 1 μm were obtained with a Cameca SX50 electron beam microprobe (EPMA).

### 25.2.4  APS coating process

Commonly, thermally sprayed Ni–Al alloys are used as bond coats (e.g. NiAl5). Commercial Ni–Al alloys for oxidation-resistant coatings have higher aluminium contents, such as NiAl18 and NiAl20. The stoichiometric composition NiAl, i.e. the NiAl31.49 alloy investigated in the present work is not a standard alloy. Its usage is

*Table 25.2*    Plasma spraying parameters

| Spraying distance, mm | Ar flow, L min⁻¹ | H₂ flow, L min⁻¹ | Current, A | Powder flow rate, g min⁻¹ |
|---|---|---|---|---|
| 100 | 45 | 9 | 600 | 50 |

limited due to the low ductility of the NiAl phase at room temperature. However, the ductility becomes better at temperatures above 600 K [26]. During the thermal spraying process, the temperature of the particles is far above this value and, consequently, one could expect dense coatings for corrosion protection by using this alloy.

Basically, thermal spraying processes are qualified for coatings, if the coating material can be melted without decomposition. Due to the relatively high melting point of the NiAl31.49 alloy (1638 °C) and its brittleness, the atmospheric plasma spraying process (APS) was used to apply the coatings on the corrosion test coupons. The principle of this process is described in DIN EN 657 [27]. This presents a wide range of plasma parameters for setting up the degree of fusion (Table 25.2). Therefore, by using suitable plasma parameters, all particles in the plasma stream are at least partially molten which results in good bonding to the substrate (adhesion) and within the coating (cohesion). The spraying system used was a conventional plasma spraying gun (SulzerMetco F4) and a Thermico control unit for adjustment of the power and gas flow. Figure 25.4 shows the APS process for simultaneous coating of the front area of six steel cylinders which are fixed in a special rotating sample holder. The side area of each cylinder had to be coated separately. Before the coating process,

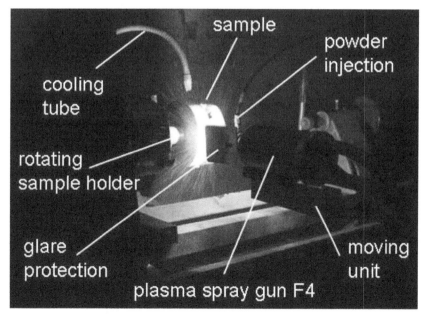

*25.4* Six steel cylinders in a special sample holder for simultaneous coating of the front areas with NiAl by atmospheric plasma spraying (APS)

the cylinder surfaces were cleaned by grit blasting with alumina F36 (size fraction 420 to 595 μm).

## 25.3    Thermodynamic calculations

The corrosion resistance of a material in $Cl_2/O_2$ environments can be defined by either its rate of mass loss or its rate of thickness loss. In industry, the value of 0.1 mm year$^{-1}$ is often taken as a limit above which corrosion of a material is considered to be detrimental. Based on these considerations, Doublet and colleagues [18,20] developed quasi-stability diagrams using a rate of thickness loss of 0.1 mm year$^{-1}$ as the criterion distinguishing between protective and non-protective situations.

Every material placed in a flowing gas mixture is covered by a boundary layer whose thickness depends on the characteristic parameters of the system including the length of the surface and the velocity and viscosity of the gas. Diagram development is based on the diffusion kinetics through this boundary layer of the species responsible for the material attack, i.e. $Cl_2$ and $O_2$, as well as on the corrosion products, i.e. metal chlorides. For the calculations, the fluxes of chlorine and metal chlorides were taken to be equal:

$$J_{Cl2} = -\Sigma J_{MClx} \qquad [25.1]$$

where $J_{Cl2}$ and $J_{MClx}$ are the chlorine and metal chloride fluxes (mol s$^{-1}$), respectively. In addition, no oxygen gradient was considered in the diffusion layer ($p_{O2}^{(B)} = p_{O2}^{(S)}$, where $p_{O2}^{(B)}$ and $p_{O2}^{(S)}$ represent the oxygen partial pressure (bar) in the furnace (B) and directly at the specimen surface (S), respectively), as the oxygen diffusion coefficient in the diffusion layer is higher than those of the chlorine and metal chlorides. At 800 °C, the calculated values of the diffusion coefficients of $O_2$, $Cl_2$ and $NiCl_2$ in argon are equal to 1.742, 1.077 and 0.834 cm$^2$ s$^{-1}$, respectively. For diagram development, three composition ranges of the environment were defined with regard to the oxygen partial pressure in the gas mixture. For each oxygen partial pressure range, typical corrosion mechanisms occur (see Fig. 25.5), and specific assumptions for the calculations were made and are detailed hereafter.

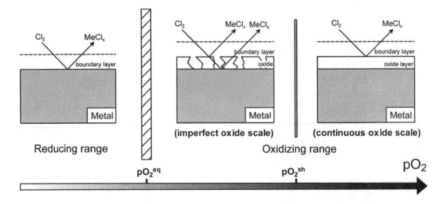

25.5 Three main types of metal corrosion by a $Cl_2/O_2$ mixture at high temperature as a function of the $pO_2$ content in the process gas. These three ranges are separated by $pO_2^{eq}$ and $pO_2^{sh}$

### 25.3.1   Reducing range (for $pO_2$ lower than $pO_2^{eq}$)

$pO_2^{eq}$ represents the equilibrium partial pressure of oxygen between the metal and its oxide for the temperature considered. In this range, the metal oxide does not form. The corrosion process is characterised by direct attack of the metal by $Cl_2$. On the basis of the work of Fruehan [28], it was shown that, in the 'reducing range', the evaporation kinetics of the metal chlorides depend only on the diffusion of chlorine to the metal surface, due to the high rate of formation of metal chlorides under these conditions. This leads to a more or less equal rate of thickness loss for all alloying elements studied, following Eq. 25.2

$$k_l = \frac{M}{RT} h_{Cl_2} pCl_2^{\ B}$$  [25.2]

where $k_l$ (g cm$^{-2}$ s$^{-1}$) is the rate of mass loss, $M$ (g mol$^{-1}$) is the molar weight of the studied species, $h_{Cl_2}$ (cm s$^{-1}$) the mass transfer coefficient of $Cl_2$ through the boundary layer and $pCl_2^{\ B}$ (bar) the partial pressure of chlorine in the bulk gas. For laminar flow, $h_{Cl_2}$ can be defined by Eq. 25.3

$$h_{Cl_2} = 0.664 \frac{D_{AB}^{2/3}}{v^{1/6}} \left( \frac{v}{L} \right)^{1/2}$$  [25.3]

where $D_{AB}$ (cm$^2$ s$^{-1}$) is the diffusivity of the reactive species in the gas phase; $v$ (cm$^2$ s$^{-1}$) is the kinematic viscosity of the gas mixture; $v$ (cm s$^{-1}$) is the average velocity of the flowing gas mixture at temperature, and $L$ (cm) is the length of the sample.

### 25.3.2   Intermediate-oxidising range (for $pO_2$ between $pO_2^{eq}$ and $pO_2^{sh}$)

In this range, the corrosion process is characterised mainly by the active oxidation process [29], i.e. complex corrosion. The oxide layer contains many cracks and pores. The development of diagrams including this range requires a combined approach of oxide formation and $Me_xCl_y$ volatilisation due to the complexity of the corrosion mechanisms which are involved.

### 25.3.3   Strongly-oxidising range (for $pO_2$ above $pO_2^{sh}$)

In this range, $pO_2$ is sufficient to allow the formation of a continuous dense oxide layer. The corrosion process is characterised by direct attack of the oxide layer by $Cl_2$. For the 'strongly-oxidising range', it has been shown, based on the work of MacNallan and Liang (for example, for chlorine attack of CoO [29]), that the rate of thickness loss of the materials is a function of diffusion through the boundary layer of both chlorine and metal chlorides. It should be pointed out that, in this oxygen partial pressure range, the formation of metal chlorides remains possible but their relative partial pressures are far below the fixed criterion ($p_{MeClx} < 10^{-4}$ bar). In this case, the growth of a dense alumina layer takes place, despite the evaporation of metal chlorides [30].

Such diagrams were developed for several pure metals (see Ref. 20 for more details). Figure 25.6 sketches the case of nickel and aluminium at 800 °C for a 1.5-cm plate with a gas velocity of 0.75 cm s$^{-1}$. This diagram also shows that the studied atmosphere (indicated by a cross) is in the 'protection' zone for aluminium and in the 'non-protective' zone for nickel. The favoured formation of a protective alumina layer will induce protection of NiAl alloys.

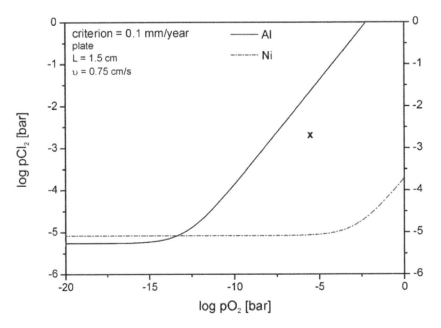

25.6 Rates of thickness loss for aluminium and nickel at 800 °C as a function of the Cl$_2$ and O$_2$ contents of the process gas. The curves correspond to a rate of thickness loss of 0.1 mm year$^{-1}$ of aluminium and nickel (solid and dotted lines), respectively. The cross indicates the atmosphere studied in this work

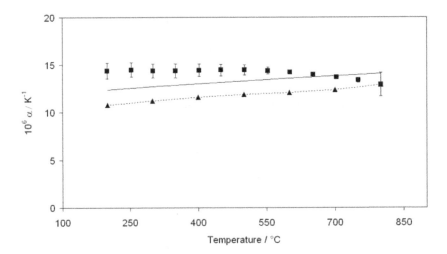

25.7 Thermal expansion coefficients of several materials as a function of temperature. NiAl (■) this work, NiAl (——) calculated values from Ref. 22, ferritic steel Fe–18Cr–1Al–1Si values (▲)

## 25.4    Results and discussion

### 25.4.1    Measurements of thermal expansion coefficients

Figure 25.7 depicts the evolution of the thermal expansion coefficient of NiAl as a function of the applied temperature. These values are compared with calculated values from the literature [22] and those for the ferritic steel used as the substrate in this study. For the thermal expansion coefficient of NiAl samples, the calculated and measured values differ by less than 10%. The difference can be ascribed to the slow formation of alumina on the specimen end surfaces in the dilatometer test despite the use of hydrogenated argon. Over the temperature range of interest (400–800 °C), the difference between the thermal expansion coefficients of NiAl and the steel remains lower than 20%. This indicates that NiAl can be used as a coating without introducing excessively high mechanical stresses at the substrate/coating interface during cooling.

### 25.4.2    Characterisation of the NiAl powder

The starting material for the spraying process was an inert gas atomised NiAl31.49 powder, classified to a size fraction ranging from 20 µm to 45 µm (Fig. 25.8). As a consequence of the production method [24], the powder particles have a spherical shape which is advantageous for the thermal spraying process with respect to the feedability and homogeneity of the powder.

The powder was analysed by differential thermal and simultaneous thermogravimetric analysis (DTA/TG). The corresponding results are shown in Fig. 25.9. The DTA signals give two weak exothermal events, starting at 513 °C and 618 °C, respectively. These might indicate phase transformations or the presence of metastable phases inside the powder which is rapidly cooled during its manufacturing process (gas atomisation). At 820 °C, an exothermic reaction occurs which corresponds to a simultaneous weight increase. Thus, at this temperature, the NiAl powder starts to oxidise. A second strong oxidation reaction starts at a temperature of approx. 1320 °C which is still far away from the melting point of the alloy (1638 °C). This thermal

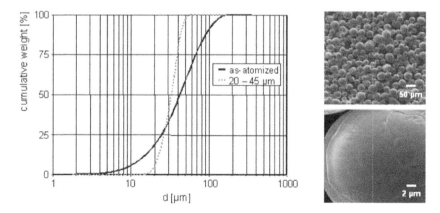

*25.8* Particle size distributions of the NiAl powder, as-atomised and classified for the thermal spraying process (left), together with SEM images of the classified fraction (right)

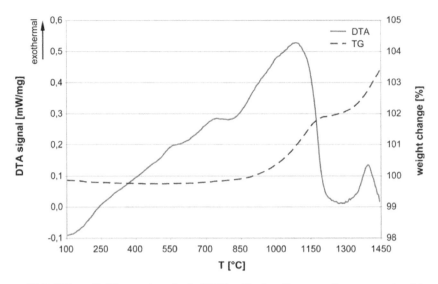

25.9 Differential thermal analysis (DTA) with simultaneous thermogravimetric analysis (TG) of the NiAl31.49 alloy powder (sample weight = 100%)

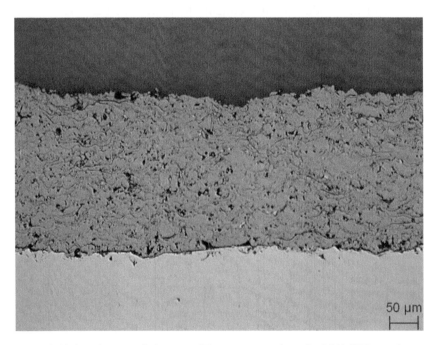

25.10 Light microscopic image of the cross-section of a NiAl APS coating

analysis was performed in an argon atmosphere. However, the residual oxygen within the measurement chamber was sufficient to show these two oxidation events. The first event is thought mainly to arise from the formation of surface oxides on the fine powder particles. At around 1150 °C, the oxidation subsides due to the self-protecting nature of the oxide films formed. Above 1300 °C, the protective character of the oxide layers starts to disappear, and non-protective oxidation occurs.

### 25.4.3   Characterisation of the APS NiAl coatings

Figure 25.10 shows the cross-section of an APS coating of the NiAl (Ni68.5Al31.5 wt.%) alloy with a thickness of approximately 300 µm before corrosion testing. The coating is characterised by a lamellar, uniform, homogeneous structure, free of macro cracks, and comparatively dense with regard to other examples of plasma sprayed coatings. Pores are identifiable as black areas, and oxides as dark grey areas (lines). No delamination occurred at the substrate/coating interface, indicating good adhesion of the coating to the substrate. Some pores arise from the voids formed during solidification of the molten material in the course of the process. Using digital image analysis, the porosity was observed to be less than 2% which is quite a good value for a plasma sprayed coating. The measurement of the oxide content within the coatings by digital image analysis is only an approximate method, since the line-shaped oxide boundaries are too fine. The oxide content estimated in this way gives values in excess of 5%.

Another characteristic of thermally sprayed coatings is the high level of mechanical stress within the coatings, which is greater the higher the difference in thermal expansion between the substrate and the coating. The X-ray diffraction analyses from the starting material (powder) and the APS coating, shown in Fig. 25.11, give an impression of this situation. The diffractogram of the NiAl coating exhibits strong broadening of the single reflections in comparison to the atomised powder. This indicates a strained crystal structure as a consequence of the presence of residual stress.

In order to study the interdiffusion stability of the coating in contact with the underlying substrate, the NiAl-coated ferritic steel was annealed in pure Ar at 800 °C for 100 h. No interdiffusion occurred at the alloy/coating interface, which contrasts with the behaviour of $NiSi_2$ coatings (Fig. 25.12). The lower melting point of the eutectic phase $NiSi–NiSi_2$ ($T_m = 992$ °C [31]) enhances the diffusion processes between coating and alloy.

The outer surface of the NiAl coating was investigated by means of SEM (Fig. 25.13). From the backscattered electron images, two different areas were identified, the pale grey areas being the NiAl matrix and the dark ones being alumina

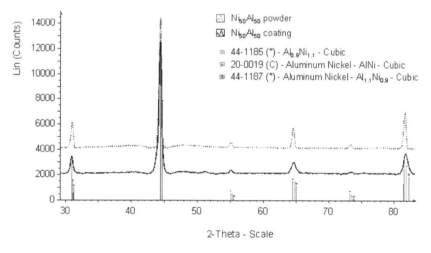

*25.11* X-ray diffractograms from the NiAl31.49 powder and from a corresponding APS coating

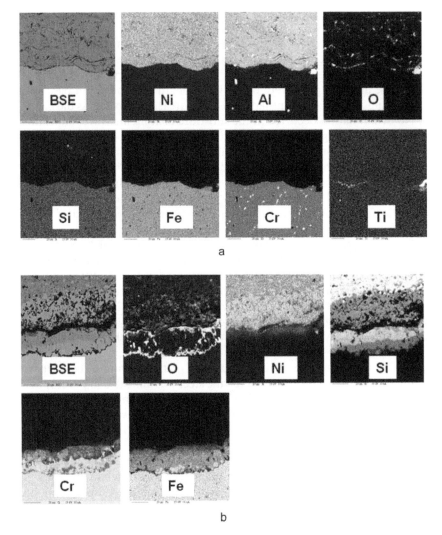

*25.12* Elemental EPMA maps of NiAl (a) and NiSi2 (b) coatings after 100 h at 800 °C in Ar

($Al_2O_3$). The same conclusions were drawn from the elemental maps determined by EPMA (Fig. 25.12) which were carried out on the cross-section. During the thermal spraying step, alumina forms due to the presence of residual oxygen in the process environment. For an aluminium activity close to $3 \times 10^{-4}$ in NiAl [32] (value given at 650 °C), the equilibrium partial pressure of oxygen for oxide formation is equal to $2.37 \times 10^{-37}$ bar at 800 °C.

### 25.4.4  Corrosion tests in chlorine-containing atmospheres

All of the corrosion tests were performed in a chlorine-containing atmosphere (0.2 vol.% $Cl_2$, 3 ppm $O_2$, bal. Ar). The corrosion tests with NiAl bulk material were carried out at 600 °C for 100 h whereas those with NiAl coatings were performed at 800 °C for up to 280 h. Microanalysis carried out on the cross-section of bulk NiAl

*25.13* SEM image of the surface of a NiAl-APS coating

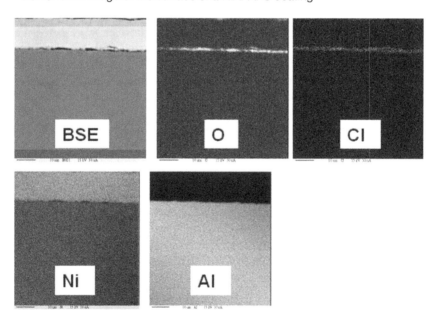

*25.14* Elemental maps of a bulk NiAl specimen after corrosion testing at 600 °C for 100 h in Ar–0.2 vol.% $Cl_2$–3 ppm $O_2$

after the 100 h corrosion tests revealed the presence of a very thin alumina layer on the sample surface with incorporated chlorine (Fig. 25.14). The thickness of the alumina layer was close to 1 μm. At the near subsurface of the alloy, nickel depletion was observed together with the growth of the alumina layer.

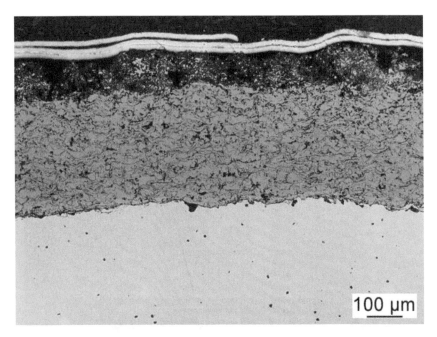

*25.15* Cross-section image of a NiAl APS coating after corrosion testing at 800 °C for 280 h in Ar–0.2% $Cl_2$–3 ppm $O_2$

Exposure tests at 800 °C with the NiAl-coated ferritic steels were performed for 280 h. The metallographic observations revealed a coating with two distinct areas (Fig. 25.15). The inner part of the coating remained unchanged whereas its outer part seemed to be less dense. Elemental maps (Fig. 25.16) identified the outer layer as almost pure alumina without incorporated chlorine. Strong depletion of nickel occurred due to its almost complete transformation into volatile nickel chlorides. This result is in agreement with the thermodynamic calculations presented above. Approximately one-quarter of the coating was transformed into alumina after 280 h. The growth kinetics of this alumina layer are quite rapid and unusual. This might be explained by nickel depletion due to its reaction with chlorine and the evaporation of the corresponding nickel chlorides. It is known that β-NiAl is an alumina former [33]. However, due to the favoured reaction between nickel and chlorine, the composition of the coating is rapidly shifted towards lower nickel contents. Therefore the activity of aluminium in the outer part of the coating increases [32], which leads to the formation of other intermetallic compounds ($NiAl_3$, $NiAl_2$, etc.). These phases have higher aluminium contents than NiAl (see phase diagram in Fig. 25.2), which enhances the formation and the 'growth' kinetics of alumina on a local scale. Furthermore, a chlorine partial pressure range may be reached in the coating imperfections (e.g. porosity, coating defects, microscopic crevices), where a gas phase transport mechanism contributes to alumina growth, similar to that postulated for TiAl or TiAl alloys with the 'halogen effect' [34]. All of these factors may accelerate the transformation of the NiAl coating into alumina. In addition, EPMA measurements have detected no chlorine in the coating, which validates the rapid evaporation of nickel chlorides through the pores and crevices of the alumina layer. At the coating/substrate interface, iron diffusion into the coating was observed whereas no chromium was detected.

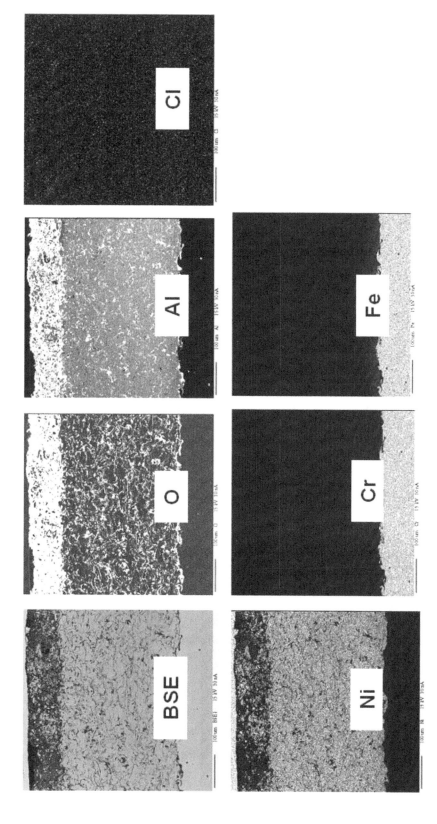

25.16 Elemental maps of the NiAl APS coating after corrosion testing at 800 °C for 280 h in Ar–0.2% $Cl_2$–3 ppm $O_2$

According to the Al–Fe–Ni and Al–Cr–Ni phase diagrams [35], chromium forms a continuous solid solution (bcc) whereas NiAl and Fe may form an equilibrium between the bcc phase and the enriched Fe-fcc phase.

## 25.5    Conclusions

From this work the following conclusions can be drawn:

- By means of combined thermodynamic and kinetic approaches, stability diagrams which take into account the gas velocity, sample geometry, etc. were found to be an efficient tool to assess the stability of materials in chlorine- and oxygen-containing atmospheres
- NiAl APS-coatings with a porosity $<2\%$ can be applied on ferritic steel for service in high-temperature environments
- Nickel from the NiAl-coating evaporates as $NiCl_2$ in chlorine-containing atmospheres
- Thick and protective alumina layers form on the top of the coatings
- One-quarter of the coating was transformed into alumina after 280 h of exposure at 800 °C in an Ar/0.2 vol.% $Cl_2$/3 ppm $O_2$ atmosphere, i.e. a gas phase transport mechanism seems to be involved in the alumina scale growth mechanism.

## Acknowledgements

The financial support of the Bundesministerium für Wirtschaft und Technologie (BMWi) via the Arbeitsgemeinschaft industrieller Forschungsvereinigungen (AiF) under contract AiF no. 13266 N is gratefully acknowledged. The authors sincerely thank Ms Ellen Berghof-Hasselbächer and Ms Daniela Hasenpflug (metallographic examinations), Ms Monika Schorr (EPMA measurements) as well as Mr Peter Gawenda (SEM analysis), for the characterisation work.

## References

1. M. Schütze, *NACE Int.*, 63(1) (2007), 4.
2. B. Aumüller, T. Weber and M. Schütze, in Proc. International Thermal Spray Conference, 23, 2002.
3. H.-W. Gudenau and S.-C. Cha, in Proc. 7th European Conference on Advanced Materials and Processes, 1, 2001.
4. F.-W. Bach and E. Lugscheider, *Schweißen Schneiden*, 54(2) (2002), 64 (in German).
5. L. Zhao, K. Bobzin, F. Ernst, J. Zwick and E. Lugscheider, *Materialwissenschaft Werkstofftechnik*, 37(6) (2006), 516 (in German).
6. J. Wilden and A. Wank, *Materialwissenschaft Werkstofftechnik*, 32(8) (2001), 654.
7. H.-D. Steffens and T. Duda, *J. Therm. Spray Technol.*, 9(2) (2000), 235.
8. B. J. Downey, J. C. Bernel and P. J. Zimmer, *Corrosion*, 25 (1969), 502.
9. P. L. Daniel and R. A. Rapp, in *Advances in Corrosion Science and Technology*, Vol. 5, 131. Plenum Press, New York, NY, 1976.
10. K. Reinhold and K. Hauffe, *J. Electrochem. Soc.*, 124 (1977), 87.
11. M. J. McNallan, W. W. Liang, S. H. Kim and C. T. Kang, *High Temp. Corros.*, 2 (1981), 316.
12. M. J. McNallan and W. W. Liang, Met. Soc. AIME-ASM, 457, ed. M. F. Rothman, 1985.
13. C. Schwalm and M. Schütze, *Mater. Corros.*, 51 (2000), 34.

14. C. Schwalm and M. Schütze, *Mater. Corros.*, 51 (2000), 73.

15. C. Schwalm and M. Schütze, *Mater. Corros.*, 51 (2000), 161.

16. R. Bender and M. Schütze, *Mater. Corros.*, 54 (2003), 567.

17. R. Bender and M. Schütze, *Mater. Corros.*, 54 (2003), 652.

18. S. Doublet, PhD thesis, ISBN 3-8322-2314-2, University of Aachen, Germany, 2006.

19. S. Doublet, P. Masset, T. Weber and M. Schütze, *Mater. Corros.*, to be submitted.

20. S. Doublet, H. Latreche, P. Masset, T. Weber and M. Schütze, *Oxid. Met.*, to be submitted.

21. R. Roy, D. K. Agrawal and H. A. McKinstry, *Annu. Rev. Mater. Sci.*, 19 (1989), 59.

22. P. Abel and G. Bozzolo, *Scripta Mater.*, 46 (2002), 557.

23. Binary alloy phase diagrams, Vol. 1, 140–142, ed. T. B. Massalki, J. L. Murray, L. H. Benett and H. Baker, 1986.

24. G. Wolf, D. Bendix and M. Faulstich, in Proc. Int. Thermal Spray Conf., 1093, Bale, Switzerland, 2005.

25. E. Berghof-Hasselbächer and M. Schütze, *Sonderbände der Prakt. Metallographie*, 31 (2000), 319–324 (in German).

26. K. Matsuura and M. Kudoh, *Adv. Eng. Mater.*, 3(5) (2001), 311.

27. N.N., DIN EN 657, June 1994, Thermal spraying – Terminology, classification, 7.

28. R. J. Fruehan, *Metall. Trans.*, 3 (1972), 2585.

29. M. J. McNallan and W. W. Liang, *J. Am. Ceram. Soc.*, 64(5) (1981), 302.

30. R. Bender, PhD thesis, ISBN 3-8265-9505-X, University of Aachen, Germany, 2006 (in German).

31. Binary alloy phase diagrams, Vol. 1, 1754–1755, ed. T. B. Massalki, J. L. Murray, L. H. Benett and H. Baker, 1986.

32. G. Rog, G. Brochardt, M. Wellen and W. Löser, *J. Chem. Thermodynam.*, 35 (2003), 261.

33. J. Jedlinski and G. Borchardt, *Oxid. Met.*, 36(3/4) (1991), 317.

34. P. Masset and M. Schütze, *Oxid. Met.*, in preparation.

35. L. Kaufman and H. Nesor, *Metall. Trans.*, 5 (1974), 1623.

# 26

# Parameters governing the reduction of oxide layers on Inconel 617 in an impure VHTR He atmosphere*

## J. Chapovaloff and G. Girardin

*AREVA NP Technical Centre, Corrosion-Chemistry Department,*
*1 rue B. Marcet BP 181, 71205 Le Creusot, France*
*jerome.chapovaloff@areva.com; gouenou.girardin@areva.com*

## D. Kaczorowski

*AREVA NP Fuel Sector, Lyon, France*
*damien.kaczorowski@areva.com*

## 26.1    Introduction

Among the Generation IV reactor concepts, the Very High Temperature Reactor (VHTR) system is designed to be a high-efficiency system using helium as the coolant. The basic technology for the VHTR has been established in earlier High Temperature Gas Reactor plants. The new concept VHTR must produce an outlet gas temperature above 1000 °C from the core and this heat is transferred to a second gas circulator through an Intermediate Heat eXchanger (IHX). This high temperature enables applications such as hydrogen production or process heat for the petrochemical industry or others.

The candidate materials for the IHX are NiCr-base alloys and two have been short-listed for this component: Haynes 230 and Inconel-617. These metallic materials must exhibit good resistance towards corrosion. As observed in previous Gas Cooled Reactors, it is expected that the VHTR's cooling gas will be polluted by air ingress, internal leakage or the degassing of adsorbed species from the large amount of graphite used. These impurities are reactive and may interact with the core graphite and with the metallic materials and may cause some loss or damage to their properties.

Typical impurities are $H_2$, $H_2O$, $CH_4$, $CO$, $N_2$, and $CO_2$ in the µbar range, which can react with the metallic components. Depending on their relative concentrations and on the helium temperature, the interaction of these impurities with the metallic materials may or may not lead to internal and surface oxidation and may cause carburisation or decarburisation of the alloy. The formation of a protective oxide on the surface of the alloy turns out to be a good solution in establishing a barrier against the impurities in the VHTR atmosphere. The oxide layer has to be thin and compact in order to avoid spallation.

---

\* Reprinted from J. Chapovaloff et al.: Parameters governing the reduction of oxide layers on Inconel 617 in an impure VHTR He atmosphere. *Materials and Corrosion.* 2008. Volume 59. pp. 584–90. Copyright Wiley-VCH Verlag GmbH & Co. KGaA.

The open literature reveals the existence of a particular reaction in which chromium oxide suffers from a destructive reaction with carbon in solution in the alloy and the protective surface layer becomes unstable. The occurrence of this reaction is indicated by a sharp rise in CO production corresponding to reduction of the oxide which no longer acts as an efficient barrier against environmental interactions. The unprotected metallic material will rapidly gain or lose carbon. These carbon transfers, called carburisation and decarburisation, respectively, can induce structural transformations that have a dramatic impact on the mechanical properties of the component.

This reaction has mainly been studied by two authors [1,2]. Their work was essentially based on NiCr-based alloys in polluted helium atmospheres. Their experimental results led to consideration of the following reaction 26.1 as being a possible explanation of the phenomenon:

$$2Cr_2O_3 + Cr_{23}C_6 \xrightarrow{T > T_a} 27Cr + 6CO \qquad [26.1]$$

This reaction occurs above a critical temperature, $T_a$, and was identified by Brenner and Graham [3] as the 'microclimate reaction'. In other words, these authors suppose that the reaction takes place in pores within oxide, allowing access for the gas to interact with the matrix. The gas phase present in the bottom of the pores establishes a bridge between the oxide and the carbides.

AREVA-NP Technical Centre has developed a specific apparatus devoted to intermediate exposure tests (up to 500 h) to determine a benign He environment (impurity levels and maximum temperature) and to optimise candidate alloys containing a range of minor elements. In this paper, the 'microclimate reaction' and the parameters which govern it are analysed.

## 26.2    Experimental procedures

### 26.2.1    Materials

The chemical composition of the tested Alloy 617 is given in Table 26.1. Specimens with dimensions of 15 mm × 15 mm were machined from sheet metal of 2 mm thickness. The specimen surfaces were ground to 2400 grit emery paper and finished with a diamond suspension of 3 μm. Before testing, they were ultrasonically cleaned in acetone, then alcohol before being dried and weighed.

### 26.2.2    Test facility

The experimental loop is illustrated in Fig. 26.1. The main idea behind its design was to allow adjustment of most of the identified parameters such as impurity levels, temperature, gas flow rate, etc.

*Table 26.1*    Chemical composition (wt.%) of the studied alloy Inconel-617

| Ni | Cr | Co | Mo | Fe | Mn |
|---|---|---|---|---|---|
| *Base* | 21.56 | 12 | 9.21 | 0.95 | 0.10 |
| Ti | Al | C | Cu | Si | B |
| 0.41 | 1.01 | 0.06 | 0.07 | 0.15 | 0.002 |

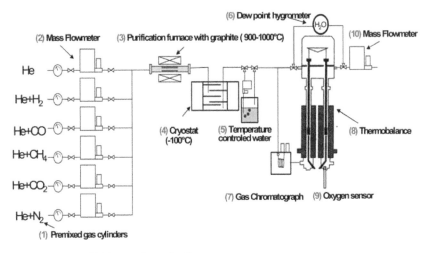

*26.1* Areva-NP Helium loop facility

To obtain low concentrations (in the range of μbar) for each impurity; gas cylinders of helium with a given level of one type of impurity and some bottles of pure helium were used. A system of mass flow controllers allowed dilution of the flow of helium plus one impurity in the pure helium. The level of each impurity could be adjusted over a wide range.

Gas was passed through graphite packed in a quartz tube and heated within a furnace up to 900 °C. The reaction of the oxygen and water contained in the gas with graphite produces CO and hydrogen. Thus, at the outlet of the furnace packed with graphite, the gas was free of oxygen and water, as observed in the real HTR.

The water level in the gas was controlled by passing it through a cryostat (−100 °C) to provide the expected low partial pressure and adjusted by passing it through deaerated water contained in a thermostat. The water content was monitored with a dew point mirror hygrometer.

Corrosion rate measurement was carried out using thermogravimetry, with an accuracy in the range of μg. The exact level of impurities was monitored by gas chromatography.

### 26.2.3   Test conditions

The corrosion tests were conducted in an impure He gas environment. Table 26.2 gives the impurity content of the He. The helium circulated in the loop (gas flow rate: 12 mL s⁻¹ at room temperature). Quadakkers and Schuster [2] developed a procedure to determine $T_a$ for Inconel 617. The oxidation experiments were carried out in two stages. The first stage was oxidation for 20 h at 850 °C. The temperature was then raised to 980 °C for a further 20 h.

### 26.2.4   Evaluation methods

After exposure, the specimens were weighed and analyses were carried out on the surface scales and cross-sections of samples using a Scanning Election Microscope equipped with an EDX-system.

*Table 26.2* Impurities levels of He (μbar) for each test. The level of oxygen is below the detection limit

|         | $H_2$ | CO   | $H_2O$ | $CH_4$ | $O_2$ |
|---------|-------|------|--------|--------|-------|
| Test 1  | 500   | 5.5  | 5      | 0      | 0     |
| Test 2  | 500   | 26.3 | 5.2    | 0      | 0     |
| Test 3  | 27    | 58.6 | 5.3    | 0      | 0     |
| Test 4  | 113   | 7    | 16     | 0      | 0     |
| Test 5  | 110   | 60   | 14.7   | 0      | 0     |
| Test 6  | 500   | 14.6 | 5.3    | 0      | 0     |
| Test 7  | 112   | 14.4 | 15.9   | 0      | 0     |
| Test 8  | 500   | 14.2 | 15.9   | 0      | 0     |
| Test 9  | 128   | 14.2 | 5.6    | 0      | 0     |
| Test 10 | 500   | 14.4 | 3.6    | 0      | 0     |
| Test 11 | 500   | 14.9 | 2.9    | 0      | 0     |
| Test 12 | 500   | 14.9 | 3.12   | 0      | 0     |
| Test 13 | 38    | 15.6 | 3.67   | 0      | 0     |

## 26.3    Results

### 26.3.1    Destruction of oxide layer in polluted He

Gas phase analysis

During heating from 850 °C to 950 °C, initial gas phase analysis showed no detectable change at the test section outlet until a sharp rise in CO production was observed. Figure 26.2 illustrates this phenomenon in the case of Test 9. The values of

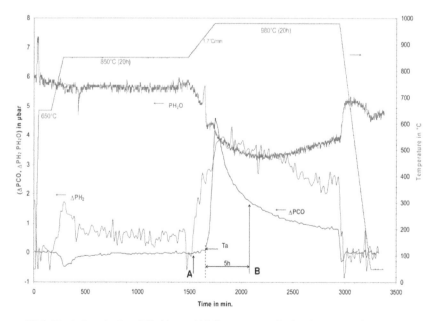

*26.2* Evolution in the CO, $H_2$ and $H_2O$ pressure during heating of Inconel 617 (Test 9). For a critical temperature, $T_a$, a CO release is detected corresponding to Reaction 26.1. Here, $T_a$ is close to 921 °C for a level of $PCO$ of 14.2 μbar

$\Delta PCO$ and $\Delta PH_2$ were determined as $PCO_{outlet} - PCO_{inlet}$ and $PH_{2outlet} - PH_{2inlet}$, respectively, by analysis of the gas at the outlet and inlet of the thermobalance during heating. The partial pressure of $H_2O$ was measured at the outlet of the thermobalance.

The temperature at which carbon monoxide production increases corresponds to the attainment of equilibrium for Reaction 26.1 and is designated $T_a$.

The first stage, exposure for 20 h at 850 °C, which is below the critical temperature, $T_a$, leads to the formation of a protective layer following Reaction 26.2

$$2Cr + 3H_2O \rightarrow Cr_2O_3 + 3H_2 \qquad [26.2]$$

As predicted by this reaction, the consumption of $H_2O$ and production of $H_2$ are experimentally observed in Fig. 26.2. Simultaneously, a second reaction of oxidation and carburisation by CO occurs and starts below 850 °C:

$$27Cr + 6CO \rightarrow 2Cr_2O_3 + Cr_{23}C_6 \qquad [26.3]$$

The second stage consists of increasing the temperature slowly until CO release is observed. As illustrated by Fig. 26.2 (Test 9), the temperature $T_a$ is around 921 °C and corresponds to a CO release of 4.5 µbar, which slowly decreases when the temperature is held constant at 980 °C. The kinetics of Reaction 26.1 are accelerated when, on the one hand, the temperature is increased and the other hand, there is CO release.

The appearance of surfaces and cross-sections of samples before and 5 h after $T_a$ was reached (Fig. 26.2 points A and B) confirmed that the release of CO had reduced the chromium oxide layer.

### Surface observations and analyses before and after $T_a$

Figure 26.3 illustrates a specimen treated for 20 h at 850 °C (point A). The nature of the scale is a nodular $Cr_2O_3$ Ti-doped. Below the nodules, the oxide is enriched in Al and Cr. The grain boundaries are also rich in Cr and Ti. Moreover, beneath the surface layer, the formation of intergranular aluminium-rich oxide is observed. Some fine Cr- and Mo-rich carbides are also observed in intra or intergranular locations.

Figure 26.4 shows a specimen after 5 h at a temperature above $T_a$ (point B). On the surface oxide, it can be observed that there is a lack of nodules representative of the reaction zone. In the near-surface region, aluminium is detected as an internal oxide. A carbide-free zone is also observed down to a depth of about 150 µm.

The release of CO reduces the chromia at the specimen surface. The occurrence of Reaction 26.1 was demonstrated experimentally, and its governing parameters are described below.

### 26.3.2    Parameters governing oxide layer reduction

#### Influence of PH$_2$

The experiments were undertaken in an atmosphere where $PCO$ and $PH_2O$ were kept constant and $PH_2$ was used as a parameter (Tests 7 to 9, 12 and 13). In each test, the critical temperature, $T_a$, was determined with the same thermal cycle and procedure as used previously. By repeating the experiment using gas compositions with different $H_2$ contents, it was possible to assess the dependence of $T_a$ on $PH_2$ for three levels of water (Fig. 26.5), as follows:

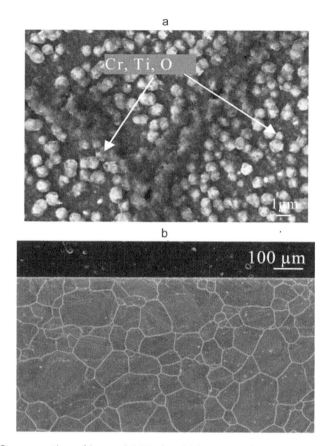

*26.3* Cross-section of Inconel 617 after 20 h at 850 °C – Fig. 26.2, point A; (a) cross-section; (b) surface

Tests 6 and 9 /atmosphere: $PCO$ $14.4\pm0.2$ µbar; $PH_2O$ $5.5\pm0.1$ µbar
Tests 12 and 13 /atmosphere: $PCO$ $15.3\pm0.3$ µbar; $PH_2O$ $3.3\pm0.3$ µbar
Tests 7 and 8 /atmosphere: $PCO$ $14.3\pm0.1$ µbar; $PH_2O$ $16\pm0.1$ µbar

It is concluded that when $PH_2$ varies between 30 and 500 µbar, $T_a$ is not affected.

### Influence of PCO

A series of tests was performed with $PCO$ as a variable (Tests 1 to 3, and 6). $PH_2$ was not constant but without influence on $T_a$, as has been demonstrated previously. Water vapour pressure was kept constant ($5.3\pm0.1$ µbar).

Figure 26.6 shows the experimental results and the dependence between $PCO$ and $T_a$. It should be noted that $T_a$ increases with the level of CO.

The comparison is done between experimental results and the theoretical values expected from Reaction 26.1 constants

$$K_x=(PCO)^3.(a_{Cr})^9=\exp[-\Delta rG/(RT_a)] \qquad [26.4]$$

$$T_a=-\Delta rG/[R(3.\ln(PCO)+9.\ln(a_{Cr}))] \qquad [26.5]$$

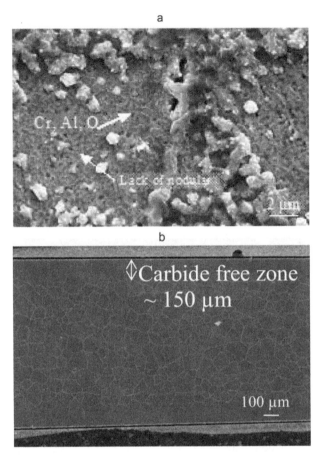

26.4 Cross section of Inconel 617; 5 h after $T_a$ – Fig. 26.2, point B; (a) cross-section; (b) surface

where

$\Delta rG$ = free enthalpy of reaction 26.1 (kJ mol$^{-1}$)
$R$ = universal gas constant (J mol$^{-1}$ K$^{-1}$)
$a_{Cr}$ = chromium activity
$PCO$ = partial pressure of CO (μbar).

The experimental results and the theoretical values are related by an exponential law, described by several authors. The best fit is obtained for $a_{Cr} = 0.76$.

The curve $T_a$ as a function of $PCO$ could define a limitation in terms of the temperature for the IHX.

## Influence of PH$_2$O

Using the same procedure as previously, water was chosen as a parameter. Hydrogen and CO pressures were the same as for Test 6 ($PCO$ 14.5 ± 0.3 μbar; $PH_2$ 40 to 500 μbar) and the water vapour pressure used varied from nearly 3 to 16 μbar.

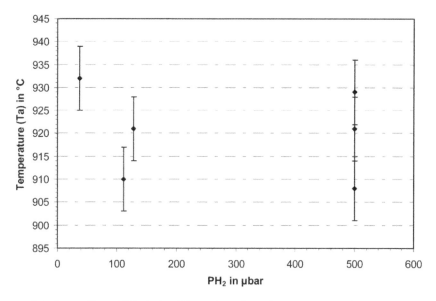

*26.5* Curve $T_a$ vs. $PH_2$ derived from the microclimate reaction for Inconel 617 in an atmosphere with $PCO$, $PH_2O$ constant

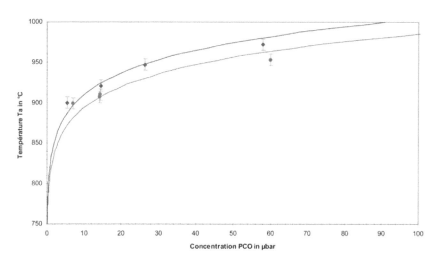

*26.6* Curve $T_a$ vs. $PCO$ derived from the microclimate reaction for Inconel 617 for several atmospheres: blue diamonds, $PH_2O \sim 5$ µbar, pink circles, $PH_2O \sim 16$ µbar; blue line, theoretical fit established from the microclimate reaction with $a_{Cr}=0.76$; pink line, theoretical fit established from the microclimate reaction with $a_{Cr}=0.69$

The results are shown in Fig. 26.7 and illustrate how $T_a$ is influenced by $PH_2O$. $T_a$ decreases by 20 °C as $PH_2O$ is increased from 3 to 16 µbar.

Consequently, curves of $T_a = f(PCO)$ can be drawn for several atmospheres with different values of $PCO$ and $PH_2O$. Figure 26.6 presents these curves of $T_a = f(PCO)$ and, in the same way, a chromium activity, $a_{Cr} = 0.69$ can be estimated.

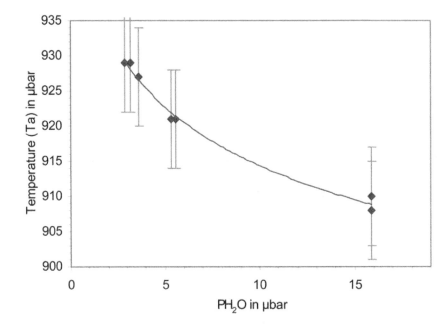

26.7 Curve $T_a$ vs. $PH_2O$ derived from the microclimate reaction for Inconel 617 in an atmosphere with $PCO$ constant and $PH_2$ variable between 40 and 500 µbar

## 26.4    Discussion

### 26.4.1    Reduction of the oxide layer

The HTR atmospheres are polluted by different impurities. This implies that carbon monoxide production can be explained by different reactions, such as

Reaction gas/gas

Depending on the impurities, the possible reactions are:

$$CO_2 + H_2 \Leftrightarrow CO + H_2O \qquad\qquad [26.6]$$

$$H_2O + CH_4 \Leftrightarrow CO + 3H_2 \qquad\qquad [26.7]$$

But considering the high gas (He) flow and the low contamination level, these reactions are not observed due to the low probability of collision between impurity particles and the very low gas/gas reactions rate.

Reaction gas/alloy

Three reactions between the atmosphere and the alloy are feasible thermodynamically

$$3CH_4 + Cr_2O_3 \rightarrow 2Cr + 6H_2 + 3CO \qquad\qquad [26.8]$$

$$Cr_{23}C_6 + 6H_2O \rightarrow 23Cr + 6CO + 6H_2 \qquad\qquad [26.9]$$

$$2Cr_2O_3 + Cr_{23}C_6 \rightarrow 27Cr + 6CO \qquad\qquad [26.1]$$

CO production is observed for Inconel 617, even without methane in the test atmosphere, and cannot be described by Reaction 26.8. Reaction 26.9 is not relevant because CO production is more important than $H_2O$ consumption. The only possible reaction is Reaction 26.1. However, this is not directly possible because it involves two solid species, and is the result of the concurrence of two other simultaneous reactions involving $H_2O$ and $H_2$. This reaction has been called by Brenner 'the microclimate reaction' because it is of major importance in the pores of the surface layers of metallic materials.

The two relevant reactions are:

1.  Reduction of chromium oxide by hydrogen:

$$2Cr_2O_3 + 6H_2 \rightarrow 6H_2O + 4Cr \qquad [26.10]$$

2.  Attack of carbides by water with the formation of carbon monoxide:

$$6H_2O + Cr_{23}C_6 \rightarrow 6CO + 23Cr + 6H_2 \qquad [26.11]$$

Adding these two reactions leads to the 'microclimate reaction'

$$2Cr_2O_3 + Cr_{23}C_6 \rightarrow 6CO + 27Cr \qquad [26.1] \ ([26.10]+[26.11])$$

Although no significant gas/gas reaction can occur, a consequence of the 'microclimate reaction' is that the local thermodynamic equilibrium of Reaction 26.1 can be reached close to the reacting metal surface, if chromium oxide and carbide are simultaneously in contact with the gaseous phase. The diffusion controlled layer of gas, which must contain $H_2$ and $H_2O$, bridges the gap between $Cr_2O_3$ and $Cr_{23}C_6$ allowing these species to be reduced to Cr and to form CO, even if they are not in close contact.

According to the experimental results, the sharp CO release is not solely due to carbides. The observations of the cross-sections show there are no 'carbides' located beneath the oxide layer to form a source of carbon. Consequently, it is supposed that it is the carbon in solution which reacts with the chromia.

The 'microclimate reaction' is

$$3C_{sol} + Cr_2O_3 \rightarrow 3CO + 2Cr \qquad [26.12]$$

The 'microclimate reaction' decreases the carbon activity near the surface and with time a gradient in carbon is established through the thickness of the sample. A carbide-free zone can be observed beneath the oxide layer (150 μm for Test 9), possibly explained by carbide dissolution (Reaction 26.13).

According to Le Chatelier's law, Reaction 26.13 takes place from left to right because of the $C_{sol}$ consumption by Reaction 26.12.

Carbide dissolution is given by

$$MC \rightarrow M + C_{sol} \qquad [26.13]$$

with $M = Cr$, Mo and $C_{sol}$ representing the carbon content in solution in the matrix.

In conclusion, the large increase in carbon monoxide production observed can be explained by the availability of the carbon close to the metal/oxide interface. The subsequent slow decrease in the CO level illustrates the slow diffusion of carbon away from the interface which impedes the reaction.

### 26.4.2   Parameters $PCO$, $PH_2O$, $PH_2$

As mentioned by Quadakkers, the thermodynamic equilibrium will never be affected for the gas/metal reactions because the flow of helium is important. The author

defines steady state rather than a thermodynamic equilibrium. From this consideration and Reaction 26.12, we can write a relationship between $PCO$, $a_{C_{sol}}$ and $a_{Cr}$ for
a particular temperature, for example $T_a$

$$K(\text{Ta}) = \frac{a_{Cr}^2 * PCO^3}{a_{C_{sol}}^3 * a_{Cr_2O_3}} \qquad [26.14]$$

It should be noted that Reaction 26.14 is defined by three variables because the activity of the oxide may be considered to be equal to 1. Every variable can be expressed
in terms of two others. For example:

$$PCO = \left( \frac{K(\text{Ta}) * a_{C_{sol}}^3}{a_{Cr}^2} \right)^{1/3} \qquad [26.15]$$

Figure 26.6 shows a plot of the predicted $T_a$ values as a function of the partial
pressure of CO. For each test, the critical temperature was determined when $\Delta PCO$
increased by 0.1 µbar.

Below $T_a$, the atmosphere is slightly carburising (without methane) and $a_{C_{sol}}$ is
considered to be constant. Furthermore, for the tests carried out (Table 26.2),
oxidation at 850 °C for 20 h may slightly affect $a_{Cr}$. Thus, in this very simplified case,
$PCO$ will be dependent only on the equilibrium constant and thus will vary with
temperature.

In different atmospheres, the activity of carbon can be modified during the oxidation treatment ($T < T_a$). Carburisation reactions by $CH_4$ or CO can increase $a_{C_{sol}}$.
Figure 26.2 shows that CO can carburise the alloy when the temperature increases
from 650 °C to 850 °C. The consumption of CO depends on the ratio $PH_2O/PCO$, as
shown in Table 26.3.

The most reasonable explanation is a competition in oxidation kinetics between
$H_2O$ and CO. It is considered that the growth of the oxide layer in atmospheres rich
in $H_2O$ no longer allows the carbon monoxide to interact with the metal surface.
Thus, carburisation by CO can be avoided.

It may be envisaged that over a long period of time, the activity of chromium in the
alloy will decrease. The oxidation of the surface implies a depletion of chromium in
the metal/oxide interface. This oxidation is not controlled by the oxidising power,
$PO_2^{eq}$, defined by the ratio $PH_2O/PH_2$ but simply by $PH_2O$ because, as Fig. 26.5
shows, $H_2$ does not modify $T_a$ when $H_2$ varies from 100 to 500 µbar.

From Fig. 26.7, the more $PH_2O$ increases, the more $T_a$ decreases for the same
partial pressure of carbon monoxide. The kinetics of oxidation in atmospheres
with 16 µbar $H_2O$ are faster than with 5 µbar $H_2O$. For example, calculations of the
parabolic rate constant for Tests 4 and 9 confirm this hypothesis

Test 4: $Kp_4 = 1.41.10^{-16}$ g$^2$ cm$^{-4}$ s$^{-1}$ for $T$: 850 °C/20 h
Test 9: $Kp_9 = 8.48.10^{-16}$ g$^2$ cm$^{-4}$ s$^{-1}$ for $T$: 850 °C/20 h

*Table 26.3*  Evolution of the ratio $PH_2O/PCO$

| Tests | $PH_2O/PCO$ | Area under curve (µbar min) |
|-------|-------------|------------------------------|
| 10    | 0.25        | 170                          |
| 6     | 0.36        | 75                           |
| 7     | 1.10        | 0                            |

It is easy to understand that the chromium activity decreases when the kinetics of oxidation are more important. Figure 26.8 shows the curve $T_a = f(PCO)$ for several chromium activities 0.8, 0.6 and 0.4 and demonstrates that $T_a$ decreases when the chromium activity decreases for a fixed $PCO$. Thus, $T_a$ decreases when $PH_2O$ increases.

By using this theoretical approach, the chromium activity for different atmospheres can be estimated. Figure 26.8 gives chromium activities of 0.76 and 0.69 for $PH_2O$ values of 15 and 5 µbar, respectively. However, this value does not represent the chromium activity of the matrix but a chromium activity in the region near the surface because estimations of the chromium activity by measurement of the saturated vapour in chromium showed that the real activity of chromium in Inconel 617 is 0.33 instead of 0.6 at 950 °C.

Furthermore, the fit obtained does not explain all of the experimental data. The model does not take into account the morphology of the oxide layer. The last possible explanation could be the effect of minor alloying additions on scale morphology. In this case, the chromia scale is doped with Ti. It is well known that the incorporation of titanium in the oxide scale increases its porosity [2].

These experiments showed that $T_a$ is influenced by several parameters such as $PCO$ and $PH_2O$. Tests of longer duration must be undertaken to take into account the parameter $a_{Cr}$. The last parameter, $a_{C sol}$, must also be studied in carburising atmospheres.

### 26.5    Conclusions

The new VHTR reactor concept uses helium as the coolant gas. But the helium is impure since it contains $H_2O$, $CO$, $CH_4$, $H_2$, and $CO_2$ in the µbar range. These impurities can induce a corrosion process in Inconel 617 used in the Intermediate Heat eXchanger (IHX). Among these oxidation reactions, reduction of the oxide layer by the so-called 'microclimate reaction' is a reaction that should be avoided in order to maintain good performance.

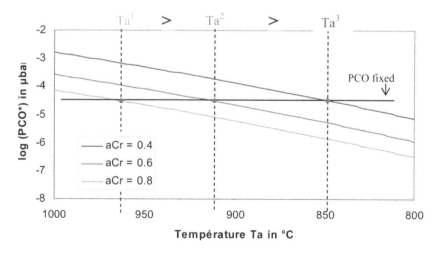

26.8 Curve $T_a$ vs. PCO for several chromium activities

It has been shown that this particular 'microclimate reaction' is controlled by $PCO$ and $PH_2O$. During operation of the power plant, it will be necessary to limit depletion of the chromium in the alloy by, for example, control of the $H_2O$ content of the helium. To maintain the optimum behaviour of the alloy, it will be necessary to choose an appropriate partial pressure of carbon monoxide to avoid destruction of the oxide layer. The carbon activity is also a very important parameter which can be influenced by the carburisation reactions. However, the existence of a critical ratio of $PH_2O/PCO$ allowing carburisation by CO has been demonstrated. Likewise, it is possible to consider a ratio $PH_2O/PCH_4$ to avoid carburisation by the methane. The presence of 20 μbar of methane in Test 6 indicated that the alloy does not carburise. Thus it does not influence $T_a$.

## References

1.  W. J. Quadakkers and H. Schuster, *Werkst. Korros.*, 36 (1985), 141.
2.  W. J. Quadakkers and H. Schuster, *Werkst. Korros.*, 36 (1985), 335.
3.  K. G. E. Brenner and L. W. Graham, *Nucl. Technol.*, 66 (1984), 401.
4.  M. R. Warren, *High Temp. Technol.*, 4 (1986), 119.
5.  K. Hilpert, H. Gerads and D. F. Lupton, *J. Nucl. Mater.*, 80 (1979), 126.

# 27

# Protective coatings for very high-temperature reactor applications*

## C. Cabet and C. Guerre

*Service de la Corrosion et du Comportement des Matériaux dans leur Environnement, DEN/DANS/DPC, CEA Saclay, 91191 Gif sur Yvette, France*

*celine.cabet@cea.fr; catherine.guerre@cea.fr*

## F. Thieblemont

*Materials and Interfaces Department, Optoelectronics Materials Laboratory, Weizmann Institute of Science, Rehovot, Israel*

*florent.thieblemont@weizmann.ac.il*

## 27.1    Introduction

Driven by world population growth, the demand for power is increasing. At the same time, the oil and gas market is experiencing tensions, and environmental concerns are more pressing. Against this background, there is a need to adapt the production infrastructure available by 2020. Sustainable nuclear power will play a role in the balanced power supply for future generations. Ten countries or so have joined their efforts to promote a 4th generation of nuclear systems that can meet the challenging objectives in the main domains of cost-efficiency, natural resource preservation, reducing radioactive waste, counteracting the risk of proliferation, safety and reliability. In this framework, important research is focusing on developing innovative Gas-Cooled Reactors (GCR). The Very High-Temperature Reactor (VHTR) is a helium-cooled, graphite-moderated, modular reactor operated in a very high-temperature mode and dedicated to supply process heat, especially for hydrogen production applications. The helium emerging from the core at temperatures as high as 1000 °C (the higher the temperature, the more efficient the energy conversion) transfers its thermal power through heat exchangers and turbo machines within the Power Conversion System vessel and the cooled helium streams back to the reactor core vessel. A special dual-section co-axial pipe, the Hot Gas Duct (HGD), links both vessels together and allows the counter flows of high-temperature helium (maximum $\sim 1000$ °C) inside and of cooled helium (maximum $\sim 450$ °C) outside. The HGD is a key component as it must retain its strength and leak-tightness over tens of years.

Considering the thermal and environmental conditions along with the extended targeted lifetime, HGD materials must fulfil high-level requirements in terms of mechanical properties, structural stability, and compatibility with the helium atmosphere [1]. Metallic materials, mainly chromia-forming nickel- or iron-base alloys, have been extensively studied for high-temperature GCR applications but their long-term durability is limited by creep as well as corrosion damage. VHTR helium, polluted by air ingress and outgassing from the large amounts of graphite, will actually contain traces of impurities such as $H_2$, CO, $CH_4$, $N_2$, etc., in quantities ranging from a few to hundreds of microbars, and a very low content of oxidising species, mainly water vapour in the microbar range [2]. These impurities are able to react with the chromia-forming alloys. Depending on the temperature, the carbon activity and the oxidising potential in the gas phase, as well as the alloy composition, the main corrosion processes are oxidation, carburisation and decarburisation [3–5]; these two last phenomena have been proven be highly detrimental and can cause rapid material failure [6,7]. While several classes of metallic materials (such as ferritic and austenitic steels or nickel-based alloys [8]) have been considered, the use of coatings has not yet been investigated for VHTR applications. However, thermal barrier coatings (TBCs), which are duplex coatings consisting of a bond coat topped with a ceramic layer, may both decrease the substrate temperature and ensure its protection against high-temperature corrosion. These systems are currently used for gas turbine materials where they encounter severe conditions including mechanical and thermal stresses and hot corrosion. In a VHTR, the environment is milder but the requirements of durability and reliability are far more demanding.

The present study has assessed the possible benefit of coating the inner side of the VHTR hot gas duct with TBCs. Two systems – Alloy 800H/NiAl(Pt)/EBPVD $ZrO_2$ (Y) and Alloy 800H/NiCrAl(Y)/CVD $ZrO_2$ (Y) – were selected and characterised in the as-received state. Then high-temperature treatments were performed in the VHTR environment for up to 1000 h. The thermal and corrosion-induced changes were studied in connection with the prospective service application.

## 27.2    Experimental procedure

### 27.2.1    Selection of the thermal barrier coating systems

The substrate is Alloy 800H, an austenitic Fe–Ni–Cr alloy, commonly used in nuclear steam-generators (ASME code case 1325-7). State-of-the-art TBCs use Partially Stabilised Zirconia (PSZ) for insulating top coats; 6–8 wt.% yttria is incorporated with zirconium oxide to stabilise partially the tetragonal phase for good strength, fracture toughness, and resistance to thermal cycling. This ceramic is relatively inert, has a high melting point, and low thermal conductivity though its properties are strongly dependent on the deposition route. The fabrication process must be able to coat the inside of a large pipe on an industrial scale. Two coating procedures have been used: liquid injection CVD (Chemical Vapour Deposition) and EB-PVD (Electron Beam – Physical Vapour Deposition). A metallic inner layer, called a bond coat, is needed between the ceramic and the substrate alloy; this aids the adhesion of the top coat, protects the substrate from corrosion and oxidation, and helps in handling expansion mismatch between the PSZ and alloy. Bond coats may be divided into two main categories:

- overlay coatings that are manufactured by the deposition of protective metallic species, generally MCrAl(Y) where M is Ni, Co, or Fe, onto the substrate surface, and

- diffusion coatings that are based on the formation of intermetallic compounds such as β-NiAl via interactions with the substrate.

Both bond coat types can be produced via several processes. It is important to go for fabrication techniques with high deposition rates convenient for industrial applications. In this study, NiCrAl(Y) sprayed via APS (Air Plasma Spray) and NiAl(Pt) were investigated. Specimens were procured by other labs with few details on the preparation technique; they had not been pre-examined.

### 27.2.2   Alloy 800H substrate

Table 27.1 gives the chemical composition of Alloy 800H. The material was machined in the shape of rectangular coupons (20 mm × 10 mm × 2 mm) and drilled with a hole ($\phi$ 3 mm) for the purpose of support.

#### Specimen Type 1 – Alloy 800H/NiCrAl(Y)/CVD ZrO$_2$ (Y)

Specimens of Type 1 were supplied by DTEN/SMP CEA Grenoble. The NiCrAl(Y) bond coat layer was formed by high-energy APS, a process that is said to provide elevated deposition rates. It is an overlay coating meaning that the adhesion between NiCrAl(Y) and Alloy 800H is mainly mechanical. The PSZ layer was deposited by liquid injection CVD at atmospheric pressure, the advantage of which is the ability to coat complex-shaped components, especially the inner surfaces of pipes; the liquid injection process allows for adequate deposition rates. No further information is available on the composition or method of fabrication of the coating.

#### Specimen Type 2 – Alloy 800H/NiAl(Pt)/EBPVD ZrO$_2$ (Y)

Specimens of Type 2 were purchased from Snecma, SAFRAN Group. The NiAl(Pt) bond coat layer was produced by vapour phase aluminisation. In this case, platinum was electrodeposited on the substrate that was then covered by aluminium and, finally, the platinum-modified intermetallic nickel aluminide was formed via interdiffusion with alloy elements, mainly Ni, at 1080 °C for 6 h. The PSZ layer was made by EB-PVD. This process achieves appropriate deposition rates. Though it is not convenient for intricate geometries, it may be possible to coat the inner surface of the Hot Gas Duct.

### 27.2.3   Corrosion testing

For this first investigation, two series of isothermal tests were carried out at 950 °C and atmospheric pressure under flowing impure helium (gas flow rate: 10 NL h$^{-1}$) for 500 and 1000 h. Runs contained only one specimen of each of the following types: substrate; specimen Type 1; specimen Type 1 without PSZ, and specimen Type 2.

*Table 27.1*   Chemical composition of Alloy 800H in wt.%

| Fe | Ni | Cr | Si | Al | Ti | C |
|------|------|------|-----|-----|-----|-----|
| 44.2 | 32.8 | 21.3 | 0.6 | 0.4 | 0.7 | 0.1 |

*Table 27.2*   Test helium composition in vppm

| H₂ | CO | CH₄ | H₂O | O₂ |
|---|---|---|---|---|
| 200 | 50 | 20 | 1–7 | <0.1 |

Table 27.2 gives the gas composition. A capacitive-probe dew point analyser monitored the water vapour content and an on-line Gas Phase Chromatograph (helium detector) analysed the permanent gas concentrations in helium. Cross-sections of specimens in the as-coated state and after the high-temperature corrosion treatments were examined using scanning electron microscopy (SEM), energy dispersive X-ray spectroscopy (EDS) and wavelength dispersive X-ray spectroscopy (WDS).

## 27.3    Examination of the as-received specimens

### 27.3.1   Specimen Type 1 – Alloy 800H/NiCrAl(Y)/CVD ZrO₂ (Y)

Figure 27.1 presents SEM cross-section images of an as-coated specimen of Type 1 with from the outside (left-hand side) to the substrate (right-hand side):

- the CVD-PSZ layer. The zirconia shows some residual closed porosity without cracks and its thickness is about 80–100 µm. Based on the relatively dense appearance of the layer, its thermal conductivity may be similar to that of bulk ceramic (approx. 2.5 W m⁻¹ K⁻¹ [9]). Figure 27.1 shows that the interface between bond coat and top coat is highly convoluted.
- the APS-NiCrAl(Y) overlay bond coat. The layer was approximately 150 µm thick and demonstrated a wavy lamellar microstructure, globally parallel to the surface. NiCrAl(Y) coatings are known to establish a dual-phase microstructure with a β-NiAl phase contained in a more ductile matrix of solid solution (γ). They are comprised of occasional dispersions of γ' (Ni₃Al), and α-Cr phase. Furthermore, some secondary phases such as Y₂O₃ and chromium carbides may coexist [10]. EDS analyses indicated an overall composition of approximately 63 at.% Ni, 17 at.% Al and 20 at.% Cr (Y concentration was too low for proper detection) that should correspond to the coexistence of γ, β, and γ' phases, according to the Ni–Cr–Al equilibrium diagram at high temperature (Fig. 27.2). It is worth noting that, along with metallic phases, a significant fraction of alumina was detected, probably due to the coating process. The precipitation of alumina within the bond coat is problematic: on the one hand, it may alter the mechanical and adhesion properties of the bond coat and on the other hand, it consumes part of the aluminium that is no longer available for intermetallic phases.
- the Alloy 800H rectangular coupon that was actually not modified by the coating process.

### 27.3.2   Specimen Type 2 – Alloy 800H/NiAl(Pt)/EBPVD ZrO₂ (Y)

Figure 27.3 shows SEM cross-section images of an as-received sample of specimen Type 2 that was composed of (from the left-hand side to the right-hand side):

- the **PSZ layer**. EBPVD zirconia, roughly 300 µm thick, was composed of highly elongated grains that had grown perpendicular to the surface. It was highly

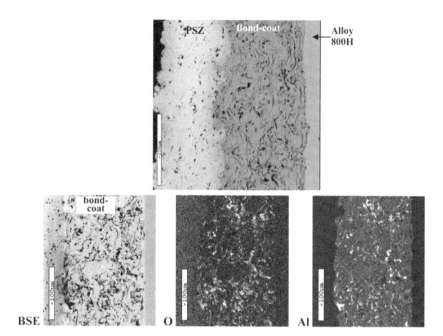

*27.1* SEM-BSE images (top and bottom left-hand side) and EDS maps of O (bottom, middle) and Al (bottom, right-hand side) – As-coated Type 1 specimen (no exposure)

porous in the bulk, but looked denser near the PSZ/bond coat interface. This typical columnar morphology is said to reduce stress build-up within the body of the coating by accommodating stress mismatch between the ceramic layer and the metallic one. The initial thermal conductivity of EBPVD-PSZ is typically 1 W m$^{-1}$ K$^{-1}$ but it increases in service up to about 1.5–2 W m$^{-1}$ K$^{-1}$ [10–12].

- the *TGO*. The image at higher magnification in Fig. 27.3 (centre) exhibits a dark film, one-tenth of a micron thick, at the PSZ layer/NiAl(Pt) bond coat interface. It was rich in oxygen and aluminium. Published TEM results [13] have reported that the bond coat, which is an alumina-former, produced α-alumina scale – a stable form of Al$_2$O$_3$ at high temperature – during the vapour phase aluminisation process. This alumina layer, named TGO (for Thermally Grown Oxide), has good adherence and a low growth rate and thus provides resistance against corrosion.

- the *NiAl(Pt) bond coat*. In agreement with other studies [13,14], the bond coat of the as-received specimen of Type 2 consisted of two zones: the initial surface was probably located at the outer zone/inner zone where small alumina particles can be observed (Fig. 27.3):

  the *outer zone* (white, single phase) arose from both diffusion of Ni out of the alloy and Al supplied by the APV process

  the *inner zone* had developed under the initial surface and was due to interdiffusion processes between the coating and Alloy 800H. It was composed of β-(Ni,Pt)Al precipitates in a γ solid solution.

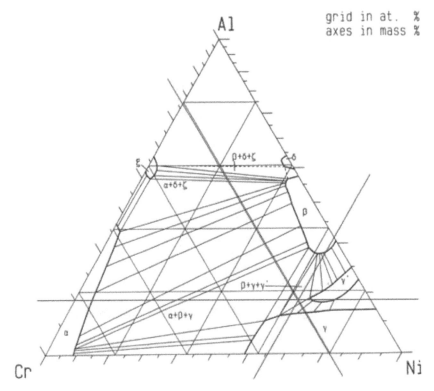

*27.2* Ni–Cr–Al phase diagram at 1025 °C – Cross lines mark initial NiCrAl(Y) overlay bond coat composition

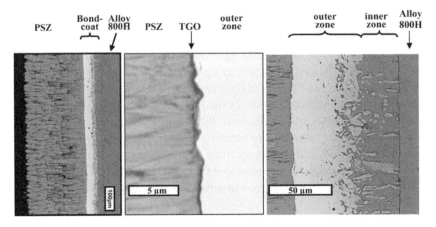

*27.3* SEM-BSE images – As-coated Type 2 specimen (no exposure) – Overall TBC (left-hand side); PSZ /bond coat interface (middle); bond coat (right-hand side)

Figure 27.4 left-hand side reports microanalyses through the different zones. In the outer zone, there were steep concentration gradients; between the PSZ/bond coat interface and the outer zone/inner zone interface, the Al and Pt contents decreased, respectively from 47 at.% to 35 at.% and from 23 at.% to 14 at.% while

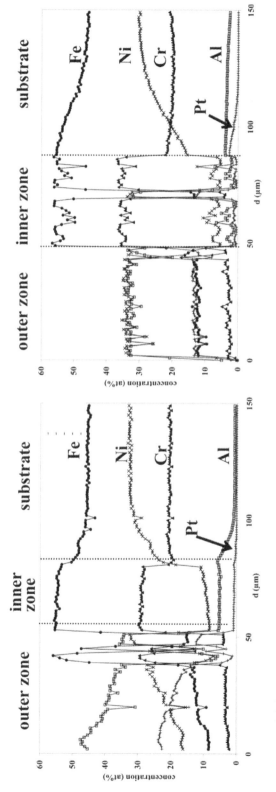

27.4 Microanalyses (at.%) of Type 2 specimen – Left hand side: as coated; right-hand side: after 1000 h at 950 °C in impure helium (d = 0 represents the PSZ/bond coat interface)

the Ni content increased from 16 at.% to 34 at.%. The outer zone also included Fe (up to 8 at.%) and Cr (up to 3.5 at.%). In the inner zone, the main γ phase was constituted of elements of the former alloy Fe, Cr and Ni but Al (up to 5 at.%) and traces of Pt from the initial coating were also detected. The chemical composition of the β-(Ni,Pt)Al inclusions followed the concentration gradients in the outer zone.

- the **_Alloy 800H substrate_**. Figure 27.4 reveals that the substrate has been significantly depleted in Ni up to around 25 μm deep while Al has diffused from the coating into the base alloy.

## 27.4    Examinations of the corroded specimens

### 27.4.1    Alloy 800H

Figure 27.5 illustrates the corrosion morphology of Alloy 800H after exposure to impure helium for 1000 h at 950 °C. The bare coupon developed a mixed surface scale made of a titanium-rich oxide (dark layer and underlying inclusions) and chromium-rich carbides (outer grey geometric compounds). Aluminium and titanium oxidised internally. It was shown [15] that chromium oxide is unstable under such conditions because it reacts with the carbon dissolved in the alloy. The mixed surface scale cannot provide any long-term protection against corrosion, and carburisation may proceed rapidly deep into the substrate.

### 27.4.2    Specimen Type 1 – Alloy 800H/NiCrAl(Y)/CVD ZrO$_2$ (Y)

The thermal and environmental conditions of the experiments produced changes in the TBC microstructure. Figure 27.6 shows a cross-section of a specimen of Type 1 after exposure to impure helium for 1000 h at 950 °C.

- Changes in the CVD-PSZ. Whatever the experimental test, the zirconia top coat cracked (through-thickness from surface to inner interface) but no spallation

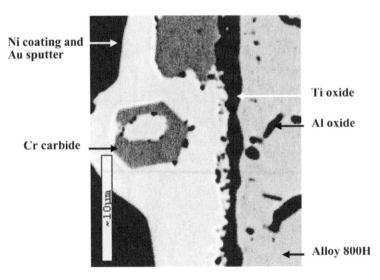

27.5  SEM-BSE image – Alloy 800H after 1000 h in impure helium at 950 °C

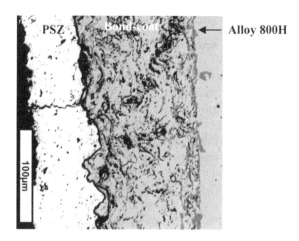

*27.6.* SEM-BSE image – Type 1 Specimen after 1000 h in impure helium at 950 °C

occurred. It is not known whether the cracking occurred during the isothermal heat treatment or during cooling (rate $\sim 2$ K min$^{-1}$).

- Growth of a TGO. An inhomogeneous alumina scale, about 2.2 μm thick, formed at the top of the bond coat. Following the shape of the surface, the TGO growth was highly convoluted. No crack developed.
- Changes in the NiCrAl(Y) bond coat. Figure 27.6 shows numerous dark phases within the bond coat that were analysed as aluminium oxide but it was practically impossible to state whether the whole amount was present initially (Fig. 27.1) or whether it had partially precipitated during the exposure (this latter assumption would imply that the TGO was not protective). Driven by the interdiffusion between the substrate and the metallic coating, the global composition of the NiCrAl(Y) phase had changed: the Ni and probably the Al concentrations had decreased whereas Fe was identified (up to 12 at.%) and the Cr content had remained fairly constant. It is also worth noticing that chromium-rich grey phases had formed within the bond coat, especially at the inner interface. EDS analyses failed to detect any element associated with the chromium; it was, therefore, assumed that these precipitates were metallic α-Cr phases. Nevertheless, complementary examinations (WDS and GDOES) are needed to prove that there was no carbon. Considering the Ni–Cr–Al diagram at high temperature (Fig. 27.2), provided that the incorporation of Fe did not interfere greatly, the depletion in Ni should involve the precipitation of α-Cr. It is known that the segregation of a separate alloy α-Cr phase with its very different coefficient of thermal expansion [16] can promote spallation.
- Changes in Alloy 800H. Interdiffusion processes occurred between the coating and the substrate that were more intense after 1000 h than after 500 h at 950 °C. It is likely that the process would have proceeded further if the corrosion test had been performed for longer times. After 1000 h, the alloy composition exhibited the following variations at depths of up to 60 μm: significant depletion in iron and slight depletion in chromium while the nickel and aluminium contents had increased.

### 27.4.3   Specimen Type 2 – Alloy 800H/NiAl(Pt)/EBPVD ZrO$_2$ (Y)

- The EB-PVD PSZ layer. There was no notable change in the zirconia scale, which exhibited good adherence (Fig. 27.7).
- Changes in the TGO. After 500 h in impure helium at 950 °C, the TBC experienced delamination. The crack occurred within the TGO (Fig. 27.7). Remains of the alumina layer are visible on the bond coat and attached to the bottom of the PSZ layer. At some spots, the adherence between PSZ and bond coat was maintained and the TGO thickness could be estimated: this was around 0.8 μm and 1.6 μm after 500 and 1000 h, respectively. Isothermal oxidation in air for 1000 h also caused TBC delamination within the TGO.
- Changes in the bond coat. Interdiffusion processes continued at 950 °C but the bond coat seemed to achieve a stationary chemical composition after 500 h. Figure 27.4 shows that the composition in the outer zone was homogeneous throughout the layer with an increase in the Ni concentration up to 36 at.% whereas the Al and Pt contents dropped down to 34 at.% and 13 at.%, respectively. Aluminium may be removed either due to growth of the alumina scale or by diffusion towards the substrate. The Cr level reached the maximum solubility in the NiAl phase but the Fe content was far below the maximum solubility in NiAl (∼45 at.%). In the inner zone, the main phase was further depleted in Ni (5 at.%). In addition, Cr-rich precipitates were observed (Fig. 27.7) that were probably due to the outward diffusion of Ni together with the low Cr solubility in Ni–Al aluminide. Microprobe WDS analyses revealed that these Cr-rich phases were carbides, although it was not possible to confirm their stoichiometry by spotting. The question was whether the carbon came from the gas phase (methane or carbon monoxide) or from the substrate. The former option would mean that, probably due to the cracking, the TGO became pervious to gases. The second alternative would imply carbon depletion in the alloy, which could be verified by complementary substrate analyses.
- Changes in the Alloy 800H. The composition of the substrate sub-surface zone has been largely altered. The Ni content was lower than the bulk value up to a depth of approximately 60 μm. On the other hand, the matrix was enriched in Pt up to a depth of approximately 15 μm and Al diffused deeply into the bulk alloy, reaching a concentration of about 3 at.%.

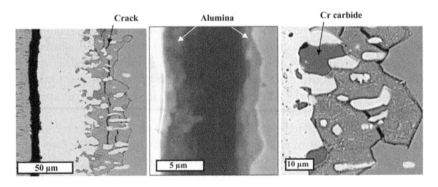

27.7 SEM images – Type 2 Specimen after 1000 h in impure helium at 950 °C – bond coat (left-hand side, BSE); cracked TGO (middle, SE); inner zone of the bond coat (right-hand side, BSE)

## 27.5    Discussion

At 950 °C in weakly oxidising helium, Alloy 800H could not develop a chromia surface scale. Instead, a transitory titanium-rich oxide formed that failed to provide
protection against carburisation. Use of TBCs dramatically improved the material
behaviour. It is worth noting that the corrosion tests were probably performed under
more severe conditions than those in a reactor. The test procedure was not actually
designed to allow for a thermal gradient through the PSZ. Considering that the
helium temperature will reach 1000 °C in the HGD, the temperature at the PSZ/bond
coat interface could be lower than during the tests at 950 °C.

Although the TBCs increased the global corrosion resistance, it appeared necessary to be more careful about the choice of bond coat and fabrication process
regarding both the as-received performance and the ageing/corrosion resistance
under the given thermal and environmental conditions. Moreover, thermal ageing
and corrosion caused intricate changes in the microstructure. Stability at 950 °C
under vacuum should be studied in order to separate the specific effects of temperature
and environment.

The NiCrAl(Y) bond coat of specimens of Type 1 was deposited by Air Plasma
Spray. The fabrication route was shown to generate a significant amount of alumina
together with the intermetallic phase. A process at low pressure should thus be more
adequate. In addition, the exposures at 950 °C in impure helium produced significant
alterations in the NiCrAl(Y) as well as in the underlying Alloy 800H. On the one
hand, nickel and aluminium diffused towards the substrate while the bond coat
appeared to be highly enriched in iron from the alloy. These chemical modifications
induced precipitation of chromium-rich phases within the bond coat. However, the
TBC demonstrated appropriate corrosion resistance over short durations with the
growth of a highly convoluted but adherent TGO at the PSZ/bond coat interface.

The bond coat of specimens of Type 2 was manufactured by platinum electro-
deposition followed by Vapour Phase Aluminisation at 1080 °C. The nickel aluminide was produced in-situ through inward diffusion of nickel from the substrate. The
as-received system also experienced dramatic changes under the given test conditions.
After a 500 h exposure, the chemical composition of the bond coat seemed to have
stabilised. The outer zone was comprised of β-Ni (34 at.%)-Al (36 at.%) doped with
13 at.% Pt, 12 at.% Fe, and 3 at.% Cr. The inner zone was a mixture of at least three
phases: β-NiAl, γ phase issued from the nickel-depleted substrate, and chromium-
rich carbide. The Alloy 800H was largely depleted in nickel and enriched in aluminium. In the as-coated state, the bond coat was topped with a thin TGO film which
continued growing during the corrosion tests. After 500 h, delamination was found
throughout the TGO. It was thus concluded that the selected bond coat, or its
deposition process, was not appropriate for coating Alloy 800H in the HGD
environment.

Regarding deposition of the ceramic, EB-PVD produced a more porous and
stress-accommodating PSZ layer than CVD. It was also faster.

## 27.6    Conclusions

Two thermal barrier coating systems: NiAl(Pt)/EBPVD-PSZ and NiCrAl(Y)/CVD-
PSZ were investigated for deposition on Alloy 800H. The corrosion tests were carried
out in impure helium at 950 °C for up to 1000 h. In the test environment, bare Alloy

800H developed a Ti-rich surface oxide that could not provide protection against carburisation. In contrast, TBCs formed alumina at the bond coat outer surface that should be far more corrosion-resistant. However, TGO of NiAl(Pt) delaminated. Under the given test conditions, NiCrAl(Y) was more appropriate for coating Alloy 800H but the APS coating process must be improved to suppress alumina formation within the bond coat.

## Acknowledgments

The authors are grateful to M. T. David of DEN/VRH/DTEC/STCF/LTIC, CEA VHR and to Mrs C. Chabrol of DRT/DTEN/S3ME, CEA Grenoble for providing the TBC systems.

## References

1. J. L. Séran, P. Lamagnère, C. Cabet, L. Guetaz, E. Wallé and B. Riou, presented at ICAPP'05, Seoul, Korea, 15–19 May 2005, paper 5419.
2. L. W. Graham, M. R. Everett, D. Lupton, F. Ridealgh, D. W. Sturge and M. Wagner-Löffler, presented at 'Gas-cooled reactors with emphasis on advanced systems', 319–352. Jülich, Germany, 13–17 October 1975.
3. W. J. Quadakkers and H. Schuster, *Nucl. Technol.*, 66 (1984), 383.
4. K. G. E. Brenner, *Nucl. Technol.*, 66 (1984), 404.
5. M. Shindo, W. J. Quadakkers and H. Schuster, *J. Nucl. Mater.*, 140 (1986), 94.
6. M. Shindo and T. Kondo, *Nucl. Technol.*, 66 (1984), 429.
7. P. J. Ennis and D. F. Lupton, presented at Int. Conf. on 'Behaviour of high temperature alloys in aggressive environments', 979–991. Petten, The Netherlands, 15–18 October 1979.
8. L. W. Graham, *J. Nucl. Mater.*, 171 (1990), 76.
9. D. Zhu, R. A. Miller, B. A. Nagaraj and R. W. Bruce, *Surf. Coat. Technol.*, 138 (2001), 1.
10. K. Fritscher and Y.-T. Lee, *Mater. Corros.*, 56(1) (2005), 5.
11. D. Zhu and R. A. Miller, Report NASA/TM-1999-2090069, USA, 1999.
12. D. Zhu, N. P. Bansal, K. N. Lee and R. A. Miller, Report NASA/TM-2001-211122, USA, 2001.
13. C. Guerre, R. Molins and L. Rémy, *Mater. High Temp.*, 17(2) (2000), 197.
14. J. H. Chen and J. A. Little, *Surf. Coat. Technol.*, 92 (1997), 69.
15. F. Rouillard, C. Cabet, A. Terlain and K. Wolski, presented at EUROCORR 2005, Lisbon, Portugal, 4–8 September 2005, paper O-358-8.
16. S. Han and D. J. Young, *Mater. Res.*, 7(1) (2004), 11.

# 28

# Oxidation of electrodeposited Cr–C at temperatures between 400 °C and 900 °C in air*

D. B. Lee

*Centre for Advanced Plasma Surface Technology, Sungkyunkwan University, Suwon 440-746, South Korea*

*dlee@yurim.skku.ac.kr*

## 28.1 Introduction

Chromium electroplating has been widely used in various industries because of its high hardness, corrosion resistance, and decorative appearance. However, the hexavalent chromium plating solution can result in serious health and environmental problems. Hence, trivalent chromium plating has been introduced as an alternative [1]. This plating is relatively non-toxic, but its industrial application has been limited because of the chemical and electrochemical problems posed by the Cr(III) solution. Hence, a number of factors are currently being investigated in an attempt to implement successfully trivalent chromium plating, including increasing the stability of the bath solution, optimising the current efficiency, and producing thick deposits [2]. The chromium salts used in trivalent chromium baths are usually chromium sulphate $(Cr_2(SO_4)_3 \cdot nH_2O)$ [1] or chromium chloride $(CrCl_3 \cdot 6H_2O)$ [3]. The organic additives containing the –CHO or –COOH group produce supersaturated carbon in the Cr deposit, resulting in it having an amorphous structure [4,5]. In this study, the amorphous Cr–C deposits were electroplated from the trivalent chromium bath, and oxidised at high temperatures of between 400 and 800 °C in air in order to study their oxidation characteristics. These characteristics constitute an important factor for practical applications, because the Cr–C deposits can be exposed to oxidising atmospheres at high-temperatures.

## 28.2 Experimental

The amorphous Cr–C deposits were electroplated using chromium sulphate, a complexing agent (HCOOK), a conductive improvement agent (KCl, $NH_4Cl$), a buffer agent ($H_3BO_3$), an anti-oxidant agent ($NH_4Br$) and an additive (polyethylene glycol) up to a thickness of $10 \sim 60$ μm. The bath composition and electrolysis conditions are given in Table 28.1. A low carbon steel substrate was cut into coupons with dimensions of 2 cm × 0.5 cm × 0.3 cm, polished to a mirror-like surface, and electroplated. The surface area of the anode used was twice that of the cathode.

The Cr–C deposited specimens were oxidised at temperatures between 400 and 900 °C in air, and investigated by scanning electron microscopy (SEM) using an

---

\* Reprinted from D. B. Lee: Oxidation of Cr -C electroplating between 400 and 900 oC in air. *Materials and Corrosion*. 2008. Volume 59. pp. 598–601. Copyright Wiley-VCH Verlag GmbH & Co. KGaA.

*Table 28.1*   Bath composition and plating conditions

| Bath composition | | Plating conditions | |
|---|---|---|---|
| Chemicals | Content | | |
| $Cr_2(SO_4)_3 \cdot nH_2O$ | 140 g L$^{-1}$ | Temperature | 30 °C |
| HCOOK | 1 M | pH | 2.0 |
| KCl, NH$_4$Cl | 1 M each | Current density | 25 A dm$^{-2}$ |
| H$_3$BO$_3$ | 0.65 M | Anode | Graphite |
| NH$_4$Br | 10 g L$^{-1}$ | Cathode | Low carbon steel |
| Polyethylene glycol | 2 g L$^{-1}$ | Agitation | Air bubbling |

instrument equipped with an energy dispersive spectroscope (EDS), an electron probe microanalyser (EPMA), an X-ray diffractometer (XRD) with Cu-K$_\alpha$ radiation, a thermogravimetric analyser (TGA), a differential thermal analyser (DTA), an X-ray photoelectron spectrometer (XPS), an Auger electron spectroscope (AES), and a transmission electron microscope (TEM operated at 300 kV) equipped with an EDS. The TEM samples were glued onto a thin Si dummy plate using epoxy resin to preserve the oxide scale, mechanically polished to a thickness of $\sim 30$ µm, and ion milled to perforation.

## 28.3    Results and discussion

Figure 28.1 shows the XRD patterns of the Cr–C electrodeposits before and after oxidation. Since the Cr–C deposit was amorphous, a diffuse diffraction pattern appeared at around 43° (Fig. 28.1(a)). The composition of the deposit analysed by EPMA was Cr–12.7 at.% C (3.1 wt.% C). Since the maximum solubility of carbon in Cr is about 0.02 at.%, the deposit was supersaturated in carbon. On the other hand, when the amorphous Cr–C deposit was electroplated from the chromium chloride bath containing Cr(III) ions, and subsequently heated under vacuum for 1 h, it completely crystallised into Cr at 400 °C, and crystalline (Cr + Cr$_{23}$C$_6$) phases appeared at 800 °C [3]. Usually, carbides did not appear in the Cr–C deposit at temperatures below 600 °C, because they did not agglomerate sufficiently to be detected by XRD [2]. The precipitation of Cr$_{23}$C$_6$ was similarly observed in the Cr–C layer deposited from the chromium sulphate bath after vacuum-annealing for 1 h at 500 and 700 °C [1]. When the current Cr–C deposit was heated in air at 400–500 °C for 5–60 h, 600 °C for 5 h, and 700 °C for 2 h (Fig. 28.1(b)), it crystallised into Cr, as has been reported previously. Further heating in air at 600 °C for 60 h, and at 700–900 °C for 5–60 h (Fig. 28.1(c) and (d)) resulted in the formation of Cr$_2$O$_3$ on the crystalline Cr. However, Cr$_{23}$C$_6$ was not found under any of the oxidising conditions used in this study, because the excess carbon atoms in the Cr–C deposit were oxidised into CO or CO$_2$, which was liberated into the air. In Fig. 28.1(c), it can be seen that Cr$_2$O$_3$ formed over the Cr deposit. The disappearance of the amorphous structure is not desirable, because amorphous metals generally show slower oxidation rates because of the absence of grain boundaries. In Fig. 28.1(d), only Cr$_2$O$_3$ was detected owing to the considerable extent of oxidation.

Figure 28.2 shows the appearance of the Cr–C electrodeposits before and after oxidation. The surface of the prepared electrodeposit was flat and semi-bright with a silver white colour (Fig. 28.2(a)). However, reduction of the hydrogen ions during electroplating resulted in hydrogen evolution, forming cracks on the surface and

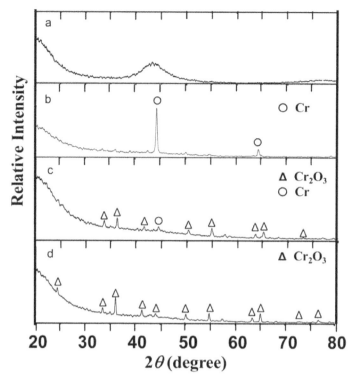

*28.1* XRD patterns of the Cr–C deposit: (a) before oxidation; (b) after oxidation at 700 °C for 2 h; (c) after oxidation at 800 °C for 5 h; (d) after oxidation at 900 °C for 60 h

along the boundaries of the nodular grains. It is known that with increasing plating time and deposit thickness, the nodular grains grow markedly and the microcracks become wider and deeper [6]. The formation of internal microcracks, the unstable bath composition, and the low current efficiency are the major drawbacks for the commercialisation of trivalent hard chromium plating. As the oxidation progressed, the cracks developed further due to the stress generated and the evolution of carbon from the coating (Fig. 28.2(b)). After oxidation at 600 °C for 5 h in air, numerous, fine oxide grains formed over the surface (Fig. 28.2(c)). They had a green colour, indicative of $Cr_2O_3$, and were non-adherent. The $Cr_2O_3$ powder formed on the surface after oxidation at 800 °C for 5 h spalled easily during subsequent handling (Fig. 28.2(d)). Dense scales cannot be obtained during the oxidation of Cr–C electrodeposits.

Figure 28.3 shows the cross-sectional image and the corresponding EPMA line profiles of the oxidised Cr–C electrodeposits. In Fig. 28.3(a), a thin oxide layer is seen on the surface of the Cr–C that has cracks. In Fig. 28.3(b), an outer, non-adherent oxide layer is seen on the Cr–C. The lower part of the Cr–C deposit consisted of $Cr_2O_3$ and retained Cr that had no solubility of oxygen. Oxygen diffused into the deposit, and iron diffused out towards the surface, according to the concentration gradient. Carbon was liberated during oxidation. It is however noted that the quantitative analysis of a light element such as carbon is notoriously difficult.

Figure 28.4 shows the XPS spectra of the Cr–C deposits before and after oxidation at 700 °C for 2 h. Figure 28.4(a) indicates that the electrodeposit consists of metallic

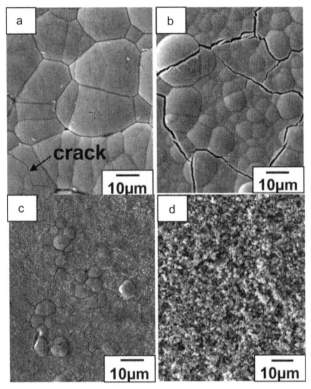

*28.2* SEM appearance of the Cr–C deposit: (a) before oxidation; (b) after oxidation at 500 °C for 5 h; (c) after oxidation at 600 °C for 5 h; (d) after oxidation at 800 °C for 5 h

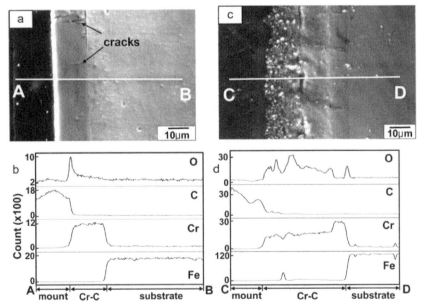

*28.3* EPMA results on the Cr–C deposit: (a) cross-sectional image after oxidation at 700 °C for 2 h; (b) line profiles across A–B; (c) cross-sectional image after oxidation at 800 °C for 5 h; (d) line profiles across C–D

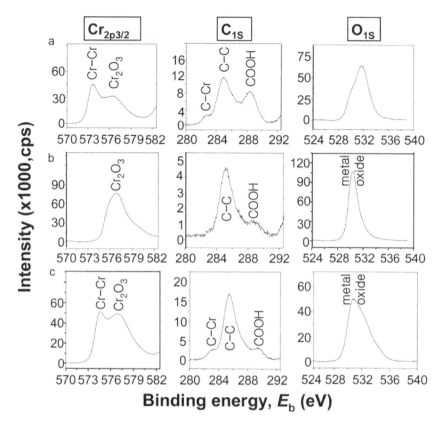

*28.4* XPS spectra of $Cr_{2p3/2}$, $C_{1s}$ and $O_{1s}$ from the Cr–C deposit: (a) before oxidation (taken from the outermost surface); (b) after oxidation at 700 °C for 2 h (taken from the outermost surface scale); (c) after oxidation at 700 °C for 2 h (taken around the interface of the scale/deposit)

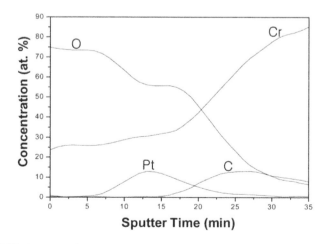

*28.5* AES depth profiles of the Cr–C deposit after oxidation at 700 °C for 2 min. The penetration rate is 200 Å min⁻¹ for the reference $SiO_2$

Cr ($E_b$ of $Cr_{2p3/2} = 574$ eV), $Cr_2O_3$ (surface oxide film, $E_b$ of $Cr_{2p3/2} = 576.9$ eV), C–Cr with a carbide bond ($E_b$ of $C_{1s} = \sim 282.6$ eV) [7], graphite ($E_b$ of $C_{1s} = 285$ eV), and probably either carboxyls (–COOH from potassium formate) [4] or C with Cl (from potassium chloride). The binding energies of –COOH and C–Cl are similar ($E_b$ of $C_{1s}$ $= 288.3$ eV). The presence of carbon in a graphite form was also previously reported in the Cr–C deposits [4]. The asymmetric, broad $O_{1s}$ spectrum peak ($E_b = 531.7$ eV) indicates that more than one oxygen-containing species are involved. Oxygen is probably in the form of –COOH, adsorbed $H_2O$, or $Cr_2O_3$ [4]. In the Cr–C deposit, there exists oxygen segregated at the outermost surface [3]. Since the oxygen content dropped rapidly to zero just beneath the outermost surface of the Cr–C deposit [3], the $Cr_2O_3$ that was detected in the XPS spectra was undetectable in the XRD pattern. Figure 28.4(b) indicates that the oxidation of the Cr–C leads to the development of $Cr_2O_3$, and the destruction of Cr–Cr bonding and carboxyls (–COOH) or C with Cl.

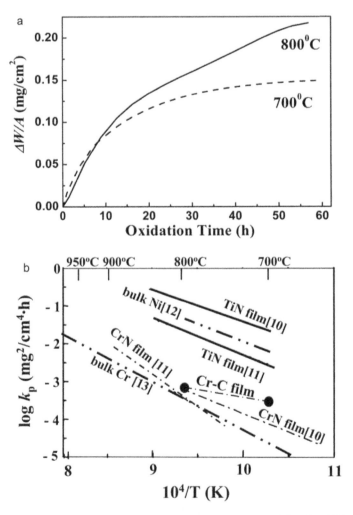

28.6 Oxidation kinetics of Cr–C deposited specimens oxidised at 700 and 800 °C in air: (a) weight gain vs. oxidation time curves; (b) Arrhenius plot of $k_p$ vs. $1/T$. The $k_p$ values of TiN and CrN films [10,11], bulk Ni [12] and bulk Cr [13] are shown

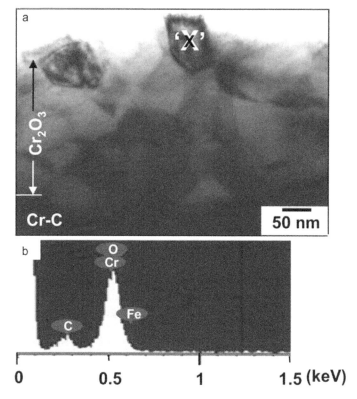

*28.7* Cr–C deposit after oxidation at 800 °C for 2 min: (a) TEM image; (b) EDS spectrum taken from 'X' in (a)

Carbon still existed mostly as graphite. The maximum binding energy of the $O_{1s}$ spectrum shifted to 530.4 eV in Fig. 28.4(b), owing to the formation of the surface $Cr_2O_3$ layer. When compared with Fig. 28.4(a), the intensity of the $O_{1s}$ spectrum increased, whereas that of $C_{1s}$ decreased in Fig. 28.4(b). Figure 28.4(c) shows the XPS data obtained after polishing off the surface oxide layer. This figure indicates the coexistence of Cr and $Cr_2O_3$, and the presence of graphite. The graphite form of C was detected in Fig. 28.4(a)–(c).

Figure 28.5 shows the AES depth profiles of the Cr–C deposit after oxidation at 700 °C for 2 min. To understand the oxidation mechanism, a thin Pt film was deposited on top of the coating before oxidation. From the location of the Pt marker, it can be seen that the outward diffusion of Cr, the inward penetration of oxygen, and the carbon loss near the surface occurred simultaneously. The rate-limiting step for polycrystalline chromia formation is generally considered to be the outward diffusion of $Cr^{3+}$ ions along the grain boundaries [8]. However, it is noted that the degree of non-stoichiometry in $Cr_2O_3$ is very small and no unequivocal explanation of the defect structure and transport properties of chromia has yet been reported [9]. In this study, the growth of $Cr_2O_3$ on the Cr–C was governed by counter-current oxygen/Cr diffusion along the short-circuit diffusion paths, together with the escape of carbon into the air.

Figure 28.6 shows the oxidation kinetics of the Cr–C deposited specimens oxidised at 700 and 800 °C in air. The weight gain curves displayed in Fig. 28.6(a) indicate that the oxidation rates increased with increasing oxidation temperature. Assuming that the parabolic rate law prevailed for the sake of simplicity, the parabolic rate constants, $k_p$, were calculated from the equation $(\Delta W/A) = k_p \cdot t$, where $\Delta W/A$ is the weight gain per surface area of the specimen in mg cm$^{-2}$, and $t$ is the exposure time in hours. In Fig. 28.6(b), the $k_p$ values of Cr–C were compared with those of other films [10,11] and bulk metals [12,13]. The Cr–C deposit oxidised more slowly than TiN films and bulk Ni [12] owing to the formation of $Cr_2O_3$, but faster than CrN films and bulk Cr [13] owing to the evaporation of carbon and the development of cracks on the surface and inside the Cr–C deposit.

Figure 28.7(a) shows the TEM image of the Cr–C deposit after oxidation at 800 °C for 2 min. Many round oxide grains with a size of 50–100 nm are seen over the Cr–C deposit. They were identified as $Cr_2O_3$ containing a small amount of Fe from the EDS analyses (Fig. 28.7(b)). The outwardly-diffused Fe ions were dissolved in the $Cr_2O_3$.

## 28.4    Conclusions

The electrodeposited Cr–C was amorphous and contained supersaturated carbon. When oxidised in air at high temperatures, it crystallised into Cr. It oxidised to fine, non-adherent $Cr_2O_3$ particles, incorporating carbon and iron ions. The oxide scales that formed were neither dense nor crack-free. Oxygen diffused into the deposit, iron diffused out towards the surface, and carbon was liberated during oxidation. The Cr–C electrodeposit oxidised faster than CrN films and bulk Cr owing to the evaporation of carbon and the development of cracks on the surface and inside the Cr–C deposit.

## Acknowledgment

This work was supported by a grant (no. R-11-2000-086-0000-0) from the Centre of Excellence Programme of the KOSEF, Korea.

## References

1. M. Takaya, M. Matsunaga and T. Otaka, *Plating Surf. Finish.*, 74 (1987), 90.
2. S. Hoshino, H. A. Laitinen and G. B. Hoflund, *J. Electrochem. Soc.*, 133 (1986), 681.
3. S. C. Kwon, M. Kim, S. U. Park, D. Y. Kim, D. Kim, K. S. Nam and Y. Choi, *Surf. Coat. Technol.*, 183 (2004), 151.
4. G. B. Hoflund, D. A. Asbury, S. J. Babb, A. L. Grogan, H. A. Laitinen and S. Hoshino, *J. Vac. Sci. Technol. A*, 4 (1986), 26.
5. D. Kim, M. Kim, K. S. Nam. D. Chang and S. C. Kwon, *Surf. Coat. Technol.*, 169/170 (2003), 605.
6. G. Hong, K. S. Siow, G. Zhiqiang and A. K. Hsieh, *Plating Surf. Finish.*, 88(3) (2001), 69.
7. A. A. Edigaryan, V. A. Safonov, E. N. Lubnin, L. N. Vykhodtseva, G. E. Chusova and Y. M. Polukaov, *Electrochim. Acta*, 47 (2002), 2775.
8. A. Ul-Hamid, *Oxid. Met.*, 58 (2002), 23.
9. P. Kofstad, *Oxid. Met.*, 44 (1995), 3.
10. P. Panjan, B. Navinsek, A. Cvelbar, A. Zalar and I. Milosev, *Thin Solid Films*, 281/282 (1996), 298.
11. H. Ichimura and A. Kawana, *J. Mater. Res.*, 9 (1994), 15.
12. D. B. Lee, J. H. Ko and S. C. Kwon, *Surf. Coat. Technol.*, 193 (2005), 292.
13. W. C. Hagel, *Trans. Am. Soc. Met.*, 56 (1963), 583.

# 29

# Experimental analysis and computer-based simulation of nitridation of Ni-base alloys – The effect of a pre-oxidation treatment*

## V. B. Trindade, B. Gorr and H.-J. Christ

*Institute for Materials Science and Engineering, University of Siegen, Siegen, Germany*
*vicente@ifwt.mb.uni-siegen.de*

## U. Krupp

*FH Osnabrück, University of Applied Sciences, D-49009 Osnabrück, Germany*

## D. Kaczorowski and G. Girardin

*AREVA-NP, Technical Center Corrosion Department, Le Creusot, France*

## 29.1    Introduction

There is an increasing demand for so-called 'clean energy', which requires a special effort to reduce the emission of gases known to be deleterious to the environment. Efforts to develop innovative new energy resources are under way, but the established actual resources, e.g. gas, coal, oil and nuclear fuel seem to be the most relevant for the coming decades. Therefore, research is ongoing to develop methods for the reduction of $CO_2$ emissions during the operation of current power plants. One practical way to achieve low emissions is to improve the efficiency of the power plant by increasing the temperature of the gas or steam flow into the turbine. The increase in temperature requires alloys to be more resistant to high-temperature degradation, e.g. corrosion, creep and thermo-mechanical load. This problem might be solved by using alloys well-designed for these new challenges, i.e. combining high chemical stability of the alloy and high creep strength in high-temperature environments as well as economical feasibility.

Ni-base superalloys are commonly used as engineering materials for high-temperature applications, e.g. as turbine blades in land-based gas turbines or in jet engines and heat exchangers. The very severe in-service conditions require both mechanical strength and oxidation resistance [1].

The design of alloys for use in corrosive atmospheres at high temperatures is usually carried out by optimisation of their chemical compositions. In this sense, chromium and aluminium are widely used as elements capable of protecting the

* Reprinted from V. B. Trindade et al.: Experimental analysis and computer-based simulation of nitridation of Ni-base alloys – The effect of a pre-oxidation treatment. *Materials and Corrosion*. 2008. Volume 59. pp. 602–8. Copyright Wiley-VCH Verlag GmbH & Co. GaA.

alloys against corrosive attack of the parent metal through the formation of a protective external oxide scale, e.g. chromia or alumina. However, internal corrosion can be observed in several alloys even underneath a dense external scale [1–4]. The ingress of corrosive species, such as oxygen, into the alloy is governed by dissociation of the oxide at the scale/metal interface and/or by penetration along defects of the scale such as cracks and pores [5]. It has been observed that internal oxidation in Ni-base alloys at moderate temperatures ($<900$ °C) occurs preferentially along grain boundaries [6–8].

The stability of the different phases such as carbides, oxides and nitrides can be calculated by their thermodynamic properties. However, due to the existence of complex solid solutions such as some carbides and oxides containing different metallic elements in their sub-lattices, a proper prediction of phase equilibrium becomes feasible only when using computational thermochemistry. Besides thermodynamics, the kinetics of the corrosion process have to be calculated by considering that most of the corrosion phenomena are controlled by solid state diffusion. Therefore, the diffusion of the different species participating in the corrosion process must be taken into account paying special attention to the high diffusivity along grain boundaries.

Internal-corrosion problems that involve: (i) more than one precipitating species; (ii) compounds of moderate/low stabilities; (iii) high diffusivities of the metallic elements or time-dependent changes in the test conditions, e.g. temperature or interface concentrations, cannot be treated by applying Wagner's original theory of internal oxidation [9]. For simulation of such a system, the combination of the finite-difference method with sophisticated computational thermodynamics might be the most promising approach.

Krupp [10], Trindade [11] and Buschmann [12] developed a computer program, InCorr, which combines diffusion and local thermodynamic calculation with the help of the thermodynamic library ChemApp [10,11].

In other studies, InCorr (a computer-based program for description of high-temperature corrosion processes) [10] has been used to calculate the inward scale formation in low-Cr steels at moderate temperatures (400–600 °C) [11], carbide formation in austenitic steel [13] and internal nitridation of Ni-base alloys [10,14]. In this paper, a further application of InCorr is demonstrated, the simulation of internal nitridation in commercial Ni-based alloys, e.g. intergranular precipitation as a consequence of high nitrogen diffusivity along alloy grain boundaries. Thermodynamic calculations using the software FactSage were shown to provide a useful tool to understand the processes at the metal–gas interface under different conditions, i.e. temperature, gas partial pressure and alloy chemical composition. Those computer tools can become powerful to carry out virtual experiments to understand high-temperature corrosion phenomena.

## 29.2    Materials and experimental procedure

Two commercial Ni-based alloys with the designations IN 617 and Haynes 230 were chosen as test materials to characterise the influence of a pre-oxidation treatment on the nitridation phenomena of those alloys. The main constituents of the materials are shown in Table 29.1.

Samples with dimensions of 10 mm × 10 mm × 3 mm were used for thermogravimetric studies. The samples were ground using SiC paper down to 1200 grit. They were finally cleaned ultrasonically in ethanol before exposure. A hole of 1 mm

*Table 29.1*  Nominal chemical compositions (wt.%) of the materials used

| Element | Haynes 230 | IN 617 |
|---|---|---|
| Al | 0.2–0.5 | 0.80–1.50 |
| Ti | – | 0.60 |
| Cr | 20–24 | 20–24 |
| Fe | Max. 3 | 3 |
| Co | Max. 5 | 10–15 |
| Mn | 0.3–1 | 0.50 |
| Mo | 1–3 | 8–10 |
| P | Max. 0.03 | Max. 0.015 |
| Si | 0.25–0.75 | 0.50 |
| S | Max. 0.015 | Max. 0.015 |
| W | 13–15 | – |
| Ni | Bal. | Bal. |

diameter served for hanging the samples in a thermobalance using a quartz thread. Isothermal thermogravimetry was carried out using a microbalance with a resolution of $10^{-5}$ g in combination with an alumina reaction chamber and a SiC furnace. After exposure, the specimens were embedded in epoxy and carefully polished using diamond paste down to 1 μm and cleaned ultrasonically in ethanol. The oxide layers were analysed using scanning electron microscopy (SEM) in combination with energy-dispersive X-ray spectroscopy (EDX).

The samples were exposed to the following conditions: (1) oxidation in a laboratory atmosphere at 900 °C; (2) nitridation in He–10 vol.% $H_2$–80 vol.% $N_2$ at 900 °C and 1000 °C; (3) nitridation in He–10 vol.% $H_2$–80 vol.% $N_2$ after the pre-oxidation treatment described in (1) above.

## 29.3    Results and discussion

### 29.3.1    Experimental results

The main surface reactions for the alloy system Ni–24Cr–1.5Al–0.6Ti–15Co–10Mo–$O_2$–$N_2$ that can occur at high temperature (1000 °C) in atmospheres containing oxygen and nitrogen are summarised in Table 29.2. According to a thermodynamic data bank tailored for the high-temperature corrosion of Ni-base alloys supplied by GTT Technologies, Figure 29.1 shows a stability diagram for the system Ni–Cr–Al–Ti–$O_2$–$N_2$ at 900 °C. With the aid of such diagrams, all of the thermodynamically stable phases can be predicted at any combination of oxygen and nitrogen partial pressures. However, information on thermodynamic equilibria alone is not fully sufficient to describe corrosion phenomena, because the kinetics of the corrosive species play a very important role during high-temperature corrosion processes. The limiting factor for the kinetics is not the chemical reaction process since the activation energy for chemical reaction at high-temperatures is very low. On the other hand, for the chemical reactions to occur, metallic elements have to diffuse from the substrate to the gas interface or non-metallic species, e.g. N and/or O, from the gas phase to the substrate interior. These diffusion processes are controlled by the defect structure of the scale (oxide and/or nitride) formed at the surface.

*Table 29.2*   Main stable compounds in the system Ni–Cr–Al–Ti–Co–Mo–O₂–N₂ at 1000 °C

| Compounds |
| --- |
| $Al_2O_3$, $Cr_2O_3$, NiO, $MoO_3$, $CoCr_2O_4$, $NiTiO_3$, $CoTiO_3$, $MoO_2$, $NiTiO_3$, $TiO_2$, $Ti_7O_{13}$, $Ti_5O_9$, $Ti_3O_5$, TiO, CrN, $Cr_2N$, $Mo_2N$, TiN, AlN, $Ni_{36}Cr_{64}N_2O$ (π-phase) |

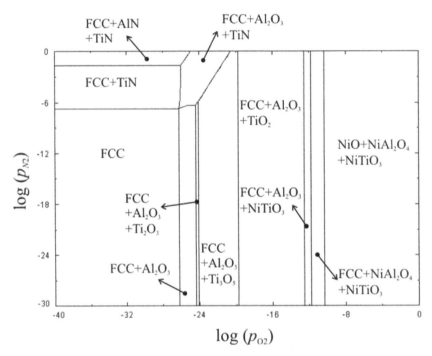

29.1 Thermodynamic stability diagram for the system Ni–Al–Ti–O₂–N₂ at 1000 °C (calculated using the FactSage commercial software)

During exposure of the alloy IN 617 to an oxygen-free nitriding atmosphere at 1000 °C, an internal nitridation zone was observed, which consists of a thick titanium-nitride zone (75 μm after 100 h and 85 μm after 250 h) and a thinner aluminium-nitride zone close to the surface. Some CrN precipitation was observed along the alloy grain boundaries, particularly within the AlN zone (Fig. 29.2).

During exposure of the alloy Haynes 230 in an oxygen-free nitriding atmosphere at 900 °C, a thin CrN layer (3 μm) was formed at the surface of the specimen followed by a zone of about 110 μm width of internal CrN precipitates along the alloy grain boundaries (Fig. 29.3a). Figure 29.3b shows the results of thermogravimetric measurements carried out for 3 days of exposure. The parabolic rate constant for the nitridation process calculated using the equation $\Delta m/A = k_p\sqrt{t}$ is $3.5 \times 10^{-4}$ mg cm⁻² h⁻½.

During oxidation in laboratory air at 900 °C for 100 h, basically only a $Cr_2O_3$ scale is formed on the surface of the Haynes 230 alloy (Fig. 29.4a). According to Fig. 29.1, other stable oxide phases can be formed at the beginning of exposure, but due to the

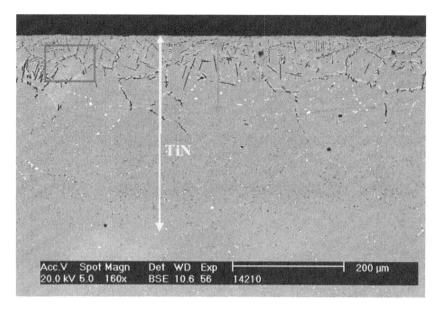

a

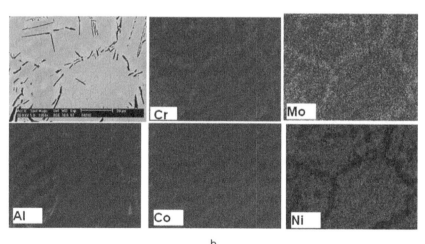

b

*29.2* (a) Typical internal nitridation zone observed in the alloy IN 617 (80 vol.% $N_2$ + 10 vol.% $H_2$ + 10 vol.% He at 1000 °C for 100 h) and (b) EDX-mapping for the area marked in (a)

low defect concentration in the $Cr_2O_3$ phase and its high thermodynamic stability, the $Cr_2O_3$ scale grows slowly (Fig. 29.4b) and prevents oxidation of the main alloying elements, e.g. Ni and Fe. The parabolic rate constant of the oxidation process was found to be $1.98 \times 10^{-3}$ mg cm$^{-2}$ h$^{-1/2}$.

The effect of pre-oxidation of the Haynes 230 alloy was investigated by measuring the kinetics of nitridation after the pre-oxidation treatment shown in Fig. 29.4. Figure 29.5 shows the nitridation curve of the specimen exposed to an oxygen-free nitriding atmosphere at 900 °C after the pre-oxidation treatment. It can be seen that the pre-oxidation treatment reduces the internal nitridation for these exposure

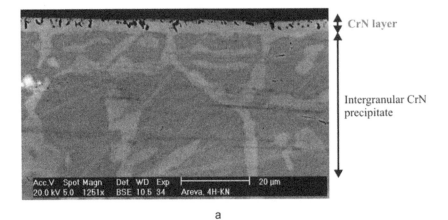

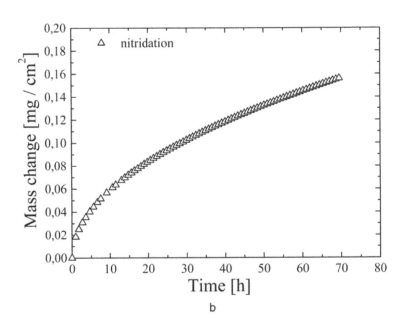

*29.3* (a) Typical internal-nitridation zone observed in Haynes 230 alloy (80 vol.% $N_2$ + 10 vol.% $H_2$ + 10 vol.% He at 900 °C for 100 h) and (b) thermogravimetric measurement

conditions. The parabolic rate constant of the nitridation process after pre-oxidation heat treatment was $2.47 \times 10^{-4}$ mg cm$^{-2}$ h$^{-\frac{1}{2}}$, which is smaller than the parabolic rate constant without pre-oxidation ($3.5 \times 10^{-4}$ mg cm$^{-2}$ h$^{-\frac{1}{2}}$). Figure 29.6 shows a cross-section of the nitrided specimen after pre-oxidation in laboratory air at 900 °C for 100 h. An intergranular chromium nitride (CrN) zone is formed below the external oxide scale.

## 29.3.2    Computer simulation

A reasonable prediction of the service life of components operating at high temperatures in aggressive atmospheres requires a full understanding of the degradation

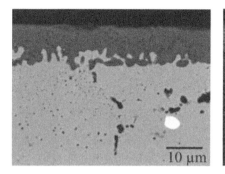
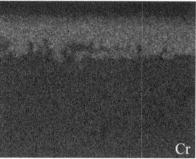

a

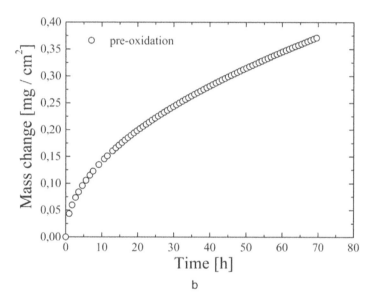

b

*29.4* Formation of Cr₂O₃ scale on Haynes 230 alloy during oxidation in
laboratory air at 900 °C for 100 h; (a) SEM cross-section micrograph and
(b) thermogravimetric measurements of the oxidation kinetics

mechanisms of the material due to mechanical loading and corrosion. To simulate
high-temperature corrosion processes under in-service conditions requires both a
thermodynamic model to predict phase stabilities for given conditions and a mathematical description of the kinetics (i.e. solid-state diffusion). To this purpose, a
computer program (InCorr – described in Refs. 11 and 15) was used. The thermodynamic program library ChemApp is integrated into a numerical finite difference
diffusion calculation to treat internal oxidation and nitridation processes in various
commercial and model Ni-base alloys.

An important parameter for the prediction of the depth of internal precipitation
in alloys at high-temperature, e.g. internal nitridation in nickel-base alloys, is the
maximum nitrogen solubility of the alloy investigated. The nitrogen solubility
depends on three factors: temperature, nitrogen partial pressure and alloy composition. Figure 29.7 shows the calculated nitrogen solubility in different alloy systems at
different temperatures in an atmosphere containing He–10 vol.% H₂–80 vol.% N₂. By

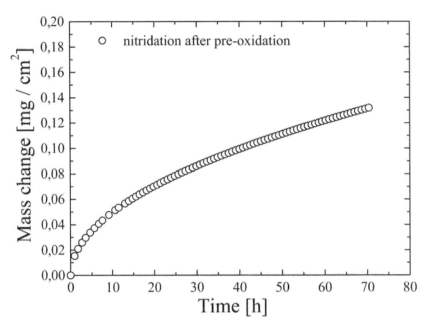

29.5 Thermogravimetric measurements for Haynes 230 alloy at 900 °C during nitridation in 80 vol. % $N_2$ + 10 vol.% $H_2$ + 10 vol.% He after the pre-oxidation treatment described in Fig. 29.4

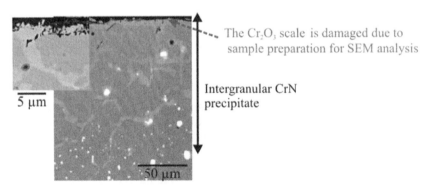

29.6 Intergranular chromium nitride formation during exposure in 80 vol.% $N_2$ + 10 vol.% $H_2$ + 10 vol.% He at 900 °C for 100 h after pre-oxidation in laboratory air at 900 °C for 100 h

comparing the solubility curve for the system Ni–1.5Al–0.6Ti–24Cr and Ni–0.5Al–0.0Ti–24Cr, it can be seen that the nitrogen solubility is slightly higher for the system containing less Al and no Ti. By adding 15 wt.% of Co to the system Ni–1.5Al–0.6Ti–24Cr, the nitrogen solubility is reduced considerably. The addition of 10 wt.% of Mo, however, increases the nitrogen solubility but does not change the nitrogen solubility if 10 wt.% Mo is added to the Ni–1.5Al–0.6Ti–24Cr–15Co alloy.

The effect of the elements Co and Mo on the nitrogen solubility in the system Ni–1.5Al–0.6Ti–24Cr is demonstrated in more detail in Fig. 29.8. The influence of these

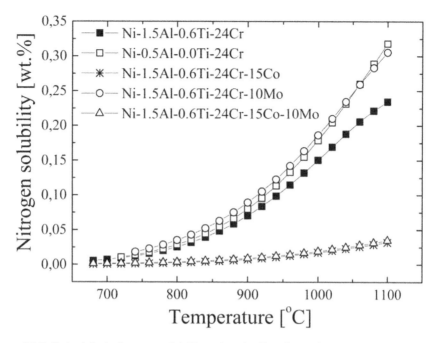

*29.7* Calculated nitrogen solubility using the FactSage thermodynamic software

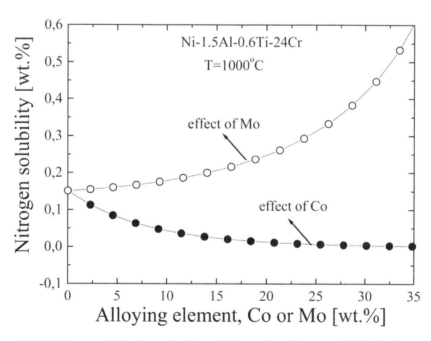

*29.8* Influence of the elements Co and Mo on the nitrogen solubility in the system Ni–1.5Al–0.6Ti–24Cr at 1000 °C in 80 vol.% $N_2$ + 10 vol.% $H_2$ + 10 vol.% He (calculated with the FactSage software)

elements, particularly Co, plays an important role on the calculation of internal nitridation of Ni-base alloys due to their influence on the nitrogen solubility.

Since internal corrosion is mainly governed by the diffusion of corrosive species, such as O, N, C, or S, and to some extent also of the reacting metallic elements such as Ti, Al, or Cr, the modelling using InCorr starts by solving the diffusion differential equations (Fick's second law) for the various elements with concentrations $C$ and diffusion coefficients $D$:

$$\frac{dC}{dt} = (D \quad C) \qquad [29.1]$$

In the model calculated with InCorr, the two-dimensional Fick's second law is solved using the implicit Crank-Nicholson technique, as described in Trindade [11] and

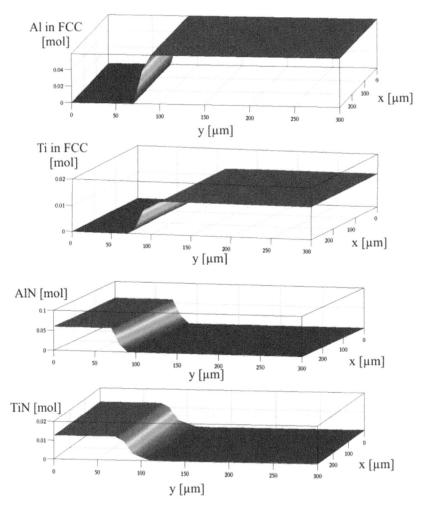

*29.9* Simulation of internal precipitation of AlN and TiN in the alloy IN 617 at 1000 °C for 100 h in 80 vol.% $N_2$ + 10 vol.% $H_2$ + 10 vol.% He

Krupp *et al.* [16]. For a better description of internal-corrosion processes, the diffusion processes have to be treated in combination with the thermodynamics of the chemical reaction between the metallic and the corrosive species. For complex systems, this can be done by a numerical thermodynamic equilibrium calculation based on the minimisation of Gibbs free energy criteria. In the InCorr model, thermodynamic calculations are performed using the commercial software package ChemApp based on specific thermodynamic data sets for the systems under consideration [11]. The lengthy computation time is drastically reduced by using the parallel

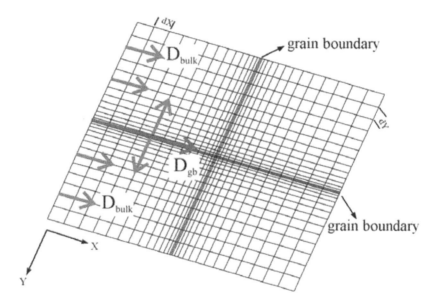

*29.10* Mesh used for two-dimensional diffusion calculation with separate diffusion coefficients for bulk diffusion and grain boundary diffusion

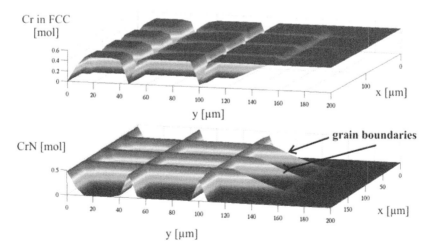

*29.11* Simulation of intergranular precipitation of CrN in Haynes 230 alloy at 900 °C for 100 h in 80 vol.% $N_2$ + 10 vol.% $H_2$ + 10 vol.% He

computing system PVM (parallel virtual machine). Using this computation methodology, the thermodynamic equilibrium calculations are distributed to individual thermodynamic workers as described elsewhere [11].

Due to the presence of Ti and Al and their high affinity to nitrogen, particularly Ti, which forms very stable stoichiometric TiN precipitates, a thick internal precipitation zone is formed after 100 h exposure in an oxygen-free nitriding atmosphere at 1000 °C. Figure 29.9 shows the simulated internal-nitridation zone formed in IN 617. The position of the reaction fronts for AlN and TiN precipitation are very well predicted using the InCorr tool (compare Fig. 29.2 and Fig. 29.9).

Since it was shown in the present study that the internal nitridation process in Haynes 230 alloy is mainly governed by grain boundary transport of the reacting species, i.e. grain boundary diffusion of Cr and N, a two-dimensional finite-difference model has been applied that distinguishes between fast diffusion along substrate grain boundaries and slow transport through the bulk (Fig. 29.10). Due to the lack of available data for interface diffusivities, on the basis of an estimated value for the grain boundary width of $\delta = 0.5$ nm [17], the grain boundary diffusion coefficient was assumed to be 500 times higher than the bulk diffusion coefficient.

The calculated intergranular precipitation depth of chromium nitrides in the Haynes 230 alloy is in good agreement with the experimental results (Fig. 29.3). Figure 29.11 shows the calculated depletion of soluble chromium in the fcc Ni substrate phase and the formation of chromium nitrides (CrN) along the alloy grain boundaries.

## 29.4    Conclusions

Exposure of the two studied Ni-base alloys shows the presence of an internal precipitation zone (AlN, TiN and CrN) after heat treatment in a nitriding atmosphere (He–10 vol.% H$_2$–80 vol.% N$_2$ at 900 °C for 100 h). The pre-oxidation treatment demonstrates that the presence of a Cr$_2$O$_3$ scale does not offer protection against internal nitridation at high temperatures. Besides the presence of a Cr$_2$O$_3$ layer, internal precipitation of chromium nitride (CrN) occurred during the nitridation heat treatment. The thickness of the CrN zone was almost the same as that formed during only nitridation heat treatment. The nitridation also took place along alloy grain boundaries.

The InCorr tool has been used to describe quantitatively the internal nitridation processes. The model is able to simulate multi-phase internal corrosion processes controlled by solid-state diffusion into the bulk metal as well as intergranular corrosion occurring in Ni-base alloys by fast nitrogen transport along the grain boundaries of the substrate.

## References

1. D. L. Douglas, *Oxid. Met.*, 44 (1995), 81.
2. W. W. Smeltzer and D. P. Whittle, *J. Electrochem. Soc.*, 125 (1978), 1116.
3. G. Böhm and M. Kahlweit, *Acta Metall.*, 12 (1964), 641.
4. L. Zhou, *Comput. Mater. Sci.*, 7 (1997), 336.
5. T. Watanabe, M. Takazawa and H. Oikawa, in *Proc. 8th Int. Conf. of Strength of Metals and Alloys* (ICSMA8), Vol. 2, 1357. Pergamon, 1988.
6. M. W. Brumm, H. J. Grabke and B. Wagemann, *Corros. Sci.*, 36 (1994), 37.

7. S. Yamaura, Y. Igarashi, S. Tsurekawa and T. Watanabe, *Acta Metall.*, 47 (1999), 1163.
8. V. B. Trindade, U. Krupp, Ph. E.-G. Wagemnhuber, Y. M. Virkar and H.-J. Christ, *Mater. High Temp.*, 22 (2005), 31.
9. C. Wagner, *Z. Elektrochem.*, 63 (1959), 772.
10. U. Krupp, PhD Thesis, Universität Siegen, Siegen, Germany, 1998.
11. V. B. Trindade, PhD Thesis, Universität Siegen, Siegen, Germany, 2006.
12. U. Buschmann, PhD Thesis, Universität Siegen, Siegen, Germany, 2006.
13. V. B. Trindade, U. Krupp, H.-J. Christ, M. J. Monteiro and F. Rizzo, *Materialwiss. Werkst.*, 10 (2005), 471.
14. S. Y. Chang, PhD Thesis, Universität Siegen, Siegen, Germany, 2001.
15. J. Crank, *The Mathematics of Diffusion*, Clarendon Press, Oxford, 1988.
16. U. Krupp, V. B. Trindade, H.-J. Christ, U. Buschmann and W. Wiechert, *Mater. Corros.*, 57 (2006), 263.
17. I. Kaur and W. Gust, *Fundamentals of Grain and Interphase Boundary Diffusion*, Ziegler Press, Stuttgart, 1989.

# 30

# Influence of alloy compositions on the halogen effect in TiAl alloys*

Patrick Masset, Hans-Eberhard Zschau and Michael Schütze

*Karl-Winnacker-Institut der Dechema e.V., Theodor-Heuss-Allee 25,
60486 Frankfurt am Main, Germany*

*masset@dechema.de*

Sven Neve

*Institut für Kernphysik der Johann Wolfgang Goethe-Universität,
Max-von-Laue-Straße 1, 60438 Frankfurt am Main, Germany*

## 30.1    Introduction

Over the last decade, interest in TiAl-based alloys has grown considerably as significant economic and environmental benefits are expected. The specific weight of this alloy family (from 3.8 g cm$^{-3}$ for $\gamma$-TiAl to 4.3 g cm$^{-3}$ for $\alpha_2$-Ti$_3$Al [1,2]) represents half that of the common Ni-based super alloys or stainless steels (8.3 g cm$^{-3}$ for single crystal alloy <001> orientation [3]). From a mechanical point of view, the inertia of rotating pieces could be decreased, and/or for a given configuration, the overall mass could be reduced significantly. In any case, this leads to a decrease in fuel consumption and/or better system efficiency. TiAl alloys are of interest in high-temperature applications for several industries: aeronautical (turbine blades, fans, etc. [1]), automotive (valves [4], turbochargers [5]), and in gas turbines for power generation. Their industrialisation has been delayed due to the production costs, which are still high, and poor oxidation resistance above 750 °C. Without surface treatment or coatings, their high-temperature oxidation in air leads to the formation of a mixed oxide scale of titania and alumina which spalls under cyclic thermal conditions and offers no long-term protection. An alternative to improve the oxidation behaviour of these alloys is to apply a surface treatment by means of halogens. Usually known as a corrosive agent [6], chlorine applied in a carefully controlled amount can enhance the growth of a pure and protective alumina layer at the surface of TiAl alloys through the so-called 'halogen effect' mechanism [7,8]. This mechanism is not based on a doping effect but on the selective formation and evaporation in the sub-surface of the alloy of gaseous aluminium halides which are oxidised during their outward diffusion because of the increasing oxygen partial pressure within the oxide scale. One of the most important features of this surface treatment is that the oxidation resistance is significantly improved without changing the basic mechanical properties of the alloy

*   Reprinted from P. J. Masset et al.: Influence of alloy compositions on the halogen effect in TiAl alloys. *Materials and Corrosion.* 2008. Volume 59. pp. 609–18. Copyright Wiley-VCH Verlag GmbH & Co. KGaA.

treated. The halogen effect was found to be effective for improving the oxidation resistance of TiAl-based alloys using different halogens (chlorine [9–11], fluorine [11], bromine [8,9], iodine [8,9]) and treatment methods (beam-line ion implantation [12,13], plasma immersion ion induction [14], dipping in F-based solutions[15]). However, the most promising results were obtained with fluorine. The effect of fluorine treatments was observed for isothermal and cyclic exposure conditions, even in wet atmospheres [16,17] at temperatures up to 1050 °C whereas chlorine-treated samples spalled during cyclic tests, especially in wet atmospheres [18]. In addition, previous investigations showed that fluorine treatments were effective with alloys containing at least 40 at.% aluminium [13]. The aim of this work was to evaluate the long-term stability of the fluorine effect for implanted alloys and the oxidation resistance of selected alloys at 900 °C in laboratory air.

## 30.2  Experimental

### 30.2.1  Materials

Sample preparation

TiAl-based alloys with different alloying elements, and mechanical and thermal treatments were used in this study. The alloy compositions and their processing are reported in Table 30.1. Samples with dimensions of 10 mm × 10 mm×1.5 mm were machined by spark erosion from the original materials. The samples were ground to remove the surface left after the cutting procedure. Before implantation, the specimens were mechanically polished down to 4000 grit SiC abrasive paper from Struers. Subsequently, they were ultrasonically cleaned in acetone and ethanol for 10 min to degrease the surface.

Sample implantation

Implantation of fluorine was carried out with the ion implanter of the Institute of Nuclear Physics (IKF) at Johann Wolfgang Goethe University in Frankfurt am Main. The implantation of $^{19}F^+$ ions was undertaken with a $CF_4$ gas source using a 20 keV acceleration potential and at a nominal dose of $1×10^{17}$ $^{19}F^+$ $cm^{-2}$. The fluorine ions were extracted from a Penning cold cathode out of the gas phase. The ions were accelerated and focused to form a beam. The focusing was adjusted to reach the maximum current at the surface of the substrate. In order to ensure homogenous implantation, the beam was scanned in front of the last aperture.

### 30.2.2  Techniques

Thermogravimetric measurements

The mass change of the samples due to oxidation was measured by a thermobalance from Setaram (France) under dry synthetic air at 900 °C for 100 h (precision: $\pm 5$ μg). For interrupted thermogravimetric measurements, the samples were oxidised in a furnace under laboratory air. The samples were removed from the furnace at regular intervals, air cooled and weighed with a laboratory balance from Satorius GmbH (Germany) within a precision of $\pm 2$ μg. After weighing, the samples were reloaded into the furnace which remained at the test temperature.

Table 30.1  Chemical compositions (at.%) of the alloys tested

| Alloy | Ti | Al | Nb | V | Cr | Mo | C | B | Ta | Zr | Sn | Si | Treatment |
|---|---|---|---|---|---|---|---|---|---|---|---|---|---|
| Ti6V4 | Bal. | 10.2 | – | 4 | – | – | – | – | – | – | – | – | Cast + heat treatment |
| IMI834 | Bal. | 10.4 | 0.4 | – | – | 0.25 | 0.2 | – | – | 1.8 | 1.6 | 0.6 | Cast + heat treatment |
| $\alpha_2$-Ti$_3$Al | Bal. | 25 | – | – | – | – | – | – | – | – | – | – | Cast + heat treatment |
| TNBV2A | Bal. | 44.5 | 8 | – | – | – | 0.2 | – | – | – | – | – | Extruded + heat treatment |
| TNBV2B | Bal. | 44.5 | 8 | – | – | – | 0.2 | – | – | – | – | – | Forged + heat treatment |
| PX3500 | Bal. | 44.5 | 2 | – | 2 | – | – | – | – | – | – | – | Fine-cast |
| TNBV3 | Bal. | 44.5 | 8 | – | – | – | 0.2 | 0.2 | – | – | – | – | Fine-cast + HiPed |
| TNBV3A | Bal. | 44.5 | 7 | – | – | 1 | 0.2 | 0.2 | – | – | – | – | Fine-cast + HiPed |
| TNBV3B | Bal. | 44.5 | 7 | – | – | 1 | – | 0.2 | – | – | – | – | Fine-cast + HiPed |
| γ-MET 100 | Bal. | 46.4 | 1.1 | – | 2.5 | – | – | 0.1 | 0.5 | – | – | – | Powder metallurgy |

## Metallographic observations

Before the metallographic investigations, the alloys were etched using procedures similar to those described in the literature [19] and the microstructures were observed with an optical microscope (Leica type DMRME) at several magnifications. The oxide scales were also post-experimentally investigated by metallographic techniques. Before the observation, the samples were gold sputtered, electrochemically coated with Ni using a Ni-Sulfamat based solution and embedded in an epoxy resin (Duro-Fast from Struers) using a Struers LaborPress3 press. The cross-sections were polished with 4000 grit SiC abrasive paper, followed by two-step polishing with 3 μm and 1 μm polycrystalline diamond suspensions from Struers. A silica-based suspension from Struers was used for the final step of polishing.

## Scanning electron microscopy (SEM) and electron microprobe analysis (EPMA)

SEM images were taken with an XL40 electron microscope from Philips (The Netherlands) coupled with an EDX probe from EDAX. WDX elemental maps with a spatial resolution close to 1 μm were obtained with an SX50 microprobe from Cameca (France).

## Proton Induced Gamma-Ray Emission (PIGE)

The non-destructive Proton Induced Gamma-Ray Emission technique was used to determine the fluorine concentration depth profile of the implanted samples before and after the oxidation tests. This method is one of the non-destructive Ion Beam Analysis (IBA) techniques whose depth resolution near the surface is about 20 nm and it allows the detection of light trace elements with concentrations in the range of tens of ppm for fluorine [20]. The PIGE measurements were performed at the 2.5 MV Van de Graaff accelerator of the Institute of Nuclear Physics (IKF) at Johann Wolfgang Goethe-University in Frankfurt am Main using the nuclear reaction $^{19}F(p,\alpha\gamma)^{16}O$.

## 30.3    Theoretical considerations

### 30.3.1    Thermodynamic calculations

Previously, it has been shown that the amount of halogen needed to obtain the 'halogen effect' must be controlled and models have been developed [7,21,22]. In the temperature range of interest (700–1100 °C), the partial pressures of the volatile species which might be involved in the halogen effect can be calculated. The sum of the partial pressures of the halides named $p(X)$ should be situated between the minimum and the maximum limit partial pressures $p(X)_{min}$ and $p(X)_{max}$, respectively, to obtain a positive effect [22]. The region where the halogen partial pressures are situated between $p(X)_{min}$ and $p(X)_{max}$ corresponds to the preferential reaction of aluminium with the implanted halogen to form volatile aluminium halides while titanium, which would form fast growing non-protective $TiO_2$, remains in the metal. The aluminium halides react with oxygen in the scale defects (crevices, nano- and micro-pores) to establish a protective and adherent (partial) $Al_2O_3$ layer almost free of $TiO_2$. The overall mechanism is described by Eq. 30.1 and Eq. 30.2

$$2\,\underline{Al}(s) + X_2(g) \rightarrow 2\,AlX(g) \hspace{2cm} [30.1]$$

$$4\,AlX(g)+3\,O_2(g)\rightarrow 2\,Al_2O_3(s)+2\,X_2(g) \tag{30.2}$$

In Eq. 30.2, $X_2$ represents a halogen gas (fluorine, chlorine, bromine or iodine) while AlX corresponds to the most stable aluminium halide. The letters 's' and 'g' refer to the solid and gaseous states, respectively. The underlined element means that the activity of the element considered in the alloy is lower than unity.

Most of the industrial alloys used in this study are composed of two phases, namely $\gamma$-TiAl and $\alpha_2$-Ti$_3$Al resulting from their processing (see microstructures of the alloys given in Fig. 30.8). For the calculations, some hypotheses and approximations were made. The solubility of oxygen in $\gamma$-TiAl varies from 350 ppm [23–25] to 3 at.% [26], whereas it reaches 7–15 at.% in the $\alpha_2$-Ti$_3$Al phase [27–30]. The calculations were performed with the assumption that the oxygen participates preferentially in the growth of alumina through the halogen effect rather than dissolving in the phases mentioned. Moreover, the Al activity may also be modified in the presence of dissolved oxygen [31]. In addition, industrial alloys contain some alloying elements, e.g. Nb, to enhance the mechanical properties. The addition of Nb was found to decrease the Al activity [32] compared to that calculated or measured in the Ti–Al binary system [33]. Although some experimental investigations were carried out in the quaternary system Ti–Al–Nb–O to define the oxide stability area versus the alloy composition [34], no thermodynamic databases were available at that time for an accurate assessment of the Al boundaries for the halogen effect in this system.

For Ti$_{1-x}$Al$_x$ alloy compositions in the range from $x=0.4$ to $x=0.5$, calculations were performed with fluorine as the halogen. The minimum $p(F)_{min}$ and maximum $p(F)_{max}$ fluoride partial pressures were calculated (see Ref. 35 for more details). The calculations were carried out using the program ChemSage [36] which takes into account the complex equilibria between the alloy phases and the fluoride-based species.

According to the simplified model proposed in Ref. 22, $p(F)_{min}$ values were calculated on the assumption that the limiting step was the alumina layer kinetics. The $p(F)_{max}$ values were fixed when the partial pressure of TiF$_3$ was higher than $p(F)_{min}$. According to Eq. 30.1 and Eq. 30.2, the partial pressures of $p(AlF)$ and $p(F_2)$ may not be fixed independently of one another. Therefore aluminium and oxygen fluxes at the scale/alloy interface should be considered. Assuming that the Al$_2$O$_3$ layer growth follows a parabolic law, the minimum halogen pressure $p(F)_{min}$ needed to supply sufficient Al via the gas phase AlF for Al$_2$O$_3$ growth was calculated from Eq. 30.3

$$p(X)_{min}=(M_{AlX}\cdot T\cdot k_p/t)^{0.5}/2126.4 \tag{30.3}$$

where $k_p$ represents the parabolic rate constant; $M_{AlX}$ is the molecular weight of the AlX compound; $T$ is the absolute temperature in Kelvin and $t$ is the time to establish

*Table 30.2* Summary of the oxidation kinetic constants of alumina vs temperature taken from the literature [37]

| Temperature, °C | $k_p$, mg$^2$ cm$^{-4}$ s$^{-1}$ |
| --- | --- |
| 700 | $1 \times 10^{-15}$ |
| 800 | $5 \times 10^{-15}$ |
| 900 | $1 \times 10^{-13}$ |
| 1000 | $1 \times 10^{-12}$ |
| 1100 | $8 \times 10^{-12}$ |

the alumina layer, which was arbitrarily fixed at 60 s in the calculations. Values of the kinetic constant $k_p$ for the formation of pure alumina were taken from the literature review of Fergus [37] and are reported in Table 30.2.

For $Ti_{1-x}Al_x$ alloy compositions of $x=0.4$ and 0.5, the upper limit $p(F)_{max}$ for the operation of the fluorine effect is defined when the partial pressure of the most stable $TiF_3$ compound reaches the value of $p(F)_{min}$ (Fig. 30.1). That is, a partial pressure window exists between $p(F)_{min}$ and $p(F)_{max}$ in which the halogen effect can exert its positive role (Fig. 30.2). The calculated values $p(F)_{min}$ and $p(F)_{max}$ for the alloys $Ti_{0.5}Al_{0.5}$ and $Ti_{0.6}Al_{0.4}$ are summarised in Table 30.3 and Table 30.4, respectively. For both compositions, the fluoride partial pressures $p(F)_{min}$ and $p(F)_{max}$ increase with increasing temperature (Fig. 30.2).

The window where a positive F-effect is expected becomes broader at higher temperatures. This results from the higher stability of AlF compared to $TiF_3$. Figure 30.3 sketches the variations of the partial pressures of the halides $MF_n$ ($M = Al,Ti$) versus temperature for the alloy $Ti_{0.6}Al_{0.4}$. The partial pressures of $MF_n$ were calculated for the upper limit where the halogen effect takes place. In the temperature range considered (700–1100 °C), the ratio $p(AlF)/p(TiF_3)$ varies between 10 and 100. This conclusion is also valid for the second composition. By lowering the aluminium content, and therefore the Al activity, the window where the halogen effect takes place is narrowed. The boundaries move towards each other and the implantation parameters may be kept identical for all alloys.

### 30.3.2   Implantation parameters

On penetrating the surface, accelerated ions lose their energy due to elastic impacts with the target nuclei (nuclear stopping) and inelastic impacts with the target

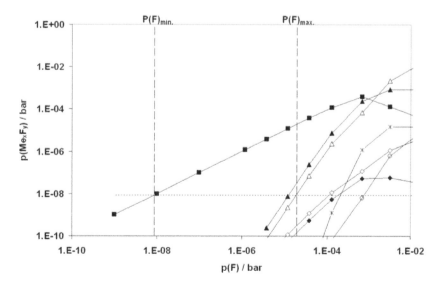

30.1 Evolution of the metal fluoride partial pressure $p(MeF_n)$ versus the fluorine pressure $p(F)$ calculated at 900 °C for the $Ti_{0.6}Al_{0.4}$ alloy. (■) AlF, (◆) $AlF_2$, (▲) $AlF_3$, (◇) $TiF_2$, (△) $TiF_3$, (○) $TiF_4$, (*) $Al_2F_6$

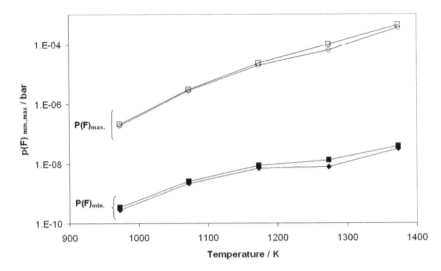

*30.2* Evolution of the metal fluoride partial pressures $p(F)_{max}$ and $p(F)_{min}$ versus the absolute temperature for the $Ti_{0.6}Al_{0.4}$ and $Ti_{0.5}Al_{0.5}$ alloys. $p(F)_{min}$: (■) $Ti_{0.5}Al_{0.5}$, (◆) $Ti_{0.6}Al_{0.4}$; $p(F)_{max}$: (□) $Ti_{0.5}Al_{0.5}$, (◇)$Ti_{0.6}Al_{0.4}$

*Table 30.3*    Calculated partial pressures $p(F)_{min}$ and $p(F)_{max}$ for the alloy $Ti_{0.5}Al_{0.5}$

| Temperature, °C | $p(F)_{min}$, bar | $p(F)_{max}$, bar |
|---|---|---|
| 700 | $3.44 \times 10^{-10}$ | $2.10 \times 10^{-7}$ |
| 800 | $2.60 \times 10^{-9}$ | $3.00 \times 10^{-6}$ |
| 900 | $8.73 \times 10^{-9}$ | $2.30 \times 10^{-5}$ |
| 1000 | $1.30 \times 10^{-8}$ | $1.00 \times 10^{-4}$ |
| 1100 | $3.86 \times 10^{-8}$ | $4.30 \times 10^{-4}$ |

*Table 30.4*    Calculated partial pressures $p(F)_{min}$ and $p(F)_{max}$ for the alloy $Ti_{0.6}Al_{0.4}$

| Temperature, °C | $p(F)_{min}$, bar | $p(F)_{max}$, bar |
|---|---|---|
| 700 | $2.84 \times 10^{-10}$ | $1.85 \times 10^{-7}$ |
| 800 | $2.15 \times 10^{-9}$ | $2.80 \times 10^{-6}$ |
| 900 | $7.15 \times 10^{-9}$ | $1.95 \times 10^{-5}$ |
| 1000 | $7.86 \times 10^{-9}$ | $6.25 \times 10^{-5}$ |
| 1100 | $3.14 \times 10^{-8}$ | $3.50 \times 10^{-4}$ |

electrons (electronic stopping). The corresponding depth profile follows a Gaussian-type distribution to a first approximation. An important parameter is the depth of maximum F concentration, δ−max. Using the software T-DYN [38], the implantation profiles have been calculated by Monte-Carlo simulations. Figure 30.4 shows the calculated depth profiles of fluorine and of the alloying elements in the γ-MET 100 alloy.

For F ions and implantation parameters of 20 keV/$1 \times 10^{17}$ $^{19}F^{+}$ $cm^{-2}$ the maximum concentration is located in a depth of about 35 nm. Except for the aluminium, the alloying component concentrations increase in the first few nanometres below the

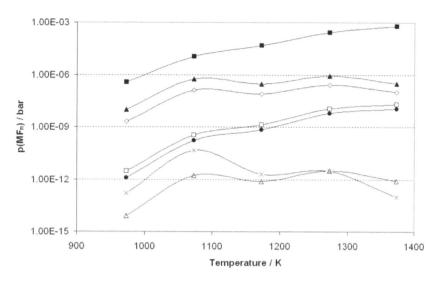

*30.3* Evolution of the partial pressures of $MF_n$ (M = Al, Ti) with temperature (alloy $Ti_{0.5}Al_{0.5}$). (■) AlF, (●) $AlF_2$, (▲) $AlF_3$, (□) $TiF_2$, (◇) $TiF_3$, (△) $TiF_4$, (×) $Al_2F_6$

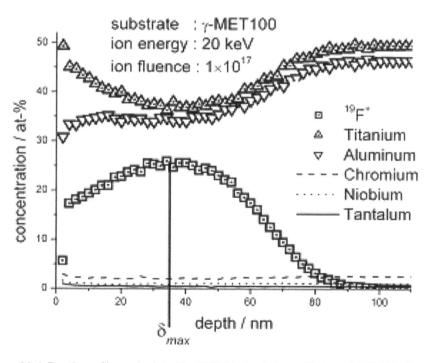

*30.4* Depth profiles calculated by T-DYN simulation with the $\gamma$-MET 100 alloy after $^{19}F^+$ implantation (ion energy 20 keV and ion fluence $1 \times 10^{17}$ F $cm^{-2}$)

surface, due to the decreasing fluorine concentration. The aluminium concentration decreases because of its comparatively high sputter yield. In the simulations, the alloying element boron has been neglected, since T-DYN does not permit more than five target components. All of the following diagrams are based on such simulated data but only that of the implanted fluorine is drawn.

### Effect of the ion fluence

The quantity of F ions (fluence) is measured by the charge accumulated at the samples during the implantation process. The relationships between the maximum F concentration, the fluence and the implantation energy are illustrated in Fig. 30.5(a) and Fig. 30.5(b), respectively. Up to $1 \times 10^{17}$ ions cm$^{-2}$, the maximum F concentration is nearly proportional to the fluence, as is apparent in Fig. 30.5(a). For higher accelerated ion values, lower fluorine concentrations are obtained due to profile broadening.

On exceeding $1 \times 10^{17}$ ions cm$^{-2}$, the depth profile becomes flat as the implanted dose is increased. A maximum concentration of approximately 82 at.% is reached at a fluence of $1 \times 10^{18}$ ions cm$^{-2}$ as depicted in Fig. 30.5(b) for the alloy $\gamma$-TiAl at 20 keV.

### Effect of ion energy

For any ion/substrate combination, $\delta$-max increases with ion energy. In addition, the maximum implanted dose is affected, due to profile broadening at higher penetration depths, as is shown Fig. 30.6(a). In Fig. 30.6(b), the evolution of $\delta$-max as a function of ion energy for a fluence of $5 \times 10^{16}$ ions cm$^{-2}$ is shown. There is a linear dependence in the energy range up to 60 keV.

### Effect of alloy composition

The alloys investigated in this work have aluminium contents which range between 10 and 45 at.%. As the aluminium is replaced by elements with high specific weight such as niobium, their density increases. Compared to stoichiometric TiAl ($\rho = 3.6$ g cm$^{-3}$), the density of Ti22Al27Nb is increased by 44% to $\rho = 5.2$ g cm$^{-3}$. From a physical point of view, higher ion stopping is expected in a denser substrate, causing a decrease in the $\delta$-max value. The shift to lower ion ranges is confirmed by the simulations shown in Fig. 30.7(a). However, there is only a slight change in $\delta$-max so that it seems appropriate to implant these alloys with the same parameters as with $\gamma$-TiAl (Fig. 30.7).

## 30.4     Results and discussion

### 30.4.1     Characterisation of the alloys

TiAl-based alloys with varying alloying elements and mechanical and thermal treatments were used in this study (Table 30.1). The microstructures of the materials were investigated by classical metallographic techniques (Fig. 30.8). The magnification was optimised for each alloy to emphasise details of the microstructure. The microstructure of the alloy $\alpha_2$-Ti$_3$Al is not depicted as it shows no specific features. Alloy

(a)

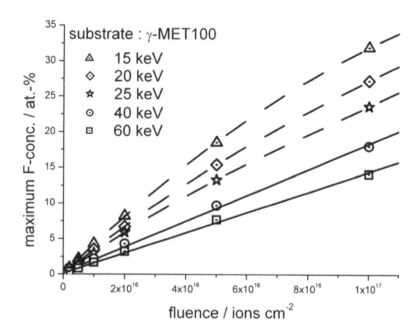

(b)

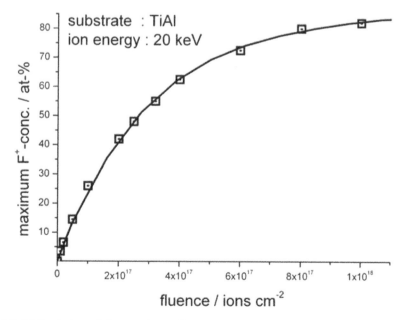

30.5 (a) Evolution of the maximum F concentration versus the fluence
($< 1 \times 10^{17}$ F cm$^{-2}$) for different ion energies ($\gamma$-MET 100 alloy), (b) Evolution of the
maximum F concentration versus the fluence at 20 keV ($\gamma$-MET 100 alloy)

(a)

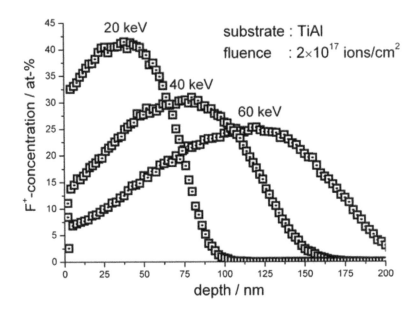

(b)

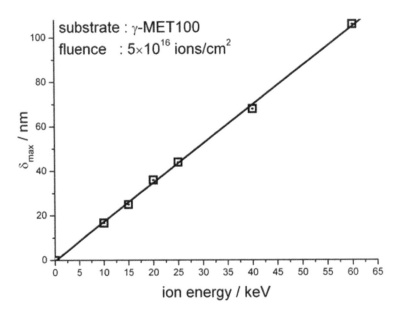

*30.6* (a) Evolution of the F concentration versus depth within the alloy for different ion energies for a fluence of $2 \times 10^{17}$ F cm$^{-2}$ ($\gamma$-MET 100 alloy), (b) Evolution of the maximum F concentration depth $\delta$–max versus the ion energy for a fluence of $5 \times 10^{16}$ F cm$^{-2}$ ($\gamma$-MET 100 alloy)

(a)

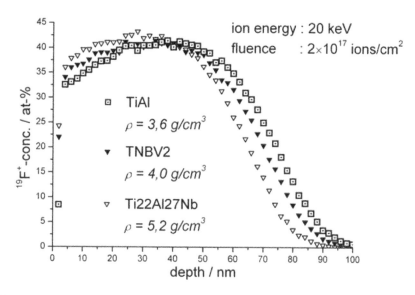

(b)

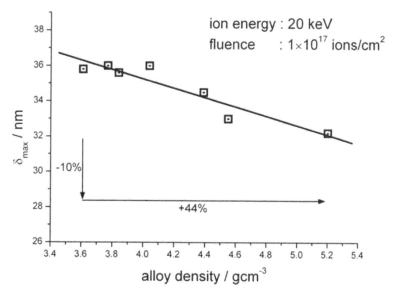

*30.7* (a) Evolution of the F concentration versus the alloy composition (influence of the density) for an ion energy of 20 keV and a fluence of $2 \times 10^{17}$ F cm$^{-2}$, (b) Change in the maximum F-concentration depth $\delta-$max versus the alloy density for an ion energy of 20 keV and a fluence of $1 \times 10^{17}$ F cm$^{-2}$

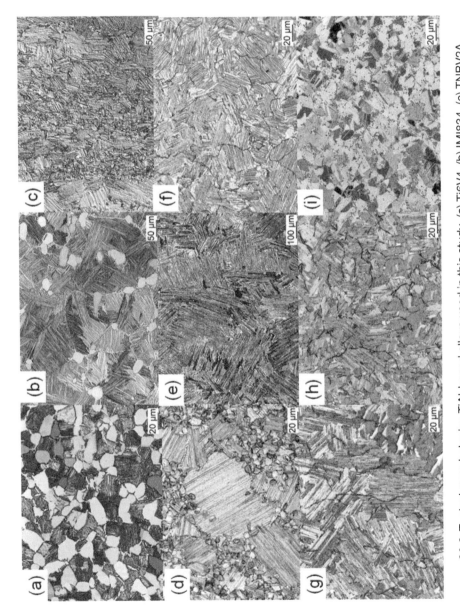

30.8 Typical morphologies TiAl-based alloys used in this study. (a) Ti6V4, (b) IMI834, (c) TNBV2A, (d) TNBV2B, (e) PX3500, (f) TNBV3, (g) TNBV3A, (h) TNBV3B, (i) $\gamma$-MET 100

IMI834 (Fig. 30.8(b)) shows a so-called bi-modal structure consisting of equiaxed primary α phase ($\alpha_p$) transformed into a lamellar matrix and a lamellar structure in which the remaining areas form the coarse β-phase. The γ-MET shows a typical near-gamma structure composed of coarse γ grains and lamellar areas of $\alpha_2 - \gamma$ grains. The fully lamellar structure consists of grains of fine $\alpha_2 + \gamma$ lamellae. The presence of dark needles is ascribed to the presence of boron in the alloy used to refine the structure during alloy casting [39]. One can observe the effect of the mechanical treatment (extrusion) and the corresponding grain size decrease for the TNBV2A alloy (Fig. 30.8(c)).

### 30.4.2    Thermobalance experiments in synthetic air

Thermogravimetric experiments over 100 h in synthetic air were carried out with unimplanted and implanted specimens (Fig. 30.9). In addition, long-term thermo-gravimetric experiments (over 2000 h) were also carried out in synthetic air, but only with some of the F-implanted samples, as this experiment is time consuming in the use of equipment. The latter were compared with interrupted isothermal oxidation experiments in laboratory air. The mass change curves were found to fit closely to each other (Fig. 30.10). From the thermogravimetric measurements, the kinetic constants of the scale were derived using Eq. 30.4

$$(\Delta m / A)^2 = k_p\, t \qquad\qquad [30.4]$$

where $\Delta m$ is the mass variation in mg, A the surface area of the sample in cm², $k_p$ the kinetic constant in mg² cm⁻⁴ s⁻¹, and $t$ the time in s. For the calculation of the $k_p$ values, a parabolic growth law was assumed. Values of $k_p$ for the specimens tested are reported in Table 30.5. For the low Al-content alloys (IMI834 and $\alpha_2$-Ti$_3$Al), the $k_p$ values were close to those obtained for pure TiO$_2$ growth. For alloys for which the Al content ranged between 40 and 50 at.%, the kinetic constants corresponded to the growth of a mixed oxide scale. The oxidation rate of alloys from the TNB family was found to be of the same order of magnitude as for Ti$_3$Al–10Nb [40] or orthorhombic Ti–22Al–25Nb [41]. The presence of Nb improves the oxidation resistance of the alloys.

### 30.4.3    Short-term exposure tests in laboratory air (100 h)

Exposure tests with F-implanted and unimplanted specimens were carried out for 100 h in laboratory air. From SEM analysis, a wave-type Al$_2$O$_3$-layer structure was observed with F-implanted samples oxidised for 100 h in laboratory air (Fig. 30.11). This structure was observed for all of the alloys studied (see example in Fig. 30.11(a), (b) and (c) at different magnifications). For comparison of the crystal size, Fig. 30.11 represents the surface of pure TiO$_2$ crystals from an unimplanted alloy, Ti$_{0.9}$Al$_{0.1}$, after 100 h exposure in laboratory air. The size of the TiO$_2$ crystals is close to 50 μm whereas Al$_2$O$_3$ crystals are smaller than 5 μm. From microanalysis of the cross-section, the wave-type structure of the alumina layer was well evidenced (Fig. 30.12). In the void formed between the surface of the original alloy and the alumina layer, fluorine was detected by EDX analysis (Fig. 30.13). It was entrapped in the course of the alumina layer growth. It also acts as a reservoir for the halogen effect. For EPMA analysis (Fig. 30.12), the signal for the Ti map was recorded with high sensitivity. The alumina layer contains a small amount of Ti. In addition, it seems that low amounts of Nb are also located in the oxide scale. At the inward surface of the original alloy, Al

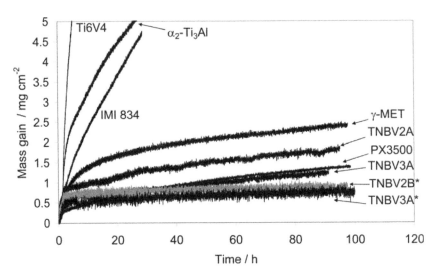

30.9 TGA curves for isothermal oxidation at 900 °C for 100 h in synthetic air of non-implanted specimens, (*) two-side implanted specimens

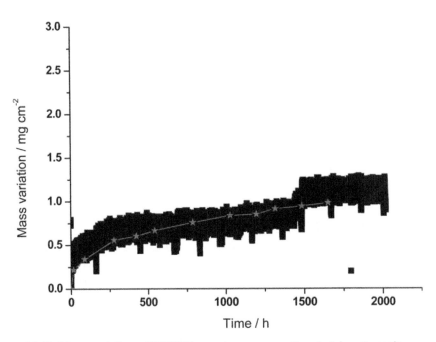

30.10 Mass variation of TNBV2A specimen versus time in laboratory air (star-shaped symbols) and synthetic air (solid line) at 900 °C

Table 30.5   Kinetic constants of oxidation of untreated alloys at 900 °C in synthetic air

| $10^{13} \times k_p$ | $\alpha_2$-Ti$_3$Al | IMI834 | TNBV2A | TNBV2B | TNBV3 | $\gamma$-MET | Al$_2$O$_3$ |
|---|---|---|---|---|---|---|---|
| $g^2\ cm^{-4}\ s^{-1}$ | 830 | 275 | 24 | 24 | 24 | 30 | 0.1–1 [42] |

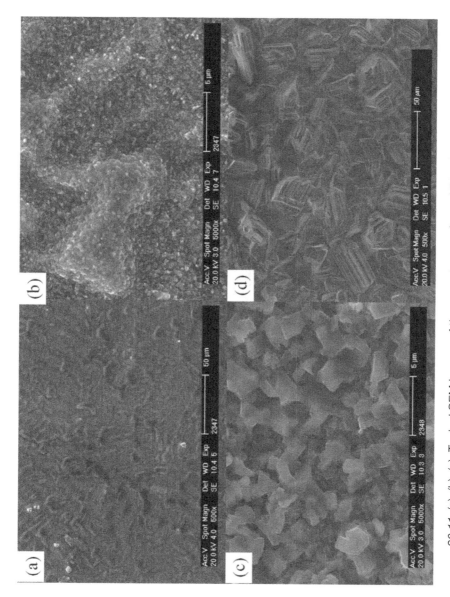

*30.11* (a), (b), (c): Typical SEM images of the wave-type surface of F implanted specimen (alloy TNBV3) after 100 h oxidation in laboratory air for different magnifications; (d) $Ti_{0.9}Al_{0.1}$ after 100 h oxidation in laboratory air

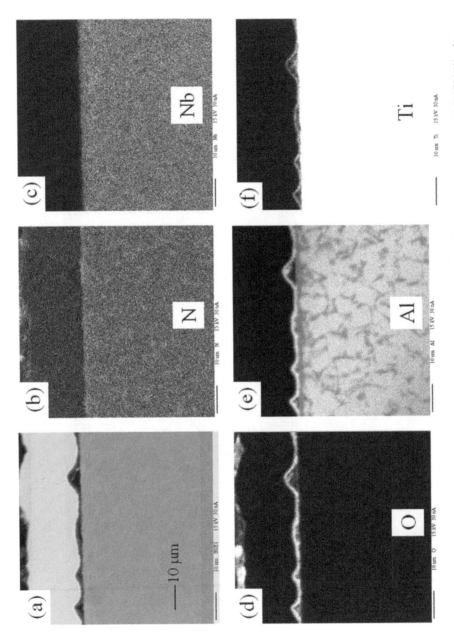

30.12 Typical composition mapping of the cross-section of the F implanted side (sample TNBV2A) after an oxidation test at 900 °C for 120 h in laboratory air

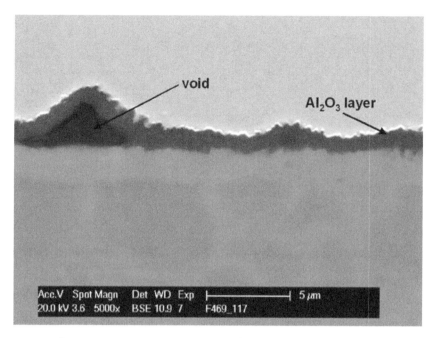

*30.13* SEM image of the cross-section of the alloy TNBV2B after 120 h of isothermal oxidation at 900 °C in laboratory air

depletion was observed due to the consumption of Al for the alumina layer growth. In this same region, oxygen was present and might be ascribed to dissolved oxygen. After 100 h of exposure in laboratory air, fluorine implantation promotes the growth of a pure and adherent alumina layer (3–4 µm thick) whereas a mixed oxide scale $Al_2O_3 + TiO_2$ (10–15 µm thick) was observed with unimplanted samples.

### 30.4.4   Long-term exposure tests in laboratory air

Isothermal oxidation tests were performed at 900 °C for over 4000 h in laboratory air using samples implanted from both sides. These long-term experiments were carried out with alloys in which the Al content was higher than 40 at.%. For Al contents between 40 and 50 at.%, the implantation parameters were kept constant according to the thermodynamic calculations. The energy and the doses were fixed at 20 keV and $1 \times 10^{17}$ F cm$^{-2}$, respectively. After cooling of the specimens, the mass changes were measured periodically. The curves of $\Delta m$ versus time are presented in Fig. 30.14 for the alloys tested. For comparison, the mass variation of the untreated $\gamma$-MET sample (in synthetic air) is also plotted. After tens of hours, the kinetic constant for the unimplanted sample is typical of those for the growth of a mixed oxide scale. For the F implanted alloys tested, all of the curves present identical features.

Assuming a parabolic growth law for the oxide scale, the kinetic constants $k_p$ were calculated using Eq. 30.4 and the values were derived from the slope of the plot ($\Delta m/A)^2$ versus time (Fig. 30.15). Values of the kinetic constant, $k_p$, are summarised in Table 30.6. These values are typical of those for the growth kinetics of a pure alumina layer. They are in agreement with previous determinations carried out with other

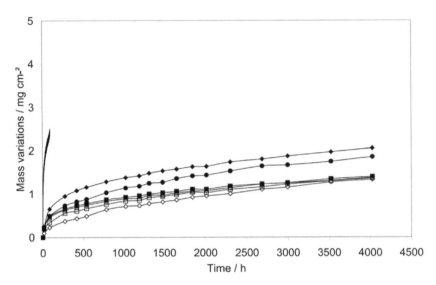

30.14 Mass variations of implanted specimens (20 keV, $1 \times 10^{17}$ F cm$^{-2}$) for i sothermal oxidation at 900 °C for 4000 h under laboratory air. Solid line, untreated $\gamma$-MET 100 (reference), and F treated specimens ($\diamond$) $\gamma$-MET 100 ($\blacklozenge$) TNBV2A, ($\bullet$) TNBV3, ($\blacksquare$) TNBV3B, ($\triangle$) TNBV3A, ($\square$) TNBV2B

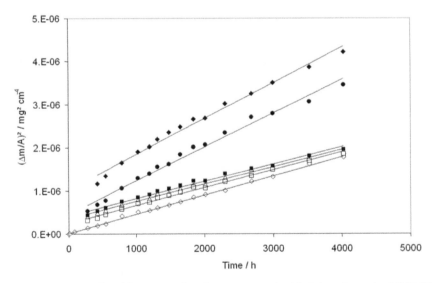

30.15 Curves $(\Delta m/A)^2$ versus time for exposure test in laboratory air at 900 °C with F treated specimens. ($\diamond$) $\gamma$-MET 100 ($\blacklozenge$) TNBV2A, ($\bullet$) TNBV3, ($\blacksquare$) TNBV3B, ($\triangle$) TNBV3A, ($\square$) TNBV2B

Table 30.6    Kinetic constants of oxidation of F implanted (energy: 20 keV, dose: $1 \times 10^{17}$ F cm$^{-2}$) alloys at 900 °C in laboratory air

| $10^{13} \times k_p$ | TNBV2A | TNBV2B | TNBV3 | TNBV3A | TNBV3B | $\gamma$-MET | Al$_2$O$_3$ |
|---|---|---|---|---|---|---|---|
| g$^2$ cm$^{-4}$ s$^{-1}$ | 2.66 | 1.25 | 2.49 | 1.21 | 1.28 | 1.23 | 0.1–1 [] |

alloys known to be alumina formers [42]. It was observed that the curves became linear after a small delay except for the $\gamma$-MET 100 alloy, and the curves did not cross the ordinate. The results of the Monte-Carlo simulations (Fig. 30.4) of the elemental profiles after implantation showed that the Al and Ti concentrations decrease in the near sub-surface of the alloy. In addition, it was found that for depths lower than 15 nm, only the Al concentration decreases, e.g. down to 27 at.% for $\gamma$-TiAl alloy, whereas the concentrations of heavier elements increase, especially Ti up to 55 at.%. The non-linearity of the curve might be ascribed to the presence of titania that formed in the earliest stage of the oxidation process due to the 'artificial' Al depletion arising from the implantation process. This might imply significant Al activity modification which impedes momentarily the formation of a pure alumina layer despite the presence of halogen. For the $\gamma$-MET 100 alloy, this effect is not that obvious as the aluminium content in the alloy is slightly higher and supports the alumina layer growth even at the beginning of the oxidation step.

In addition, PIGE measurements were carried out after 4000 h of exposure to measure the fluorine profile with the $\gamma$-TiAl alloy (Fig. 30.16). The F concentration maximum is located deeper in the oxide scale. A residual F concentration of 0.02 at.% was detected in the area which corresponds to the scale. At the oxide/metal interface, the maximum concentration reached 1.2 at.%. In Fig. 30.17, the maximum F concentration is presented as a function of the exposure time. This shows that the maximum F concentration remains approximately constant with exposure time after 500 h. Evidently, the possible fluorine losses during long-term oxidation remain very low and the remaining amount of fluorine serves as reservoir for extended use of the halogen effect.

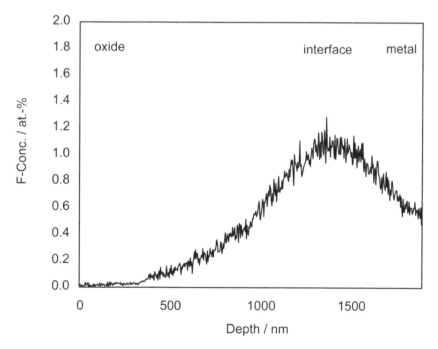

*30.16* Fluorine concentration in the alloy ($\gamma$-MET 100) versus the depth after 4000 h exposure in laboratory air at 900 °C

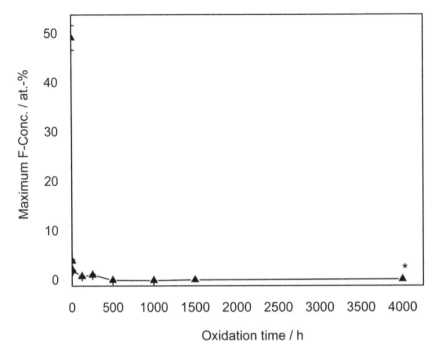

*30.17* Maximum value of fluorine concentration in the implanted alloy γ-MET 100 (fluence: $2.04 \times 10^{17}$ F cm$^{-2}$ / 20 keV) with the exposure time in laboratory air, (*) fluence: $1 \times 10^{17}$ F cm$^{-2}$ / 20 keV

## 30.5    Conclusions

The main conclusions of this work are detailed in three parts.

### 30.5.1   Thermodynamic calculations

- The fluoride partial pressures characterising the halogen effect are similar for the alloys $Ti_{0.6}Al_{0.4}$ to $Ti_{0.5}Al_{0.5}$
- The minimum and maximum fluoride partial pressures characterising the halogen effect increase with temperature
- The width of the fluorine partial pressure window for the halogen effect increases with temperature
- The implantation calculations imply that parameters should be kept constant for alloys of compositions $Ti_{0.6}Al_{0.4}$ to $Ti_{0.5}Al_{0.5}$.

### 30.5.2   Concentration profile simulations

- For ion fluences lower than $1 \times 10^{17}$ F cm$^{-2}$, a linear dependence was observed between the maximum F concentration depth, δ-max, and the ion fluence
- δ-max varies linearly with the ion energy up to 60 keV
- The alloy density does not significantly affect the F concentration profile
- δ-max decreases slightly with increase in the alloy density.

### 30.5.3  Exposure tests at 900 °C

- For F-implanted specimens (20 keV, $1 \times 10^{17}$ F $cm^{-2}$), the halogen effect was sustained for at least 4000 h in laboratory air at 900 °C
- The oxidation kinetic constants of F implanted samples of the alloys $Ti_{0.6}Al_{0.4}$ and $Ti_{0.5}Al_{0.5}$ range between 1.2 and $2.7 \times 10^{-13}$ $g^2$ $cm^{-4}$ $s^{-2}$, which is typical for alumina layer growth
- PIGE measurements showed the presence of approximately 1 at.% fluorine at the alloy/scale interface which acts as reservoir for the halogen effect
- The growth of alumina might be delayed at the beginning of the exposure test due to the decrease in the Al content in the alloy subsurface from the implantation process as evidenced by the concentration profile simulation.

### Acknowledgements

The authors acknowledge the financial support of the Bundesministerium für Wirtschaft und Technologie (BMWi) via the Arbeitsgemeinschaft industrieller Forschungsvereinigungen (AiF) under contract AiF–Nr 177 ZN. The authors sincerely thank Ms Ellen Berghof-Hasselbächer, Ms Daniela Hasenpflug (metallographic examinations) and Ms Monika Schorr (EPMA measurements) as well as Mr Peter Gawenda (SEM analysis) for their characterisation work. The authors are grateful to the ion implanter team of the Institute of Nuclear Physics (IKF) at Johann Wolfgang Goethe University in Frankfurt am Main for their technical support. The companies GfE Gesellschaft für Elektrometallurgie GmbH/Nuremberg, MTU Aero Engines/Munich, Royce-Royce/Dahlewitz are thanked for providing TiAl-based alloys used in the frame of this project.

### References

1. C. Leyens and M. Peters (eds.), *Titanium and Titanium Alloys: Fundamentals and Applications*. Wiley-VCH, ISBN 3-527-30534-3, 2003.
2. M.-R. Yang and S.-K. Wu, *Bull. Coll. Eng., N.T.U.*, 89 (2003), 3.
3. D. M. Dimiduk and D. B. Miracle and C. H. Ward, *Mater. Sci. Technol.*, 8 (1992), 367.
4. T. Noda, *Intermetallics*, 6 (1998), 709.
5. T. Tetsui and S. Ono, *Intermetallics*, 7 (1999), 689.
6. M. McNallan, *Oxid. Met.*, 46 (1996), 559.
7. M. Schütze and M. Hald, *Mater. Sci. Eng.*, A239–240 (1997), 847.
8. M. Kumagai, K. Shibue, M.-S. Kim and M. Yonemitsu, *Intermetallics*, 4 (1996), 557.
9. M. Schütze, G. Schumacher, F. Dettenwanger, U. Hornauer, E. Richter, E. Wieser and W. Möller, *Corros. Sci.*, 44 (2002), 303.
10. H.-E. Zschau, M. Schuetze, H. Baumann and K. Bethge, *Mater. Sci. Forum*, 461–464 (2004), 505.
11. A. Donchev, H.-E. Zschau and M. Schütze, *Microsc. Oxid.*, 6 (2006), 1.
12. A. Donchev, E. Richter, M. Schütze and R. Yankov, *Intermetallics*, 14(10/11) (2006), 1168.
13. P. Masset, H.-E. Zschau and M. Schütze, in Proc. 8th Conf. on Materials for Advanced Power Engineering, Part 2, 783, ed. J. Lecomte-Beckers *et al.*, Liège, Belgium, 2006.
14. U. Hornauer, E. Richter, E. Wieser, W. Möller, A. Donchev and M. Schütze, *Surf. Coat. Technol.*, 173–174 (2003), 1182.
15. H.-E. Zschau, V. Gauthier, M. Schütze, H. Baumann and K. Bethge, *Oxid. Met.*, 59(1/2) (2003), 183.

16. A. Zeller, F. Dettenwanger and M. Schütze, *Intermetallics*, 10 (2002), 59.
17. P. Masset, A. Donchev, H.-E. Zschau and M. Schütze, in Proc. Eurocorr 2007, Maastricht, The Netherlands, 2006.
18. A. Donchev and M. Schütze, *Mater. Sci. Forum*, 461–464 (2004), 447.
19. E. Copland, PhD Thesis, University of New South Wales, Australia, 1995.
20. J. R. Tesmer and M. Nastasi (eds.), *Handbook of Modern Ion Beam Materials Analysis*. Materials Research Society, Pittsburgh, PA, USA, 1995.
21. G. Schumacher, PhD Thesis, University Aachen, Germany, 2001, in German.
22. A. Donchev, B. Gleeson and M. Schütze, *Intermetallics*, (2003).
23. A. Menand, A. Hugent and A. Nérac-Partraix, *Acta Mater.*, 44(12) (1996), 4729.
24. A. Huguet and A. Menand, *Scripta Mater.*, 76/77 (1994), 191.
25. A. Nérac-Partraix and A. Menand, *Scripta Mater.*, 35(2) (1996), 199.
26. A. Kußmaul, PhD Thesis, University of Stuttgart, Germany, 1998, in German.
27. X. Li, R. Hillel, F. Teyssandier, S. K. Choi and F. J. J. Van Loo, *Acta Metall. Mater.*, 40(11) (1992), 3149.
28. B. Lee and N. Saunders, *Z. Metallkd.*, 88(2) (1997), 152.
29. M. X. Zhang, K. Hsieh, J. DeCock and Y. Y. Chang, *Scripta Metall.*, 27 (1992) 1361.
30. F. Dettenwanger, PhD Thesis, University of Stuttgart, Germany, 1997, in German.
31. E. Opila, in Proc. 29th Int. Conf. Adv. Ceram. Comp., 311, 2005.
32. M. Eckert, H. Nickel and K. Hilpert, in Proc. 5th Eur. Conf. Adv. Mater. Proc. Appl., 303, 1997.
33. A. Rahmel and P. J. Spencer, *Oxid. Met.*, 35(1/2) (1991), 53.
34. M. Allouard, Y. Bienvenu, I. Nazé and C. B. Bracho-Traconis, *J. Phys. IV, Colloque C9*, supplement to *J. Phys. III*, 3 (1993), 419.
35. P. Masset and M. Schütze, *Oxid. Met.*, in preparation.
36. ChemSage program, Version 4.22, GTT-Technologies, Herzogenrath.
37. J. W. Fergus, *Mater. Sci. Eng.*, 338(1–2) (2002), 108.
38. J. Biersack, *Nucl. Instrum. Meth. Phys. Res. B*, 153(3–4) (1999), 398.
39. Y.-W. Kim and D. Dimiduk, in *Structural Intermetallics 2001*, ed. K. J. Hemker, D. M. Dimiduk, H. Clemens, R. Darolia, H. Inui, J.M. Larsen, V. K. Sikka, M. Thomas and J. D. Whittenberger, The Minerals, Metals & Materials Society, 2001.
40. C. Leyens, *J. Mater. Eng. Perform.*, 10(2) (2001), 225.
41. C. Leyens, *Oxid. Met.*, 5/6 (1999), 52.
42. G. Schumacher, C. Lang, M. Schütze, U. Hornauer, E. Richter, E. Wieser and W. Möller, *Mater. Corros.*, 50 (1999), 162.

# 31

# Modelling of the long-term stability of the fluorine effect in TiAl oxidation*

## H.-E. Zschau and M. Schütze

*DECHEMA e.V., Karl Winnacker Institut, Theodor-Heuss-Allee 25,
D-60486 Frankfurt am Main, Germany*
*zschau@dechema.de*

## 31.1    Introduction

The increasing interest in $\gamma$-TiAl-based alloys is motivated by their excellent specific strength at high temperatures which offers high potential for applications in the aerospace and automotive industries.

Due to the reduction by about 50% in density compared to the currently-used Ni-based superalloys, a lower moment of inertia is achieved for rotating parts such as turbine blades, exhaust valves and turbocharger rotors. This leads to lower mechanical stresses and reduced fuel consumption. The so-called 'halogen effect' offers an innovative way to improve the poor oxidation resistance at temperatures above 800 °C [1–7]. For example, by applying fluorine to the TiAl surface by ion implantation [8] or treatment with diluted HF solution [9] followed by oxidation at 900 °C in air, the formation of a dense alumina scale is achieved which protects the TiAl against corrosion. A theoretical model was developed which explains this halogen effect by the preferred formation and transport of volatile aluminium halides through pores and microcracks within the metal/oxide interface and their conversion into alumina, forming a protective oxide scale on the surface [4]. The maximum of the fluorine depth profile was found to be located at the metal/oxide interface [8,9].

However, the amount of halogen and, in particular, fluorine before and after oxidation shows a distinct difference. After fluorine ion implantation, the maximum amount of fluorine near the surface measured by PIGE (Proton Induced Gamma-ray Emission) is reduced significantly after only 1 h of oxidation at 900 °C. Similar behaviour was observed after treatment with 0.11 mass% HF [10]. Industrial applications will only be possible if the fluorine effect can be stabilised for a time of at least 1000 h. Thus, the aim of this work was to study the time dependence of fluorine depth profiles in TiAl during isothermal and cyclic oxidation at 900 °C in air.

## 31.2    Materials and methods

Cast $\gamma$-TiAl (Ti–50 at.% Al) and the industrial alloy $\gamma$-Met (46.6 at.% Al) were prepared as coupons with dimensions of 8 mm × 8 mm × 1 mm and polished with SiC paper down to 4000 and 1200 grit, respectively. Microstructural investigations showed minor amounts of the $\alpha_2$-Ti$_3$Al phase (lamellar structure) within the $\gamma$-TiAl phase.

---

\* Reprinted from H.-E. Zschau and M. Schütze: Modelling of the long-term stability of the fluorine effect in TiAl oxidation. *Materials and Corrosion.* 2008. Volume 59. pp. 619–23. Copyright Wiley-VCH Verlag GmbH & Co. KGaA.

Following previous work [8], fluorine ions of 20 keV were implanted into TiAl with a fluence of $2 \times 10^{17}$ F cm$^{-2}$. The chosen energy corresponds to a projected range of 34 nm due to the stopping power of fluorine ions in TiAl as calculated by the software package TRIM [11]. In the case of oxidation up to 500 h at 900 °C in air, only one side of the specimen was implanted whereas for oxidation times > 500 h, both sides and the edges were implanted. All implantations were performed using the 60 kV implanter at the Institute of Nuclear Physics (IKF) at Johann Wolfgang Goethe-University in Frankfurt/Main.

Other TiAl specimens were treated on one side with a droplet of 0.11 mol% HF solution as reported in Refs. 8 and 9. After drying, the samples were oxidised for 1, 12, 120 and 250 h at 900 °C in air.

The non-destructive PIGE-technique was used to determine the fluorine concentration depth profiles within the first 1.4 µm of the HF-treated and of the ion-implanted samples before and after oxidation. A detailed description of the technique can be found in Ref. 12. The PIGE measurements were carried out at the 2.5 MV van de Graaff accelerator of the IKF using the nuclear reaction $^{19}$F(p, $\alpha\gamma$)$^{16}$O at a resonance energy of 340 keV. The information depth of the PIGE depth profiling technique using this resonance is limited to 1.4–1.5 µm.

Finally, all samples were inspected by metallographic methods and SEM.

## 31.3.    Results and discussion

### 31.3.1    Isothermal oxidation at 900 °C in air

Isothermal oxidation of implanted samples

The initial fluorine depth profiles, as calculated by the code T-DYN [13] after ion implantation, are illustrated in Fig. 31.1 for several fluences and an ion energy of

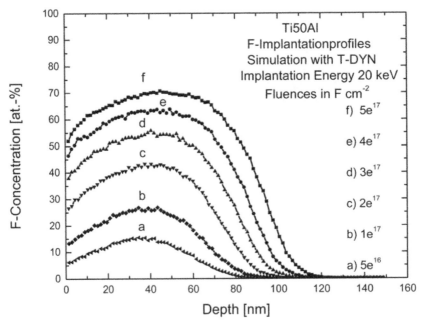

31.1 Fluorine depth profiles in TiAl for several fluorine fluences and an ion energy of 20 keV (T-DYN calculation)

20 keV. For the chosen fluence of $2 \times 10^{17}$ F cm$^{-2}$ (curve c), a maximum fluorine concentration of about 44 at.% was obtained.

The behaviour of the fluorine content during isothermal oxidation was studied at 900 °C for oxidation times of 1, 12, 120, 250, 500 and 1000 h at 900 °C in air. The metallographic inspections and measurements of mass gain also revealed the formation of a thin alumina layer on the surface in all cases.

From the corresponding fluorine profiles summarised in Fig. 31.2, it can be concluded that the alumina scale also acts as a diffusion barrier for longer oxidation times at 900 °C and prevents significant fluorine loss through the surface. This is an essential condition for the stability of the fluorine effect which acts at fluorine concentrations as low as 1–5 at.% in the profile maximum. The development with time of the integral and maximum fluorine concentrations summarised in Figs. 31.3 and 31.4, respectively, shows a rapid decrease within the initial hours of oxidation followed by a very slow decrease (Fig. 31.3) or a nearly asymptotic behaviour (Fig. 31.4). In order to describe the experimental data by a quantitative function, they were fitted by a 'decay curve'. The part of the curve with a slow fluorine decrease in the maximum amount starts at an oxidation time of about 12 h and follows the exponential fit up to 1000 h

$$c_F^{impl}(t) = c_0 + A \exp\left(-\frac{t}{t_1}\right)$$  [31.1]

where $c_0 = 2.71$; $A = 1.10$ and $t_1 = 148.58$. This function describes the behaviour of the fluorine maximum which is located at the metal/oxide interface. The dramatic

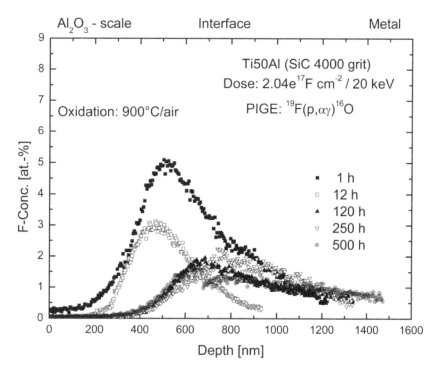

31.2 Fluorine depth profiles of implanted TiAl-samples ($2.04 \times 10^{17}$ F cm$^{-2}$/ 20 keV) after isothermal oxidation at 900 °C in air (obtained by PIGE)

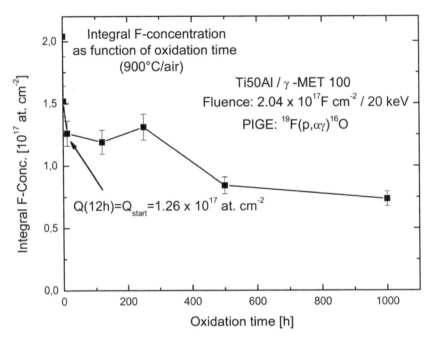

*31.3* Time dependence of integral fluorine concentration of implanted samples (2.04 × 10¹⁷ F cm⁻²/20 keV) during isothermal oxidation at 900 °C in air

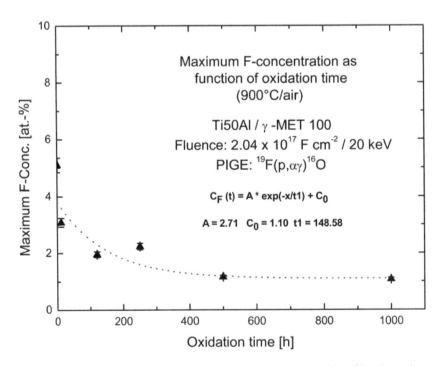

*31.4* Time behaviour of the maximum fluorine concentration of implanted samples (2.04 × 10¹⁷ F cm⁻²/20 keV) during isothermal oxidation at 900 °C in air)

fluorine loss from the surface within the first hour was not taken into account by fitting because this is regarded as the result of the evaporation of fluorine species before a dense oxide scale has been formed [10].

The fluorine effect is connected to the existence of a fluorine reservoir at the metal/oxide interface [8–10]. The analytical formulae describing the time behaviour of this fluorine maximum reveal that the local fluorine enrichment should be nearly stable for a much longer timescale than 1000 h. However, the validity for longer oxidation times at a temperature of 900 °C is a field for further investigation.

## Isothermal oxidation of HF-treated samples

The fluorine depth profiles in Fig. 31.5 indicate the stability of the fluorine content, at least up to oxidation times of 500 h at 900 °C in air. The near-surface region with negligible fluorine content marks the alumina scale again, whereas the maximum is located at the metal/oxide interface followed by a metal region influenced by fluorine diffusion. In all cases, the alumina scale could be observed by metallographic techniques.

The maximum fluorine content illustrated in Fig. 31.6 falls within the first hour of oxidation to values <5 at.%, which is more rapid than for the implanted samples. After an oxidation time of about 12 h, it is characterised by relatively stable behaviour, comparable to the results obtained for the implanted samples. It can again be described by the same exponential fit

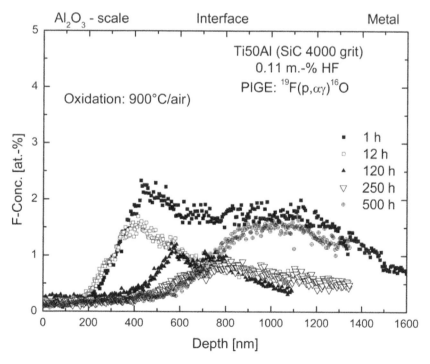

*31.5* Fluorine depth profiles of HF-treated (0.11 mass% HF) TiAl samples after oxidation at 900 °C in air (obtained by PIGE)

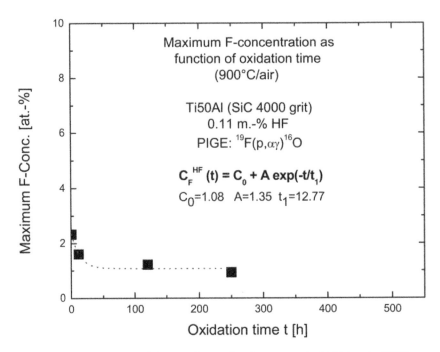

*31.6* Time behaviour of the maximum fluorine content of HF-treated TiAl samples (0.11 mass% HF) during oxidation at 900 °C in air)

$$c_F^{HF}(t) = c_0 + A \exp\left(-\frac{t}{t_1}\right)$$  [31.2]

with $c_0 = 1.08$, $A = 1.35$ and $t_1 = 12.77$.

The sample oxidised for 500 h was treated twice before oxidation and showed a slightly higher fluorine content. Therefore, this sample was not considered for fitting by Eq. 31.2. Similar to the implanted case, the fluorine loss from the surface within the first hour was not considered by the fitting procedure. After formation of the alumina scale, the fluorine loss from the surface stopped and the remaining fluorine was sufficient to establish a long-term fluorine effect at 900 °C.

### 31.3.2   Thermocyclic oxidation of implanted samples at 900 °C in air

The service conditions of components manufactured from gamma-TiAl alloys are characterised by thermocyclic processes between 900 °C and ambient temperature. Therefore, the material was tested using 24 h cycles in air at 900 °C and 1 h cycles in air at 900 °C to study the fluorine content up to an oxidation time of 1500 h. The behaviour of the integral fluorine content and the maximum fluorine content is illustrated in Figs. 31.7 and 31.8. In order to establish a thin alumina scale on the surface, the samples were pre-oxidised for 168 h in air at 900 °C air before starting the cyclic oxidation (1 h cycles in air at 900 °C).

From the metallographic inspections and mass gain measurements, it was clear that, even in the case of rapid cyclic oxidation, a protective alumina scale existed on the surface. From the analytical point of view, the integral fluorine content – after

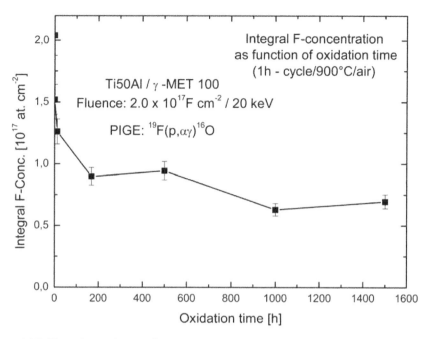

*31.7* Time dependence of integral fluorine concentration of implanted samples (2 × 10¹⁷ F cm⁻²/20 keV) during cyclic oxidation (1 h cycles at 900 °C in air)

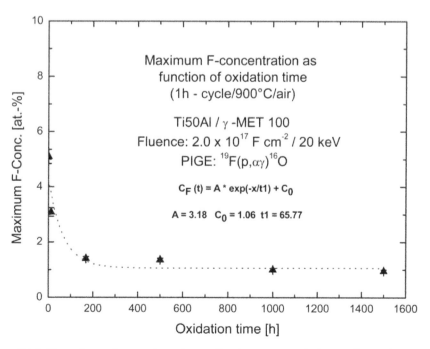

*31.8* Time dependence of maximum fluorine content during cyclic oxidation (1 h cycles at 900 °C in air). Implantation as in Fig. 31.7

decreasing rapidly during the initial hours of isothermal oxidation – showed only a slight further decrease and stabilised at a value of about $7 \times 10^{16}$ F cm$^{-2}$ (Fig. 31.7). As in the case of isothermal oxidation, the fluorine maxima can be fitted by an exponential function according to

$$c_F^{impl}(t) = c_0 + A \exp\left(-\frac{t}{t_1}\right)$$    [31.3]

with the parameters $c_0 = 3.18$, $A = 1.06$ and $t_1 = 65.77$. For oxidation times $> 500$ h, the fluorine maxima attained values of about 1 at.%, which offers long-term stability for thermocyclic oxidation.

### 31.3.3    Diffusion coefficient of fluorine in TiAl at 900 °C

Data for the diffusion coefficient of fluorine in TiAl at 900 °C are not yet available. However, the measured fluorine depth profiles allow the diffusion coefficient of fluorine in TiAl at 900 °C to be estimated. The method suggested in Ref. 14 starts with the standard diffusion equation

$$\frac{\partial c}{\partial t} = D\frac{\partial^2 c}{\partial x^2}$$    [31.4]

where $c$ is the fluorine concentration in atoms cm$^{-3}$ and $D$ is the diffusion coefficient in cm$^2$ s$^{-1}$.

The analytical solution is given by the formula

$$c(x,t) = \frac{Q}{2\sqrt{Dt\pi}}\exp\left(-\frac{x^2}{4Dt}\right)$$    [31.5]

where $Q$ denotes the number of atoms cm$^{-2}$ deposited at the metal/oxide interface. The diffusion coefficient $D$ can be calculated using the logarithmic form of Equation 31.4

$$\ln c = \ln\left(\frac{Q}{2\sqrt{Dt\pi}}\right) - \frac{x^2}{4Dt}$$    [31.6]

which is a linear function with respect to $x^2$.

The diffusion coefficient was calculated from the isothermal oxidation data by using the depth profiles at oxidation times of 120 h, 250 h and 500 h for the implanted sample (Fig. 31.2). The value of $Q$ was set at $Q = 1.26*10^{17}$ atoms cm$^{-2}$, according to Fig. 31.3. We assume that, after oxidation for 2 h, no significant fluorine loss from the surface takes place and the inward diffusion dominates. After transforming the depth profile and fitting according to Eq. 31.6, the resulting diffusion coefficient was calculated using the linear term. The results are summarised in Table 31.1.

The scatter is within one order of magnitude. The fluorine depth profiles were determined without removing the initially formed oxide layer and without sample destruction, which may explain the scatter. Also, irregularities at the metal/oxide interface may have influenced the scatter. The fluorine maxima are located within a depth region around 700–800 nm. Nevertheless, the diffusion coefficients can be

*Table 31.1*   Diffusion coefficients of fluorine in TiAl at 900 °C as calculated from the fluorine depth profiles of the implanted TiAl samples

| Oxidation time, h | Diffusion coefficient, $10^{-15}$ cm$^2$ s$^{-1}$ |
| --- | --- |
| 120 | 2.55 |
| 250 | 0.72 |
| 500 | 1.41 |

used to estimate the decrease in the fluorine content due to inward diffusion into the metal.

## 31.4   Conclusions

The establishment of the fluorine microalloying effect during the oxidation of TiAl in air requires a surprisingly high fluorine enrichment on the surface, which is much higher than predicted by the thermodynamic model [3,4]. This was observed after the application of fluorine both by ion implantation and by treatment with diluted HF. On the other hand, after oxidation for only 12 h in air at 900 °C, the amount of fluorine was significantly lower compared to the initial application.

The main fluorine loss occurs during heating the TiAl and within the initial hours of oxidation at 900 °C. From this and from the shape of the fluorine profiles, it can be concluded that in a second step, the fluorine diffuses into the material leading to a flatter depth profile. A thin alumina layer – once formed – acts as a diffusion barrier to prevent further loss of fluorine through the surface.

This behaviour was observed both after ion implantation and after HF treatment. During isothermal oxidation for up to 1000 h at 900 °C, the remaining fluorine content is characterised by a slowly decreasing behaviour for ion implantation as well as for HF treatment because of the strong diffusion barrier formed by the alumina scale. Even for cyclic oxidation for up to 1500 h at 900 °C, there is a stable amount of fluorine at the metal/oxide interface to provide reliable protection against oxidation.

Empirical equations have been derived to describe the time behaviour of the fluorine maximum during oxidation after both implantation and HF application. The diffusion coefficient for fluorine diffusion in TiAl at 900 °C has also been calculated. Investigations with increased oxidation times at a temperature of 1000 °C are in progress to obtain more data which are required for the development of a lifetime model of the fluorine effect. In both cases of fluorine application – ion implantation and HF treatment – the residual amount of fluorine is sufficient to establish the fluorine effect.

The results offer the potential for technical exploitation of the fluorine microalloying effect for oxidation protection of TiAl alloys at temperatures above 800 °C in air.

## Acknowledgement

The work has been funded to a large extent by Deutsche Forschungsgemeinschaft (DFG) under contract no. SCHU 729/12-2, which is gratefully acknowledged by the authors.

**References**

1.  A. Rahmel, W. J. Quadakkers and M. Schütze, *Mater. Corros.*, 46 (1995), 271.
2.  M. Kumagai, K. Shibue, K. Mok-Soon and Y. Makoto, *Intermetallics*, 4 (1996), 557.
3.  M. Schütze and M. Hald, *Mater. Sci. Eng.*, A239–240 (1997), 847.
4.  A. Donchev, B. Gleeson and M. Schütze, *Intermetallics*, 11 (2003), 387–398.
5.  G. Schumacher, F. Dettenwanger, M. Schütze, U. Hornauer, E. Richter, E. Wieser and W. Möller, *Intermetallics*, 7 (1999), 1113.
6.  G. Schumacher, C. Lang, M. Schütze, U. Hornauer, E. Richter, E. Wieser and W. Möller, *Mater. Corros.*, 50 (1999), 162.
7.  U. Hornauer, E. Richter, E. Wieser, W. Möller, G. Schumacher, C. Lang and M. Schütze, *Nucl. Instrum. Meth. Phys. Res.*, B 148 (1999), 858.
8.  H.-E. Zschau, V. Gauthier, M. Schütze, H. Baumann and K. Bethge, in Proc. Int. Symp. Turbomat, 210, Bonn, Germany, 17–19 June 2002.
9.  H.-E. Zschau, V. Gauthier, G. Schumacher, F. Dettenwanger, M. Schütze, H. Baumann, K. Bethge and M. Graham, *Oxid. Met.*, 59 (2003), 183.
10. H.-E. Zschau, M. Schütze, H. Baumann and K. Bethge, *Nucl. Instrum. Meth. Phys. Res.*, B240 (2005), 137–141.
11. J. Ziegler, J. Biersack and U. Littmark; *The Stopping and Range of Ions in Solids*, Version 95, Pergamon Press, New York, 1995.
12. J. R. Tesmer and M. Nastasi (eds.), *Handbook of Modern Ion Beam Materials Analysis*, Materials Research Society, Pittsburgh, PA, USA, 1995.
13. J. Biersack, *Nucl. Instrum. Meth. Phys. Res.*, B153 (1999), 398.
14. J. Philibert, S. J. Rothman and D. Lazarus, *Atom Movements, Diffusion and Mass Transport in Solids*, EDP Science, Les Ulis, 1991.

# High-temperature oxidation of $Ti_3Al_{0.7}Si_{0.3}C_2$ in air at 900 °C and 1000 °C*

## D. B. Lee and Thuan Dinh Nguyen

*Centre for Advanced Plasma Surface Technology, Sungkyunkwan University,*
*Suwon 440-746, South Korea*

*dlee@yurim.skku.ac.kr*

## S. W. Park

*Multifunctional Ceramic Research Centre, KIST, Seoul 136-791, South Korea*

## 32.1 Introduction

The nanolaminated ternary compound, $Ti_3SiC_2$, consists of $Ti_3C_2$ octahedrons stacked along the c-axis which are separated by layers of Si atoms. This compound has been extensively studied for a number of applications, due to its unique combination of both metallic and ceramic properties [1]. It is electrically and thermally conductive, easily machinable, ductile with a high stiffness-to-hardness ratio, thermally stable up to at least 1700 °C in an inert atmosphere, damage tolerant, maintains its strength at high temperatures, and is resistant to thermal shock and chemical attack including high-temperature oxidation. The scales formed on $Ti_3SiC_2$ are typically composed of an outer $TiO_2$ layer and an inner $(TiO_2 + SiO_2)$ mixed layer [2–4]. $Ti_3SiC_2$ displays good oxidation resistance because of the presence of $SiO_2$.

$Ti_3AlC_2$ is another nanolaminated ternary compound that also exhibits superior metallic and ceramic properties [5]. However, it has been less well studied, owing to the fact that it is a relatively new member of the ternary carbides and because of the difficulties involved in preparing high purity, fully dense specimens [1]. The machinable, layered $Ti_3AlC_2$ is isostructural with $Ti_3SiC_2$. Depending on the investigators, inconsistent oxidation resistance and scale morphologies have been reported. For example, Wang and Zhou reported that $Ti_3AlC_2$ had better oxidation resistance than $Ti_3SiC_2$, because an adherent, continuous inner $Al_2O_3$ layer formed below the outer $TiO_2$ layer [6,7]. In contrast, Barsoum reported that $Ti_3AlC_2$ had much poorer oxidation resistance than $Ti_3SiC_2$ because the scales that formed were highly striated, consisting of three repeating layers: an $Al_2O_3$ layer; a $(Ti_{1-y}Al_y)O_{2-y/2}$ layer where $y < 0.05$; and a porous layer [8,9].

$Ti_3SiC_2$ and $Ti_3AlC_2$ form a complete range of solid solutions. Recently, Zhou and colleagues synthesised a $Ti_3Al_{1-x}Si_xC_2$ ($x = 0.05 \sim 0.25$) solid solution using the solid–liquid reaction synthesis and simultaneous densification processes [10,11]. A

* Reprinted from D. B. Lee and T. D. Nguyen: High-temperature oxidation of Ti3Al0.7Si0.3C2 compound at 900 and 1000 °C in air. *Materials and Corrosion*. 2008. Volume 59. pp. 624–8. Copyright Wiley-VCH Verlag GmbH & Co. KGaA.

mixture of Ti, Al, Si and graphite was used as the starting powder. The synthesis temperature used was 1500 °C, with a soaking time of 1 h, and the hot pressing pressure was 38 MPa of Ar. The $Ti_3Al_{0.75}Si_{0.25}C_2$ that was synthesised displayed excellent oxidation resistance up to 1200 °C for at least 20 h, due to the formation of a continuous $\alpha$-$Al_2O_3$ layer, which formed below a thin, discontinuous $TiO_2$ layer. Silicon existed not only in the $TiO_2$ layer, but also in the $\alpha$-$Al_2O_3$ layer, as $Al_2(SiO_4)O$. A small amount of $Ti_5Si_3$ precipitated from the $Ti_3Al_{0.75}Si_{0.25}C_2$ solid solution [11].

In the present study, a powder mixture of $TiC_x$ ($x=0.6$), Al and Si was used as the starting powder, and a hot pressing method was used to synthesise highly pure, dense $Ti_3Al_{0.7}Si_{0.3}C_2$ compounds. This new process benefits from simultaneous reaction and densification at a relatively low processing temperature for a short reaction time. Changing the synthesising process and the compound composition would be expected to affect the sample purity, sample density, and the sizes of the matrix grains, which could influence the oxidation kinetics and scale structures significantly. The aim of this study was to describe the high-temperature air-oxidation behaviour of $Ti_3Al_{0.7}Si_{0.3}C_2$ synthesised by the newly developed process. The characteristics of the oxides formed, the distribution and roles of Ti, Al, Si and C in the scale, and the oxidation mechanism are discussed based on the experimental results.

## 32.2   Experimental

Ti ($<45$ μm, 99.9% purity) and C ($\sim 10$ μmφ, 99.95% purity) powders were mixed at a molar ratio of Ti/C$=3:0.67$, and pressed at 1500 °C for 3 h under a vacuum of 1.3 Pa. The $TiC_x$ ($x=0.6$) pellets synthesised in this way were ground using a SPEX™ shaker mill, and sieved to a diameter of $<45$ μm. Powders of $TiC_x$, Al ($<45$ μmφ, 99.9% purity), and Si ($<70$ μmφ, 99.9% purity) were mixed at a molar ratio of 3:0.75:0.25 in a SPEX shaker mill for 10 min, and hot pressed at 1400 °C under a pressure of 25 MPa for 60 min in flowing Ar gas. During the hot pressing, a reaction occurred between the $TiC_x$, Al and Si powders. The synthesised $Ti_3Al_{0.7}Si_{0.3}C_2$ pellets were cut into specimens with dimensions of 10 mm × 5 mm × 5 mm, which were then ground to a 1000 grit finish, ultrasonically cleaned in acetone and methanol, and oxidised isothermally at 900 and 1000 °C in atmospheric air for up to 50 h. The weight changes were continuously monitored as a function of time using a thermogravimetric analyser (TGA). The specimens were investigated by means of a differential thermal analyser (DTA), an Auger electron spectroscope (AES), a scanning electron microscope (SEM) equipped with an energy dispersive spectrometer (EDS), an electron probe micro-analyser (EPMA), and an X-ray diffractometer (XRD) with Cu-K$\alpha$ radiation.

## 32.3   Results and discussion

Figure 32.1 shows the SEM microstructure of the synthesised $Ti_3Al_{0.7}Si_{0.3}C_2$. The fully compact specimen has lamellar grains about 10 μm in length and 4 μm in thickness.

Figure 32.2 shows the XRD pattern of $Ti_3Al_{0.7}Si_{0.3}C_2$, along with those of $Ti_3AlC_2$ and $Ti_3SiC_2$. These three compounds all have a layered hexagonal structure. Si atoms partially substituted into the Al sites in $Ti_3AlC_2$ to form $Ti_3Al_{0.7}Si_{0.3}C_2$. Monolithic $Ti_3Al_{0.7}Si_{0.3}C_2$ was successfully synthesised without any impurities. When changing from $Ti_3AlC_2$ to $Ti_3Al_{0.7}Si_{0.3}C_2$, and then to $Ti_3SiC_2$, the diffraction angles such as those of the (006), (008), and (104) planes shifted to larger values because the

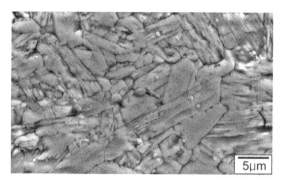

*32.1* SEM image of etched $Ti_3Al_{0.7}Si_{0.3}C_2$

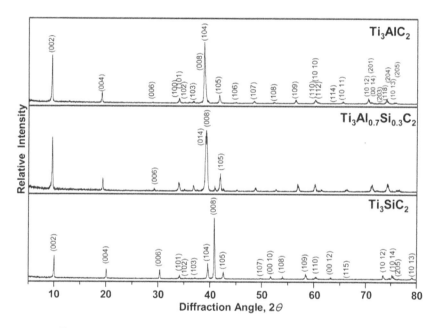

*32.2* XRD patterns of $Ti_3Al_{0.7}Si_{0.3}C_2$, $Ti_3SiC_2$ and $Ti_3AlC_2$

magnitude of the lattice parameter, $c$, decreased more significantly than that of $a$ [10].

Figure 32.3 shows the TG-DTA analytical results for $Ti_3Al_{0.7}Si_{0.3}C_2$. The oxidation rate increased gradually with increasing oxidation temperature. From around 450 °C, the endothermic reaction due to the heating of the sample changed to an exothermic reaction, because $Ti_3Al_{0.7}Si_{0.3}C_2$ began to be oxidised noticeably.

Figure 32.4 shows the XRD patterns of $Ti_3Al_{0.7}Si_{0.3}C_2$ after oxidation at 900 and 1000 °C for 30 h. The oxide scales formed consisted primarily of $\alpha$-$Al_2O_3$ and rutile–$TiO_2$. The $SiO_2$ that also formed was not detectable using XRD owing to its amorphous structure. The matrix peaks were strong in Fig. 32.4(a), because of the small extent of oxidation. As the oxidation progressed, the scale thickened, and the matrix peaks disappeared. Since the semi-protective $TiO_2$ grew much faster than

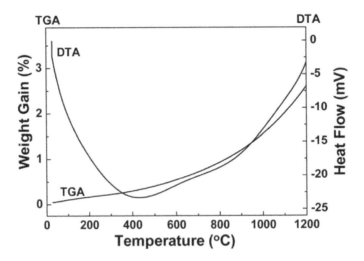

*32.3* TG-DTA curves of $Ti_3Al_{0.7}Si_{0.3}C_2$, which were obtained during heating from room temperature to 1200 °C in air with a heating rate of 10 K min$^{-1}$

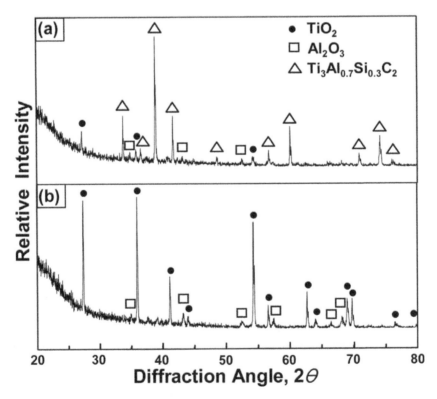

*32.4* XRD pattern of $Ti_3Al_{0.7}Si_{0.3}C_2$ after oxidation for 30 h in air: (a) at 900 °C; (b) at 1000 °C

the highly protective Al$_2$O$_3$, the TiO$_2$ peaks were stronger than the Al$_2$O$_3$ peaks in Fig. 32.4(b). Here, the oxide surface was covered with TiO$_2$.

Figure 32.5 shows the AES depth profiles of Ti$_3$Al$_{0.7}$Si$_{0.3}$C$_2$ after oxidation at 900 °C for 7 min in air, which were obtained in order to understand the oxidation mechanism during the initial oxidation period. This inert Pt marker experiment was performed by sputter-depositing a thin Pt film on top of Ti$_3$Al$_{0.7}$Si$_{0.3}$C$_2$ before its oxidation. From the location of the Pt film, it is seen that oxygen diffused inwards, while Ti and Al tended to diffuse outwards. Carbon tended to escape from the surface, but Si simply stayed in the Ti$_3$Al$_{0.7}$Si$_{0.3}$C$_2$ sample. It is noted that TiO$_2$, being an n-type semiconductor, grows primarily by either the outward diffusion of interstitial Ti$^{4+}$ ions or the inward diffusion of O$^{2-}$ ions via oxygen vacancies, depending on the defect concentrations [12]. On the other hand, α-Al$_2$O$_3$ is generally known to grow very slowly by the inward diffusion of oxygen [13]. However, the non-stoichiometry in Al$_2$O$_3$ is very small, and the defect structure of α-Al$_2$O$_3$ is not yet known unequivocally [14].

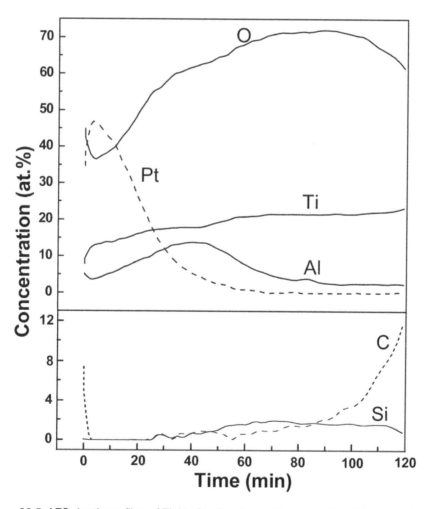

32.5 AES depth profiles of Ti$_3$Al$_{0.7}$Si$_{0.3}$C$_2$ after oxidation at 900 °C for 7 min in air. The penetration rate is 18 nm min$^{-1}$ for reference SiO$_2$

To understand the oxidation mechanism in the later stages of oxidation, another marker test was performed by spraying fine Pt particles onto the matrix surface before oxidation, as shown in Fig. 32.6. The scale formed after oxidation at 1000 °C for 50 h in air is divided into three layers, namely, an outer $TiO_2$ layer containing $Al_2O_3$ particles, an intermediate $Al_2O_3$ layer, and an inner mixed layer that is rich in $TiO_2$, but deficient in $Al_2O_3$ and $SiO_2$. Platinum particles are located at the top of the intermediate $Al_2O_3$ layer, indicating that the outer $TiO_2$ layer grows by the outward diffusion of $Ti^{4+}$ ions, while the inner $(TiO_2 + Al_2O_3 + SiO_2)$ mixed layer grows by the inward diffusion of $O^{2-}$ ions. $TiO_2$ formed by the outward diffusion of $Ti^{4+}$ and the inward diffusion of $O^{2-}$ ions (Fig. 32.6). On the other hand, silicon ions in oxides are relatively immobile, because of the higher bonding energy of $Si^{4+}-O^{2-}$ (465 kJ $mol^{-1}$). Hence, the Si in the inner mixed layer was oxidised more or less *in situ* by the inwardly diffusing $O^{2-}$ ions. $^{18}O$ tracer diffusion studies have confirmed that

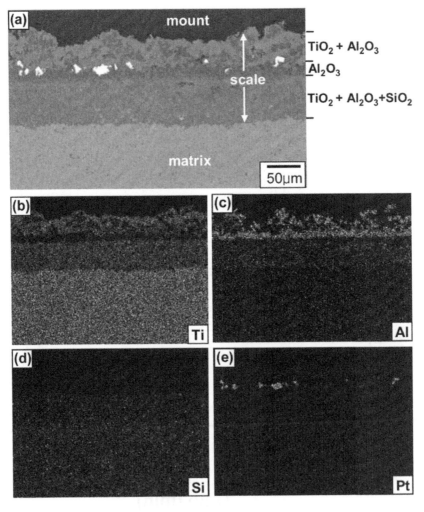

*32.6* SEM/EDS analysis on $Ti_3Al_{0.7}Si_{0.3}C_2$ after oxidation at 1000 °C for 50 h in air: (a) cross-sectional back-scattered electron image; (b) Ti map; (c) Al map; (d) Si map; (e) Pt map

the growth of $SiO_2$ is dominated by inward oxygen diffusion [14]. The $Al_2O_3$ particles embedded in the outer $TiO_2$ layer may either be formed by the outward diffusion of $Al^{3+}$ ions (Fig. 32.5), or be the result of the upward migration of $Al_2O_3$ particles that were originally formed around the intermediate $Al_2O_3$ layer. The outwardly diffusing $Ti^{4+}$ ions may have induced an upward plastic flow of $Al_2O_3$ particles. Since the growth rate of $TiO_2$ is much faster than that of $Al_2O_3$ and $SiO_2$ owing to the higher lattice defect concentration, the protectiveness of $Ti_3Al_{0.7}Si_{0.3}C_2$ would depend strongly on the continuity and compactness of the intermediate $Al_2O_3$ layer. Another barrier to oxidation is the inner mixed layer, containing $Al_2O_3$ and $SiO_2$.

Figure 32.7 shows the typical structure of the oxide scale formed on $Ti_3Al_{0.7}Si_{0.3}C_2$. The mature scale consisted of an outer $TiO_2$ layer containing $Al_2O_3$ particles, an

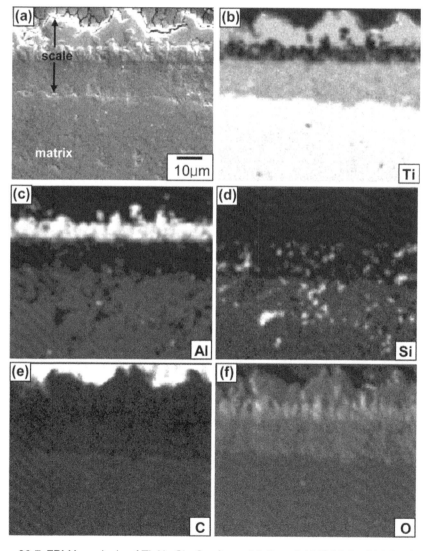

*32.7* EPMA analysis of $Ti_3Al_{0.7}Si_{0.3}C_2$ after oxidation at 1000 °C for 30 h in air: (a) cross-sectional image; (b) Ti map; (c) Al map; (d) Si map; (e) carbon map; (f) oxygen map

intermediate $Al_2O_3$ layer, and a $TiO_2$-rich, inner mixed oxide layer. Carbon is hardly seen in the scale, owing to its liberation as CO or $CO_2$ gases. Upon its exposure to oxygen, the surface of $Ti_3Al_{0.7}Si_{0.3}C_2$ would be covered with $TiO_2$, $Al_2O_3$, and $SiO_2$ crystallites, because no equilibrium exists between the gas phase and the matrix. Rutile, having the highest growth rate, soon overgrows alumina and silica and progressively covers the entire surface, forming coarse outer oxide grains. The outwardly moving Ti ions may carry a certain amount of Al ions with them towards the surface, forming $Al_2O_3$ particles inside the outer $TiO_2$ layer. The consumption of Ti at the outer surface depletes the Ti immediately below and thereby enriches the Al enough to form an intermediate $Al_2O_3$ layer. This layer contains interdispersed $TiO_2$ particles, so that oxygen can diffuse through the intermediate $Al_2O_3$ layer down to the matrix. $Al_2O_3$ itself is highly stable, and exhibits low diffusivities for both cations and anions. Since the amount of Si in $Ti_3Al_{0.7}Si_{0.3}C_2$ is small, the $SiO_2$ that is formed exists as scattered particles in the inner mixed oxide layer. Here, rather strong Al-depletion occurs owing to the consumption of the Al above. Since there is still enough Ti below the intermediate $Al_2O_3$ layer, the inner mixed oxide layer is rich in $TiO_2$.

Figure 32.8(a) shows the appearance of the oxide scale formed on $Ti_3Al_{0.7}Si_{0.3}C_2$ that was oxidised at 1000 °C for 5 h in air. Initially formed tiny oxide crystallites were

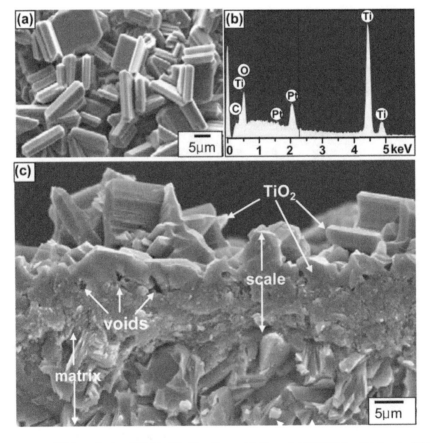

*32.8* SEM analysis of $Ti_3Al_{0.7}Si_{0.3}C_2$ after oxidation at 1000 °C for 5 h in air: (a) top view; (b) EDS spectrum of (a); (c) fractograph

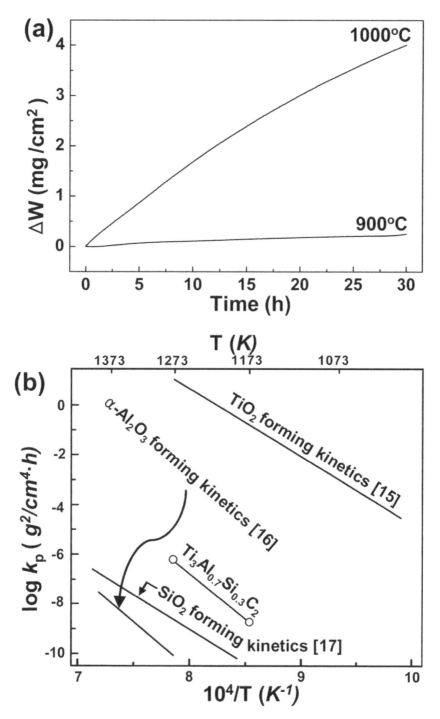

*32.9* (a) Isothermal oxidation kinetics of $Ti_3Al_{0.7}Si_{0.3}C_2$ at 900 and 1000 °C in air; (b) temperature dependence of the parabolic rate constants for $Ti_3Al_{0.7}Si_{0.3}C_2$. The $k_p$ values of $TiO_2$ [15], $\alpha$-$Al_2O_3$ [16], and $SiO_2$ [17] are superimposed

grown into the characteristic pillar-like, coarse rutile–$TiO_2$. The corresponding EDS spectrum shown in Fig. 32.8(b) indicates that the whole surface was covered with $TiO_2$. The fractograph shown in Fig. 32.8(c) displays the $TiO_2$ grains that formed at the outer surface. Partial sintering of the coarse $TiO_2$ grains is evident. The intermediate $Al_2O_3$ grains and the inner mixed oxide grains are inevitably fine, because alumina and silica grow quite slowly and suppress the grain growth of the intermixed titania. Microscopic voids are seen at the interface between the outer $TiO_2$ grains and the intermediate $Al_2O_3$ grains. Voids form due to the Kirkendall effect caused by the outward diffusion of Ti ions, the escape of carbon in the form of CO or $CO_2$, and the mismatch in the volume expansion between the $TiO_2$ and $Al_2O_3$ grains. Such voids provide the channel for oxygen diffusion and act as stress concentrators and, therefore, adversely affect the scale integrity or adherence. The lamellar $Ti_3Al_{0.7}Si_{0.3}C_2$ grains were disintegrated into fine oxide grains at the oxidation front. This oxidation front is not ideally uniform, owing to the anisotropy of the matrix grains.

Figure 32.9(a) shows the weight gain vs. oxidation time curves of $Ti_3Al_{0.7}Si_{0.3}C_2$ at 900 and 1000 °C. $Ti_3Al_{0.7}Si_{0.3}C_2$ oxidises almost parabolically, indicating that the oxidation process is diffusion controlled. The weight gain increased with increasing oxidation temperature. The scale thicknesses measured after oxidation at 900 and 1000 °C for 30 h were 5 and 42 μm, respectively. The curves shown in Fig. 32.9(a) were fitted to the equation, $\Delta W^2 = k_p \cdot t$, where $\Delta W$ is the weight gain per unit area, $t$ is the oxidation time, and $k_p$ is the parabolic rate constant. From Fig. 32.9(b), it is seen that $Ti_3Al_{0.7}Si_{0.3}C_2$ oxidises considerably more slowly than the $TiO_2$-forming kinetics [15], but faster than the $\alpha$-$Al_2O_3$-forming [16] and $SiO_2$-forming [17] kinetics. The activation energy for the oxidation of $Ti_3Al_{0.7}Si_{0.3}C_2$ obtained from Fig. 32.9(b) was 706.5 (kJ mol$^{-1}$).

## 32.4    Conclusions

Fully dense, monolithic $Ti_3Al_{0.7}Si_{0.3}C_2$ compounds having lamellar grains were oxidised in air. $Ti_3Al_{0.7}Si_{0.3}C_2$ oxidised almost parabolically at 900 and 1000 °C according to the equation

$$Ti_3Al_{0.7}Si_{0.3}C_2 + O_2 \rightarrow \text{rutile–}TiO_2 + \alpha\text{-}Al_2O_3 + \text{amorphous } SiO_2 + (CO \text{ or } CO_2)$$

The oxidation proceeded primarily by the simultaneous inward diffusion of oxygen ions to form an inner mixed oxide layer that was rich in $TiO_2$, but deficient in $Al_2O_3$ and $SiO_2$, and the outward diffusion of Ti ions to form an outer $TiO_2$ layer. These oxide layers were separated by an intermediate $Al_2O_3$ layer. The silicon in the inner mixed oxide layer was less mobile, and was oxidised more or less *in situ* by the inwardly diffusing oxygen ions. The $Al_2O_3$ particles embedded in the outer $TiO_2$ layer may be formed by either the outward diffusion of $Al^{3+}$ ions or the upward plastic flow of $Al_2O_3$ particles owing to the outwardly diffusing $Ti^{4+}$ ions.

## Acknowledgments

This work was supported by a grant (no. R-11-2000-086-0000-0) from the Center of Excellency Program of the KOSEF, and by a grant (no. 05K1501-00610) from the 'Center for Nano-Structured Materials Technology' under the 21$^{st}$ century frontier R&D programme of the MOST, Korea.

## References

1. M. W. Barsoum, *Prog. Solid State Chem.*, 28 (2000), 201.
2. M. W. Barsoum, T. El-Raghy and L. U. J. T. Ogbuji, *J. Electrochem. Soc.*, 144 (1997), 2508.
3. M. W. Barsoum, L. H. Ho-Duc, M. Radovic and T. El-Raghy, *J. Electrochem. Soc.*, 150 (2003), B166.
4. Z. Sun, Y. Zhou and M. Li, *Corros. Sci.*, 43 (2001), 1095.
5. N. V. Tzenov and M. W. Barsoum, *J. Am. Ceram. Soc.*, 83 (2000), 825.
6. X. H. Wang and Y. C. Zhou, *Corros. Sci.*, 45 (2003), 891.
7. X. H. Wang and Y. C. Zhou, *Acta Mater.*, 50 (2002), 3141.
8. M. W. Barsoum, *J. Electrochem. Soc.*, 148 (2001), C551.
9. M. W. Barsoum, *J. Electrochem. Soc.*, 148 (2001), C544.
10. Y. C. Zhou, J. X. Chen and J. Y. Wang, *Acta Mater.*, 54 (2006), 1317.
11. J. X. Chen and Y. C. Zhou, *Oxid. Met.*, 65 (2006), 123.
12. Y. M. Chiang, D. P. Birnie III and W. D. Kingery, *Physical Ceramics*, 109. John Wiley & Sons, NY, 1996.
13. R. Prescott and M. J. Graham, *Oxid. Met.*, 38 (1992), 233.
14. P. Kofstad, *Oxid. Met.*, 44 (1995), 3.
15. P. Kofstad, *High Temperature Oxidation of Metals*, 175. John Wiley & Sons, NY, 1966.
16. M. W. Brumm and H. J. Grabke, *Corros. Sci.*, 33 (1992), 1677.
17. I. C. I. Okafor and R. G. Reddy, *JOM*, 51(6) (1999), 35.